TJ 808 .K75 1986
Kristoferson, Lars A.
Renewable energy technologies : their
applications in developing countries

TJ 808 .K75 1986
Kristoferson, Lars A.
Renewable energy technologies : their
applications in developing countries

RENEWABLE ENERGY TECHNOLOGIES

Their Applications in Developing Countries

L. A . KRISTOFERSON
and
V. BOKALDERS

A Study by the Beijer Institute
The Royal Swedish Academy of Sciences

PERGAMON PRESS

OXFORD · NEW YORK · BEIJING · FRANKFURT
SÃO PAULO · SYDNEY · TOKYO · TORONTO

U.K.	Pergamon Press, Headington Hill Hall, Oxford OX3 0BW, England
U.S.A.	Pergamon Press, Maxwell House, Fairview Park, Elmsford, New York 10523, U.S.A.
PEOPLE'S REPUBLIC OF CHINA	Pergamon Press, Room 4037, Qianmen Hotel, Beijing, People's Republic of China
FEDERAL REPUBLIC OF GERMANY	Pergamon Press, Hammerweg 6, D-6242 Kronberg, Federal Republic of Germany
BRAZIL	Pergamon Editora, Rua Eça de Queiros, 346, CEP 04011, Paraiso, São Paulo, Brazil
AUSTRALIA	Pergamon Press Australia, P.O. Box 544, Potts Point, N.S.W. 2011, Australia
JAPAN	Pergamon Press, 8th Floor, Matsuoka Central Building, 1-7-1 Nishishinjuku, Shinjuku-ku, Tokyo 160, Japan
CANADA	Pergamon Press Canada, Suite No. 271, 253 College Street, Toronto, Ontario, Canada M5T 1R5

First edition 1986
Reprinted 1987

Library of Congress Cataloging in Publication Data

Kristoferson, Lars A.
Renewable energy technologies.
1. Renewable energy sources. 2. Renewable energy
sources—Developing countries. I. Bokalders, Varis,
1944-   . II. Beijer Institute. III. Kungliga
Svenska vetenskapsakademien. IV. Title.
TJ808.K75   1986   621.042   86-766

British Library Cataloguing in Publication Data

Kristoferson, Lars A.
Renewable energy technology : a study by the
Beijer Institute, the Riyal Swedish Academy of
Sciences.
1. Renewable energy sources
I. Title II. Bokalders, V. III. Beijerinstitutet
621.042   TJ163.2
ISBN 0-08-034061-X

In order to make this volume available as economically and as rapidly as possible the authors' typescripts have been reproduced in their original forms. This method unfortunately has its typographical limitations but it is hoped that they in no way distract the reader.

Printed in Great Britain by A. Wheaton & Co. Ltd., Exeter

CONTENTS

PREFACE AND ACKNOWLEDGEMENTS

This Study arose from discussions with Karin Wohlin at the Energy Desk of SIDA´s Industry Division. Following the growing awareness of the seriousness of the energy situation in developing countries and the often confusing signals about the supply/demand potential, SIDA felt the need for a review and assessment of the prospects for new and renewable energy technologies and their possible role in energy planning. In order to be better able to assess requests and proposals for energy-related projects and their relevance in development terms, SIDA has supported a number of studies on energy planning and energy technologies. This is one of them. It is an overview and introduction to the subject meant for non-specialists, and its aim is to summarize and assess present experiences of a number of new and renewable energy sources in a development context. The study focusses on renewable energy technologies for the rural sector, which is, of course, only one limited part of the complex problem of energy in developing countries. Thus, we do not treat issues like fossil fuels, nuclear energy, conventional hydropower, industrial energy supply, electrification or energy conservation. Moreover, matters of general policy are not discussed in detail; only insofar as those matters are directly connected to development of renewable energy are they treated in this study, for example in connection with certain aspects of fuelwood and forestry. Detailed treatment of such issues can be found in many other studies, e.g. by the World Bank.

Obviously, a book that tries to cover a subject as wide and diverse as Renewable Energy in Developing Countries cannot be written without the substantial input of many specialists. Much of this input was received as reviews of specific technologies, commissioned by us for the first version of this study: "Prospects for New and Renewable Energy Sources in Developing Countries" (Volumes A-E), Beijer Institute, 1984 (see list in Appendix 3). We therefore would like to thank the following for their valuable contributions:

- N.K. Bansal, Indian Institute of Technology, New Delhi, India
- Jackie Bergman, Energikonsult, Angpanneföreningen, Stockholm, Sweden
- Dennis Costello, Littleton, Colorado, U.S.A.
- Staffan Engström, National Energy Administration, Stockholm, Sweden
- B. van Gelder, KWDP/ICRAF, Nairobi, Kenya
- Michael R. Goe, International Livestock Centre for Africa, Addis Ababa, Ethiopia
- Guido Gryseels, International Livestock Centre for Africa, Addis Ababa, Ethiopia
- Stig Gummesson, United Stirling AB, Malmö, Sweden
- Thomas Harris, Clark Univ., Worcester, Mass., U.S.A.
- Intermediate Technology Industrial Services (ITIS), London, U.K.
- Kirsten Johnson, Clark Univ. Worcester, Mass. U.S.A.
- Cindi Katz, Clark Univ., Worcester, Mass., U.S.A.
- Björn Kjellström, Exergetics AB, Trosa, Sweden
- Kathryn Lawrence, Littleton, Colorado, U.S.A.
- Björn Lundgren, ICRAF, Nairobi, Kenya
- Christer Nordström, Askim, Sweden
- Jan Erik Nylund, Swedish University of Agricultural Sciences, Uppsala, Sweden
- Britt Sahleström, National Energy Administration, Stockholm, Sweden
- M.S. Sodha, Indian Institute of Technology, New Delhi, India
- Per Johan Svenningsson, Beijer Institute, Stockholm, Sweden
- Torgny von Wachenfeldt, Dept. of Marine Botany, Lund University, Lund, Sweden

Their background reports have been published separately in a special volume of commissioned reports (Vol. E of the above study).

We warmly acknowledge B. van Gelder and Björn Kjellström for their special efforts in updating and revising Chapter 1 and Section II of this book respectively.

We are also indebted to the following for taking time to review the first version of this study and suggest improvements:

- David Hall, Imperial College, London, U.K.
- Haile Lul Tebicke, Science and Technology Commission, Ethiopia
- Armando Caceres, CEMAT, Guatemala
- Derek Lovejoy, NRED/TC, United Nations
- F.J. Mouttapa, FAO, United Nations
- P.J. Johnston, Pacific Energy Development Programme, ESCAP, United Nations.

In addition, we would like to thank the following for their input: Mark Newham, for his editorial work on an earlier version of the text; Visvaldis Bokalders for preparing the figures; Solveig Nilsson and Blanche Debenham for patiently working at their word processors; Margaret Goodman for many useful suggestions for the improvement of the text; Janos Pasztor for his helpful specialist advice in many Chapters; Phil O´Keefe for many valuable suggestions and other substantial input; Gordon Goodman for constant support throughout the work on the study, and Julian Dison for the time and effort he put in as the Technical Editor of this book.

We have relied on input from all those persons mentioned above. However, the responsibility for all errors and misunderstandings is completely our own.

Several other in-depth studies and technical monographs of specific renewable technologies and their application in developing countries have been undertaken recently, and more are certain to come. A great number of articles and reviews have also appeared in the recent literature. In drawing on the information from a large number of such publications, this Study tries to assess briefly as many of the important technologies as possible from a perspective not only covering technology, but also adding comments on economics, environment, and social issues. Emphasis has been placed on a breadth rather than depth, and on trying to give a sense of the complexity and variety of the role of new energy technologies in developing countries. We hope that this approach will make it easier to identify and understand the prospects and limitations of technology in the social problem of energy and development.

A few words of caution to the reader may be added regarding the level of technical detail in the text, in particular for those of our prospective readers who are more scientifically inclined or experts in any one of the technologies treated. This is not a scientific textbook, but rather meant as an introduction and overview for laymen, aimed at a general audience. Figures and data on costs etc. are therefore given more for illustrative and comparative purposes, than for an attempt to give exact information to be used for commercial investments in various technologies, for example. Cost data for new and renewable energy are known to be very site specific and notoriously uncertain. We have also avoided giving detailed references on all aspects, but have rather chosen to give a basic overall bibliography for each technology, usually 5-15 references per Chapter.

The ground covered in this Study is enormous, and it is impossible to be aware of all progress in this rapidly changing field. Gaps and mistakes are therefore unavoidable. We shall be very pleased to receive any comments and constructive criticisms.

Finally, we are happy to acknowledge the generous funding of this study from SIDA (the Swedish International Development Authority) and, especially, the support and encouragement of Karin Wohlin.

The present energy problems of developing countries will not be solved by renewable energy sources. But they cannot be solved without them.

Lars Kristoferson
Varis Bokalders

INTRODUCTION

Over-optimism about technological development is the hallmark of energy conferences. Exactly 20 years before the Nairobi conference on New and Renewable Sources of Energy (NARSE) in 1981, a similar conference on "New Sources of Energy" was held in Rome, again by the United Nations, again covering developing country issues. This first conference presented a largely optimistic and conflict-free picture of solar and wind energy products. In the period between the first and second conference, however, little was done to further the diffusion of solar and other renewable energy technologies. Oil was cheap and fuelwood issues had not yet become critical. The limited research and development work that took place, focussed on technical issues. Social, economic, environmental and policy considerations seldom arose.

As part of the change in energy policies which followed the "oil crisis", many industrialized countries expanded their development efforts in renewable energy, and substantial technical progress was made. A small fraction of these efforts is devoted to Third World applications, and these may well expand as new markets for energy technologies become established in the developing countries. A major problem, therefore, is to screen, select, adapt and manage emerging energy technologies for the Third World. International aid agencies have a crucial role in this area. They may also assist the build-up of a local energy R & D and manufacturing capability in developing countries, which is the long-term but necessary basis for a balanced energy development.

Of course, there is a strong interaction between renewable energy sources and commercial energy, such as oil. The current slackening of the oil market will certainly make oil more affordable to some and thus make many renewable energy projects less interesting. Many renewable energy issues, such as fuelwood provision, are, however, largely unaffected by oil price fluctuations. Oil is no alternative for the rural poor, being an unaffordable fuel compared to "free" fuelwood. This is the more tragic, since a global switch from fuelwood to petroleum fuels for cooking would correspond to some extra 150 Mtoe per year, a mere 5% of current world oil consumption (equal to the 1984 oil production of Mexico). Thus it is poverty and not the lack of physi-

cal resources that is the limiting factor in solving the fuelwood problem of the developing countries.

Too often in the past there has been a facile assumption that the Third World can rapidly bypass an era of fossil fuel and go straight into a solar future. The optimists argued that solar energy is an abundant free resource, especially in the Third World, and that the lack of an existing infrastructure, including administration, is an advantage because there is little technological inertia.

In general terms the opposite is nearer to the truth. Lack of infrastructure is a major obstacle to change, especially managed change. The weak economies of the Third World, which are characterized by a lack of foreign exchange and limited household purchasing power, slow down the rate of technical change. Any large-scale implementation of new and renewable energy technologies, like most features of industrial development, will therefore first occur in the advanced economies. This is clearly shown by today's development and trends. In the near short the role of such technologies in the Third World will therefore be limited, especially in the rural areas. There, until economic development has brought the majority of farmers above the subsistence level, wood and agricultureal wastes will continue to be the dominating source of energy for the foreseeable future.

There is also a growing awareness that New and Renewable Sources of Energy (NARSE) should not be used as a catch-word for Third World energy development. Such programmes must not lump together widely different technologies and methods in different states of research, development and testing. This must be paralleled by an increased understanding of the complex social and economic issues that influence these technologies. The problem is not whether "NARSE" or "conservation" or anything else is a strategy to follow, but whether some specific technological improvement or method makes sense in social and economic terms in a specific situation. Only a careful, site-specific evaluation can determine the "appropriateness" of a technology. This need for careful evaluation should not, however, be taken as a "clarion call to inaction". Time does not wait for large-scale, detailed evaluations to be completed. In many cases, the problems are already apparent, or the market is ready to accept new products. In both cases, appropriate action can start immediately. The need for rapid tree planting is an example the spread of micro-hydro plants in several countries another. The precise methods for successful implementation (e.g. the degree of social participation in reforestation efforts) are not, however, easy to establish without the evaluation of local conditions.

It is important to realize, after all, that it is the low level of economic activity and the inelastic household budgets that create barriers to NARSE diffusion, particularly in rural areas. This ineffective demand precludes rapid change. Only when economic development has created a more substantial diversity of end-use demands, such as water pumping for irrigation and power for food processing, will NARSE (and, in fact, any other new technologies) enjoy widespread diffusion.

As studies and tests obtain results and in-the-field experience is accumulated, much of the earlier uncritical embracing of new and renewable sources of energy (NARSE) for developing countries now seems to be replaced by a guarded optimism founded on more critical lines. As hopes, expectations and general goodwill face technological, economic and social realities, actual problems and prospects are becoming more clear. Consequently, clearer time scales for introduction are usually stretched out, and it is realized that the viability of most renewable energy projects also critically depends on where they are set up. NARSE is always site-specific.

The energy problem of the Third World is not one single, simple problem. It is a complex and multi-faceted problem experienced in modern industry and the traditional households. Effective energy demand, which is not to be confused with energy need, covers a wide range of end-uses. One conclusion that can be drawn is this, because there is not one single, simple problem, there cannot be one single, simple solution.

The flexibility thus required to achieve progress with NARSE, also requires that attention be paid to questions of both supply and demand. Fuel and technologies will have to be matched to the task, which also means that NARSE and conventional energy must be seen as complementary to each other. The need for such a complementary strategy is now being increasingly recognized by Third World governments. Current energy plans in, for example, China and India explicitly state that neither conventional nor renewable energy can solve the problems of rural energy development on their own. For the foreseeable future, it will not be possible to supply rural areas with oil or grid power in large enough quantities to support the desired economic growth. This means an increased reliance on local and therefore renewable energy sources. Also in the longer run, NARSE can be seen as a promising and, indeed, necessary complement in the national energy mix. Strong emphasis on NARSE development has therefore been given by both those countries, not least in institutional and funding terms. Many other countries, such as Brazil and the Philippines, have also tried to launch ambitious programmes for NARSE development, and several others are in various stages of implementing such programmes.

The identification of specific fuel-technology combinations to certain tasks, is something that is associated more with modern than traditional forms of energy. Consequently, the successful development of NARSE will require a focus on modern, not traditional, sector problems. Such a focus carries certain implications. Firstly, there is a need to move away from the present dominating craft tradition of alternative technology production to the modern design building, and testing of NARSE technologies. NARSE will only be successful if it is mass consumed. The most effective way to achieve mass consumption is mass production. Such production systems have little to do with the tradition of alternative technology. Secondly, the problems with the diffusion of NARSE are above all social and not technical ones. In assessing the site-specificity of NARSE, close attention must be paid to the cultural milieu in which NARSE falls.

It is important to realize that NARSE will not make a spectacular contribution to the total energy requirements of developing countries in the short and medium term. With the total energy budgets dominated by woodfuel in the household sector, the short-term impact of NARSE on energy _per se_ will be negligible. NARSE can have a substantial strategic and catalytic impact on development where the right combination of supply and demand is present, for example agricultural residues as fuel for agricultural processing industris. In many cases, NARSE will therefore accelerate economic growth. Even if NARSE may not be able to make a substantial impact on national energy balances in the near future, their impact on development can thus be crucial, since NARSE will make possible many new ways to energize agricultural and industrial developments.

NARSE is thus no panacea for the energy problems of the developing world. On the other hand, it is also clear that several of those new technologies will have a great importance for the future. Many of them are, in fact, badly needed if problems are going to be solved. The main question is to sift the wheat from the chaff. There is obviously a need for more dispassionate analysis, even though this may sometimes be difficult in a field where many of the basic data are still lacking or very uncertain.

It is useful, bearing in mind the implications of this argument, to distinguish between two kinds of energy technologies: household improvement technology and energy appliance technology. The first category must be cheap, produced from locally-available material and easily replicated without sophisticated infrastructure; the main target group is the rural poor. The second category is comparatively expensive, manufactured, and requires a sound support structure; the target group is the middle and upper levels of the monetarised economy. A third category, only briefly treated in this Study, is industrial technologies. All of these categories represent segments on a sliding scale, and clear-cut distinctions are therefore often difficult to make. It is becoming clear, however, that different implementation strategies will have to be pursued in the different categories if widespread dissemination is going to become possible. The role of cheap and inventive financial schemes is obviously crucial.

Judging from the situation in the industrialized countries, general electrification is one of the main keys to industrialization and development, this is also true in the rural areas of the developing world. In this context, new and renewable energy technologies may also have an important role to play, sometimes as a crucial first step. The introduction of local electrification may be considerably helped in many areas if local energy resources are used in the build-up of local or regional grids, which are then gradually interconnected as demand grows and infrastructure is developed over the years. This was, in fact, the way most grids in Europe were developed some hundred years ago. Small scale hydro developments and the introduction of biomass-fuelled power plants, which are often very competitive with an expensive and sometimes unreliable diesel supply, are obviously attractive in this respect.

Energy is thus not simply a commodity – it has become a critical factor of production in addition to land, labour and capital. A vigorous informed effort is needed to promote NARSE wherever feasible, for, in the energy futures of developing countries, every contribution is critical. Local, small-scale, decentralised initiatives are indispensable in this respect as the ongoing depletion of woodfuel threatens even the subsistence base of the Third World population.

During the last four or five years, as a result of the extensive studies and field trials, much useful experience has, however, been gained, and we are now in a much better position than in 1981 to express well-founded opinions about the viability of various technologies and to assess their prospects for the future.

For example, to take some obvious examples on the solar energy side, solar cookers and solar thermal pumps seem to have lost much of the attraction such programmes enjoyed a few years ago. On the other hand, photo-voltaic cells (not least for pumping) and solar crop-driers continue to develop, and seem to be able to spread gradually into larger and larger market niches. This is also the case with solar water-heaters for urban and institutional use. The revival of traditional methods for the passive solar heating and cooling of buildings is another promising area, especially when combines with the use of modern civil engineering practices. Small-scale hydropower and windpumps have become very successful in many applications. In some places, they may be very profitable compared to, say, diesel, while in other locations they may be a very bad choice.

On the bioenergy side, a better understanding of the prospects and problems of tree planting is now rapidly emerging, with an emphasis on local, replicable, village or on-the-farm agroforestry methods. Similarly, concerning improved woodstoves, several promising designs have now emerged and started to disseminate, following a period of slow progress a few years ago. Regarding biomass engines and biomass engine-fuels the situation has also become much clearer in the last few years. Ethanol and producer gas seem viable under certain conditions only, whereas methanol and vegetable oil as engine fuels are currently not competitive with conventional solutions (e.g. diesel engines). Similar examples can be found regarding most other types of NARSE in the following Chapters.

Briefly, it is clear that the so called new and renewable sources of energy (NARSE) will have a very useful role to play in developing and powering the Third World, although they will not play the dominant role suggested by some a few years ago. There is nothing like a "NARSE strategy" for the energy development of the Third World. Instead, as research and commercialization continue, various energy sources will slowly find and occupy their respective market niches, and constitute an integral part of the energy systems of the developing world.

It is against this general background that this book has been written. We hope that the perspective chosen, - basically a technological one, but as far as possible integrated with social and economic issues - will provide useful background information to practitioners, policymakers and others involved or interested in the very active field of energy and development.

ABSTRACTS

Fuelwood: Forestry and Agroforestry (Chapter 1). Fuelwood, mainly for cooking, is still the dominating energy source for most people in developing countries and is likely to remain so for the foreseeable future. However, serious deficits are now appearing in many places and show signs of worsening and becoming much more widespread. The only way to stop this is by intensive tree-planting on an unprecedented scale. Ongoing developments in forestry for energy includes tree-planting methods (including agroforestry) species selection etc., all of which has to be organised differently in different socio-economic circumstances. Local participation is essential, in order to achieve economic and social viability in fuelwood supply in rural areas. Urban and industrial demand can however, often be met competitively by large-scale plantations. One of the main problems in tree-planting is to find the right species, methods and organisational approach for different social circumstances. The impact on the national energy balance is often crucial, and in local terms it is sometimes a question of survival. Development support could be substantially increased, and may be channelled into carefully selected, self-replicating rural and urban tree planting and agroforestry schemes.

Energy crops (Chapter 2) The interest in energy crops have increased greatly during the last decade, particularly in industrialized countries. Research programmes are under way in e.g. USA, Sweden, New Zealand and Brazil. Ongoing development is for example concerned with species selection, yield improvement and marginal land utilization, and it focusses on production for industrial use. The cost-effectiveness of energy crops is extremely site- and crop-specific. A major problem, particularly in developing countries, is land use competition between food and fuel production (cash-crops). The impact on the national energy balance is small. Development support may be channelled into pilot trials and economic/environmental evaluations.

Agricultural residues and organic wastes, (Chapter 3). Although extremely important as cooking fuel in many areas in developing countries, these have been largely neglected by energy planners and aid agencies until very recently. As wood-fuel supplies are depleted, agricultural residues are increasingly being diverted for use as domestic fuel. Agricultural residues appear in very different forms, and ongoing development includes resource mapping, soil effects and suitable end-use and conversion technologies. The fuel source is suitable for local use only, since transport adds highly to the cost. The economics of their use are site-specific, but can often be very good. The main problems include seasonality of supply, competing end- uses (e.g. as fodder or fertilizer) and increased burdens for the poor created by the commercialization of earlier free goods. The impact on the national energy energy balance is very large, and is locally crucial. Development support may be channelled into resource mapping, ecological issues and support of the establishment of systems for the rational use of residues for energy.

Peat (Chapter 4) has a long history as a fuel in several northern countries. However, peat resources are widespread also in developing countries. Ongoing development includes industrial scale operations in industrialized countries, and a few technology transfer projects now also started in developing countries. Peat is a relatively localized fuel, and cannot carry large transport costs. The economics can be favourable in both large-scale industrial and small-scale manual operations, depending on the cost of alternatives. The main problems concern de-watering, the environmental effects of exploitation and combustion (particularly on a small scale), since peat is a smoky and difficult fuel. Impact on energy balances is limited. Development support may be channelled into resource mapping, technology transfer and adaptation for specific projects.

Briquetting of biomass (Chapter 5) of biomass for the production of more versatile fuels is a well-developed technology for large-scale use. Ongoing development includes cost minimizing, applications in agricultural systems, and small-scale applications using simple equipment. The economics are good, but very sensitive to transport costs and the regular availability of suitable biomass. In addition, the burning of briquettes or pellets may cause technical problems requiring specific equipment in certain applications. The impact on national energy balances is small, but can be locally important. Development support may be channelled into integrated pilot projects, probably on a large to medium scale, involving the simultaneous development of supply and end use technologies.

Charcoal production. (Chapter 6) Improved charcoal kilns come in many different designs, from improved simple earthen kilns to expensive brick and metal constructions. The technology is well developed. Ongoing development includes local adaptation, efficiency improvements and cost-minimizing. High efficiency through modern design could however, be achieved at the high social cost of displacing large numbers of existing small charcoal makers. The introduction of efficient metal and brick kilns would require major institutional change, and involve the establishment of a new commercial infrastructure. Support for the introduction of improved earthen kilns is another possibility, increasing efficiencies moderately while keeping employment intact. Major positive impact on deforestation is an attractive "by-product". The cost-effectiveness of most types of improved kilns is good, and most designs can be produced locally or by local industry, although investments are a problem for small producers. The impact on national energy balances can be significant. Major programmes sponsored by development support must be carefully evaluated from a social and resource availability point of view, and support may be channelled into the introduction of more effective kilns, evaluation of the establishment of large-scale periurban plantations for charcoal making etc.

Wood and Charcoal Stoves (Chapter 7). Today, most cooking in developing countries is done on the simple, traditional "three-stone" fireplace, or on simple charcoal stoves. These are usually neither energy-efficient, safe nor convenient. Improved stoves come in many varieties, most of which, however, have failed to be accepted by local users. Many different approaches have been tried, and following the intensive research and field trials in recent years, several promising prototypes and stove programmes are now being developed. Ongoing development includes local adaptation and efficient means for production and dissemination. The overall economics depend on a variety of factors such as the evaluation of health, comfort and other benefits in addition to energy efficiencies, as well as the actual or perceived cost for fuelwood now used for cooking and heating. (In fact, health problems associated with smoke from open fires may be a major cause of disease in developing countries.) The limited resources spent so far on most stove programmes indicate that stove cost may be one limiting factor in dissemination, and that the time-scale for wide acceptance of new woodstoves is long. The impact on the energy balance may become very significant at both the national level and the household level. Development support may be channelled to the local testing or implementation of adapted stoves; stove programmes integrating energy, health, comfort and development aspects, and improved contacts between the local and international level of stove development.

Biomass combustion in small scale industry (Chapter 8) is a well established technology though with plenty of room for improvement. Ongoing developements, mostly done in industrialized countries, are mainly concerned with improving fuel handling systems and energy efficiencies. Much of the equipment needed could be produced cheaply by local industry in developing countries. The overall cost-effectiveness is site- and application specific, but seems very favourable in many cases where an adequate biomass supply is available. The impact on the national energy balance is small, but the local impact on productivity may be of considerable interest. Development support may be channelled into pilot projects and information transfer to local entrepreneurs and organisations in developing countries.

Biogas (Chapter 9). This is a well developed technology used extensively in a few developing countries to promote health and produce fertilizer and energy. Ongoing developments include cost minimizing and increased efficiency and reliability. The economics are site-specific, depending on i.a. local construction costs, the combined availability of animal/human wastes and water, and the combined demand for all benefits delivered. The main problems concern the lack of infrastructure and technical back-up, as well as the use of non-reliable designs. The potential impact on the energy balance is small in most countries, but can be locally and (sometimes) regionally important. Development support may be channelled into dissemination programmes, including infrastructure build-up.

Draught animal power (Chapter 10) for transport, agriculture and energy is an ancient technology, used globally. However, in many countries this technology has been neglected, due to over-optimistic expectations for mechanization. Ongoing developments include adaption of implements, breeding programmes and infrastructure build-up. It is often very cost- effecitve provided that problems with animal disease and the need for an adequate infrastructure can be solved by suitable institutional and scientific back-up. The impact on the national and local energy balance is significant. Development support may be channelled into the build-up of local infrastructure, including training.

Alcohol fuels (Chapters 11, 12 and 14), ethanol and methanol, are well-established as commercial engine fuels. Fuel ethanol programmes exist or are under way in several developing countries, with varying economic results. Methanol for fuels is still limited to experiments in a few industrialized countries. Ongoing ethanol development is mostly concerned with industrial scale research on new manufacturing processes, the use of new raw materials and small scale processes. Most developing countries have to import the technologies for industrial ethanol production, which (with a few exceptions) is only economically made on a large scale. The economics are site and country specific, and depend on such factors as land use competition, government policy and subsidies, petrol prices, choice of technology, and the cost of using alternative resources.

Vegetable oils (Chapters 11 & 12) as fuel for engines is a well-known technology. Ongoing development is concerned with a number of prototype and field tests, mostly in industrialized countries, focussing on engine problems, fuel preparation and crop selection. Fuel production and engine modification can be made with local resources, although advanced technologies would yield better results. The overall economics are very site- and application-specific, and in most cases are not favourable compared to standard fuels. Other problems concern land use competition and serious maintenance problems with engines, if vegetable oils are used for a long period. The esterification of these might be one way out of those problems, but adds to cost. The impact on the national energy balance is limited. Development support may be channelled to selected pilot projects.

Producer Gas (Chapters 12 and 14) made from wood and charcoal was used extensively during World War II in both industrialized and developing countries. The technology is now being actively revised for the purpose of replacing oil for transport, shaft power and process heat. Ongoing development includes pilot projects, technology development and the adaptation of gasifiers to other fuels. Local manufacture occurs in Brazil, Philippines and India, and experiments are taken place in a large number of other countries. The cost-effectiveness is very sitespecific, depending on factors such as local manufacture and the costs of fuel, manpower, maintenance and of available alternatives. The impact on the national energy balance is small, but can be important for local productivity. Development support may be channelled to information exchange, technology development, pilot projects etc.

Steam engines and steam turbines. (Chapters 13 and 14) have a long history as a reliable but not very energy-efficient source of small scale power. Ongoing development includes a renewal of interest in developing countries, not least due to the flexibility of steam engines with respect to fuels and their general reliability. Steam engines are locally manufactured in Brazil and India, for example. The economics are often favourable, but depend a great deal on the cost of wood for fuel and of available alternatives. The impact on national energy balance is small, but can be important for local productivity. Development support may be channelled into hardware development, economic evaluations and pilot projects.

Stirling engines (Chapters 13 and 14) is an old, well established technology, that never became really widespread due to competition from internal combustion engines. Ongoing development includes new advanced designs as well as renewed interest in simpler designs. Stirling engines are reliable and adaptable to most local fuels. Local manufacture might be difficult. The economics are is usually unfavourable, but may be attractive in special applications. Problems include high costs for advanced models and lack of commercial hardware for simpler designs. The impact on the energy balance is insignificant. Development support may be channelled to prototype development and tests.

Solar water heaters (Chapter 15), mostly for household use, is an old widespread and well-developed technology. Ongoing development is mainly concerned with cost-minimizing and an increase in durability and quality, aimed at urban populations and institutional buildings, hotels, etc. The economics are often very attractive, especially for replacing electricity, when electricity prices are high (as they often are in developing countries). Energy impacts, mainly for replacing oil and electricity, may become significant even though costs will preclude access to large sectors of the population, especially in rural areas where there is little demand for hot water. Development support may be channelled to encourage local industry by using its products in development programmes.

Photovoltaics (Chapter 16) is a well-developed technology in widespread use for many special applications. Ongoing development includes cost reductions through higher efficiency, better manufacturing methods, etc. Manufacturing is advanced and on a large scale and is dominated by the U.S.A, Japan and Europe. Only a few developing countries (e.g. India) have production or assembly factories. Today, it is only cost-effective for special applications, such as remote communications, lighting, pumping, health stations, etc. Costs are generally expected to continue dropping to a level where it may become increasingly competitive with other power production. If sufficiently low costs can be achieved for photovoltaic systems, the potential market and the impact on energy balances may become significant. Development support may be channelled into pilot and demonstration projects.

Advanced solar heating systems (Chapter 17) for the production of low-grade process heat in industries are in the prototype and testing stage. Only a few installations exist. Ongoing development is mainly concerned with cost-minimizing and systems development, and most of it is done outside developing countries. The economics are favourable only in special cases and designs are very site and process-specific.

Solar water distillation (Chapter 18) is an old application of solar energy, and several designs have been tested for long periods for the production of potable water from salt, brackish and polluted water. The water produced is costly, and the overall economics depend on whether other ways of water supplying are possible. Ongoing development mainly concerns minimizing costs for large-scale units and increasing yields in smaller units through the development of new concepts. Energy impacts are insignificant but the availability of potable water can become extremely important locally.

Solar driers (Chapter 19) are still largely in the prototype stage, but some practical successes have already been achieved. Interest is growing due to the prospect of their improving crop quality and minimizing post-harvest losses, particularly for many cash-crops where the economics are sometimes very favourable. Intensive development and tests are now taking place in many developing countries. Solar driers are climate and crop specific, and only a few parts have to be imported. Impact on national energy balances is small but local impacts may become very significant. Problems arise from the low durability of materials in low cost designs, and the fact that performance is limited in the absence of electric fans. Development support may be channelled into continuing efforts to develop economic and climate- and crop-specific designs and support pilot programmes.

Solar cookers (Chapter 20) have been tried for decades as a substitute for fuelwood, but have never become widely accepted, mainly due to their impracticability and cost. Solar cookers with heat storage are even being developed to allow cooking after sunset. Interest is now generally declining except for a few special situations. Their cost-effectiveness is doubtful and their impact on future national energy balances insignificant.

Passive heating and cooling of buildings (Chapter 21). This uses old, traditional and well-developed techniques combined with modern engineering principles. Ongoing development is concerned with the adaptation of general principles to local building practices. The economics are usually favourable, compared to the cost of using commercial fuels for heating and cooling. Development support may be channelled into institutional build-up, together with information and training programmes, as well as applying those methods to aid financed buildings.

Small-scale hydropower (Chapter 22) is a well-established, widespread technology for local power production. Ongoing development is concerned mainly with cost reductions through local adaptation and simplified designs. Its main use would be in non-grid areas with adequate water resources. Other promising uses include water powered mills and pumps. The potential impact for local energy development is site-specific but can be significant, as shown by recent experience. The impact on national energy balance in most countries is, however, more limited. Problems concern partly lack of information on available concepts. The economics are very attractive if suitable sites are available and alternative energy costs high. Development support may be channelled to resource mapping, implementation and the support of local industry, for example.

Tidal power (Chapter 23) is a well-established technology for power production. The economics are very site-specific and few good sites may exist in developing countries. Exploration is done on commercial terms.

Wave power and OTEC (Ocean Thermal Energy Conversion) (Chapter 23) are both in the research and prototype stages. Ongoing development includes research and pilot tests on cost-effective designs and systems analysis. Current high costs and lack of experience have prevented any commercial applications so far, although several research projects, also in the waters of developing countries, have been suggested.

Wind power (Chapter 24) is an established technology for small-scale power production (and in the prototype stage for large-scale units). Grid connected small units are already commercially viable in some countries, and thousands have been installed. On the small-scale, up to some 100 kW, ongoing development includes the development of reliable and cost-effective stand-alone units (with diesel backup) as well as other very small systems for applications such as battery charging. The economics are site-specific, depending on wind resources and cost of energy alternatives. Impact on national energy balances is limited but can be locally important. Development support may be channelled into the introduction of well-tested systems along with resource mapping, for example.

Wind pumps (Chapter 26) is a well-established technology in widespread use. Ongoing development is concerned with modern, reliable light-weight systems adapted for developing countries. The economics are site-specific but generally good, given adequate wind resources. The prospects are usually better for the supply of drinking water than for irrigation, and also in areas without power grids. Local production is already taking place in a number of countries. The impact on the energy balances is insignificant on a national scale, but can become critically important locally. Problems include the difficulties of the commercialization of new products. Development support may be channelled into resource mapping, the support of local industry, prototype testing and the support of financing schemes for local farmers.

Hand pumps (Chapter 27) have a long history of use in developed countries, but have only recently been tried in developing countries in large numbers. Most trials have, however, met considerable problems, since designs were not adapted to the hard conditions met in rural villages but rather to family use in industrialized countries. The main problems are cost and ease of local maintenance. Development support may be channelled into development and commercialization of sturdy and easily maintained pumps as well as to the organisation of local maintenance.

Section I

BIOENERGY: PRODUCTION, CONVERSION, UTILIZATION

1 FUELWOOD: FORESTRY & AGROFORESTRY

Introduction

Until recently firewood and other plant fuels have been a prime source of energy for people. Industrialization used additional sources also, first wind and water power, then hydrocarbons to provide mechanical energy. While fossil fuels, themselves of organic origin, presently run most of the industrialized world, firewood is still the main source of domestic energy in most non-industrialized countries. Wood for energy also seems to be making a comeback in the industrialized world, being a conditionally renewable resource produced within the country of consumption in contrast to imported fossil fuels.

Only in the last decade has the role of fuelwood been fully recognised. Despite this recognition estimates of the production and consumption of fuelwood remain crude. On a global scale, fuelwood and other biomass account for some 15% of total energy consumption. In contrast, the cumulative share of hydro and nuclear power is six times smaller. In developing countries, biomass is the predominant fuel accounting for 43% of total energy consumption. But even in developed countries, fuelwood accounts for a significant proportion of the energy budget. The potential of fuelwood as a renewable energy resource is not recognised largely because market mechanisms currently undervalue the benefits obtained from renewable energy. The potential of forest energy resources can be gauged by comparing these resources with global hydro-electric potential; the estimated annual energy increment from designated forest areas is five times the capacity of the world´s hydro-electric potential.

Forest and biomass resources are often seen as the fuels of poor rural dwellers in the Third World. Increasingly, however, these fuels are the urban fuels preferred even in high income households in developing countries because supplies can be guaranteed. Additionally, fuelwood is increasingly used in the commercial sector, particularly for curing tobacco and tea, and in industry. It is also often used in processed forms such as charcoal, distillates and gas.

A number of developed countries derive a considerable proportion of their energy consumption from fuelwood: Sweden obtains 14% of its energy requirement, Canada 8% and Australia 3%. In the United States over 4%

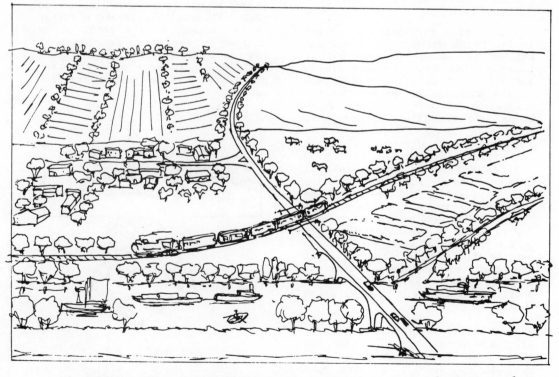

The Chinese "Four round-abouts": trees around houses and along roads, railways and channels.

of the total energy consumption is derived from fuelwood and several million households depend on fuelwood for domestic heating.

In many developing countries and in remoter areas of the developed world, wood is frequently the cheapest available fuel not only per unit weight but also per unit of heat. No special facilities are required for wood storage and, in addition, it is perfectly safe to store the energy supplies for long periods. When dried, fuelwood burns safely, although, with the latest technologies skilled labour is frequently required for efficient operation. Wood-fuelled power stations are also feasible, even utilizing advanced technologies like fluidised bed combustion.

Careful planning must be done if conventional forestry is to provide part of the future fuel requirements. Particular attention has to be paid, within the wood monoculture systems to obtaining a sustained off-take from the forest throughout the crop cycle. Above all, the low priority which is accorded to conventional fuelwood production, a low priority that is mirrored by the lack of invest-ment, should be reversed. There is no reason why the criteria, if not the actual methods applied to forest production for poles and industrial feedstock cannot be applied to for-estry production for energy.

Firewood and Charcoal Consumption: Use and Prospects

The estimated firewood consumption in the Third World in 1950 was 1.5 billion m^3, which was some 83% of total timber use in these areas, and about 90% of all firewood used in the world. The amount has in-creased by 26% over ten years, which keeps pace with the 25% growth in population during the same period. The total growth of the world's trop-ical forests is between 10 and 20 billion m^3/yr. In theory this is more than enough production capacity, but, in reality, the supply is ex-tremely uneven, and in vast areas firewood shortages are growing alarm-ingly. This is because forests per se are rarely used for energy - trees outside the forest are the principal source of fuelwood, although charcoal is often made from larger scale clear-cutting, particularly close to urban markets.

Figures on the magnitudes and trends of fuelwood supply and demand are of such an alarming character that most international agencies have given urgent priority to the prob-lems. For example, in a recent study of 15 developing countries, FAO found that the present rate of fuelwood

plantation establishment would have to be increased by more than 10 times to 70 000 hectares per year to avoid predicted shortages at the turn of the century.

The world's tropical forests and woodlands are currently being de-stroyed at an alarming rate. Of 1.2 x billion ha of closed forest and 0.7 billion ha of open woodland, making up 40% of the total tropical land mass, the area felled yearly is estimated to be 7.5 million ha and 3.8 million ha respectively, or an annual regression rate of 0.58%. In individual countries this rate can be much higher, e.g. 4.44% in the Ivory Coast or 2.73% in Costa Rica. Some of this area (possibly one third) regenerates to secondary forest or bush, but most is taken over by agriculture. Replanting does not keep pace with the deforest-ation: less than 1 million ha are planted annually, and only in tropical Asia is there an effective push for reforestation.

The average per capita consumption according to several different esti-mates is 0.5-0.8 m^3 per person each year. In Africa, firewood accounts for 58% of the total energy consump-tion. In Asia the corresponding value is 17%, which is a consequence of a severe shortage except in the humid forest zone. In Latin America, the low value, 8%, reflects a much heavier reliance on fossil fuels, and the more advanced industrializa-tion. In many poor countries, up to 90% of the energy comes from wood fuel. Charcoal represents a rather small volume, with a total of 16 million t per year, but plays an important role in urban areas. Char-coal will grow at a rate that ap-proximates to the pace of urbani-zation.

THIS IS 20 KG OF WOOD; ONE FAMILY'S ANNUAL SUPPLY IS TYPICALLY 200 SIMILAR LOADS

5

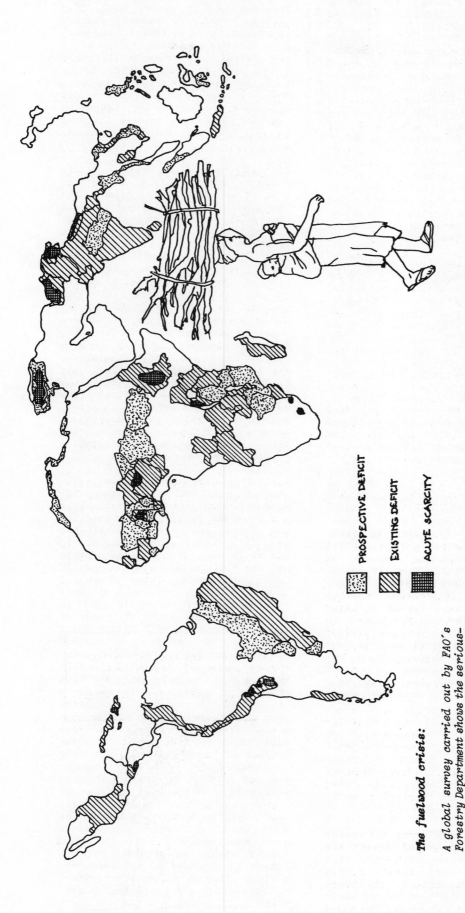

The fuelwood crisis:

A global survey carried out by FAO's Forestry Department shows the seriousness of the fuelwood situation in developing countries. The map shows the extent of the affected areas and the scale of the scarcity problem. Women in the Sahel region (far right) have to trek many kilometres in order to gather enough fuel for cooking and heating. It may take 300 man-days of work to satisfy the annual fuelwood needs of a single household.

PROSPECTIVE DEFICIT

EXISTING DEFICIT

ACUTE SCARCITY

Source: FAO (1981) Map of the Fuelwood Situation in the Developing Countries.

Leucaena plantation.

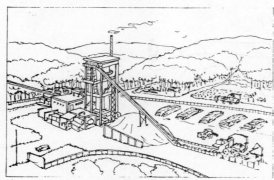

Outline of woodfuelled power station.

The emphasis on traditional uses of wood for energy demand should not be allowed to distract from the potential for industrial application of energy forestry. In many industrialized countries, like the Netherlands and Sweden, the potential of intensive low forest farming has been explored for electricity generation and production of petrol substitute. In the Philippines large scale Leucaena plantations are projected to feed several power plants and in southern Brazil, extensive Eucalyptus plantations (a total of 1.5 x 10^6 ha) provide charcoal required for steel mills. In other places, like Zambia, tobacco growers maintain their own plantations to produce fuel for the drying and curing process. The technology employed may be highly specialized, or common plantation technologies of pulpwood production as in the Brazilian case.

General strategies have been proposed to deal with the problem of increasing fuelwood scarcity. These include:

Reducing consumption: Most firewood today is burned in simple, three stone fireplaces with limited efficiency. The large-scale introduction of improved stoves is therefore one possibility to reduce consumption. Stove programmes are described in Chapter 7.

Making better use of available timber. While areas around the urban centres regularly face shortages,

remote forest areas may have vast potentials in the form of logging residues and other waste. One bottleneck is usually transport, another is management and marketing. Charcoal production brings down haulage costs considerably and urban dwellers are often accustomed to this handy fuel. The inefficient kilns now used, however, make cooking with charcoal much more wood-demanding than using fuelwood directly. Improved charcoal methods are described in Chapter 6.

Plantations. In a natural mature forest, usually only a few m^3/ha. This is an evolutionary climax, the "purpose" being to perpetuate the biotope, not to produce for the benefit of man. On the other hand, in plantations and managed natural forest where species and edaphic conditions make it possible, the growing potential of the site can be exploited and substantially higher yields obtained. In very dry areas, it may never exceed 2-5 m^3/ha, but where water is abundant and soil is good, almost a tenfold increase in production can be achieved. Plantations are simple to manage and yield only the desired product. High technology energy farming and conventional forestry plantations have much to offer. Before proceeding to a discussion of other wood growing opportunities it is worth considering plantation possibilities in some depth.

While tree plantations for the provision of industrial wood, e.g. pulpwood and sawmill timber, are familiar all over the world, plantations for energy purposes, even household use are, in general, of a recent origin except in some North-western Europe and Far East farming systems. Basically, an energy forest does not have to differ at all from a pulpwood plantation: the target in both is to maximize bulk growth with heavier trees, and operate with fairly short rotations compared to stands aimed at producing quality timbers. For energy purposes, smaller sizes can be used than for pulp where the wood has to be debarked. It is worth stressing that fuelwood production does not necessarily have to be separated from other wood production: first thinnings from industrial plantations will make fuelwood, and light construction timber for domestic use may well be extracted from most energy plantations. At the end of a rotation, much of the logging residues can be either collected for fuel directly, or be carbonized. The important thing is to allow leaves, twigs and possibly bark to return to the soil to maintain fertility. Removal of the low nutrient wood is of little ecological significance.

A more important thing to bear in mind when planning fuelwood provision through tree plantation is the social aspect; who owns and controls the plantation? Where will it be located? etc. The way of handling and managing a ten-thousand hectare plantation differs radically from the procedure where ten thousand farmers grow their one-hectare wood-lots!

Other Opportunities

Shelterbelts, greenbelts and planta-tions along roads and water are, along with the new push for social (community) forestry, a less stringent version of the plantation strategy rather than being a development of agroforestry. In many senses this is a functional definition – foresters are rarely involved in agroforestry but practic a whole range of con-ventional forestry programmes from plantations to community woodlots.

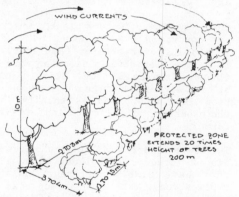

Tree shelter-belts can be used as windbreaks.

Many agricultural areas suffer from wind erosion and it has long since been an established practice to plant trees as windbreaks between fields. In modern times, this has been done on a large scale, for example, in China where, for example, the formerly deforested Shantung province has changed into wooded farmland. These plantations are at the same time a very important source for small timber and firewood. Depending on land tenure systems, these plantations can be run by landowners, cooperatives or government authorities; usually the actual work is done by local farmers.

Greenbelts designate larger zones of tree plantations or other permanent vegetation. One famous greenbelt has been planted in Algeria against the desert; another, 1,500 km long, has been established in China like a Great Wall through a giant operation of mass mobilization. The size of these undertakings usually require central planning and management, while the execution necessarily will involve much local participation.

In China, as in other countries, however, there seem to be serious difficulties in maintaining efforts on this scale.

A very interesting example of a green-belt established by private initiative for basically economic incentives is the Eucalyptus zone around Addis Ababa in Ethiopia. The whole city, situated in a landscape which at the turn of the century was totally without forest cover, is now embedded in forest, with streets and gardens looking like clearings in a dense woodland, according to recent aerial photographs. This was created to supply fuelwood for the growing im-perial capital, starting with some hectares established by private land-owners, later encouraged by tax relief and free seed distribution, but never with aid of, or organized by, any Forest Service. Following this suc-cess, Eucalyptus is now grown every-where in the countryside and contri-butes decisively to the wood supply.

Initially in Gujarat state and then in other parts of India, the Forest Service took the initiative to utilize roadsides for plantations. In 1978, one third of the State´s 17 000 km of suitable roads and canals were planted with a (mostly) triple line of trees on each side, and the work has continued to the extent of 1500 km per year. The local people are entitled to mow fodder grass under the trees, helping to keep down weeds, and will ultimately have access to the timber. In China, the "four around" programme (trees around houses, between fields, along roads and streams) has achieved much in some places: in one county alone, Chunhua in Shaanxi, 15 million trees were planted. Other related efforts have, however been less successful.

As a response to the difficulty in forestry of establishing tree planta-tions under many land tenure condi-tions, the concept of village forestry or community forestry has taken form during the last decade or so. The important factor in this forestry model is popular participation. This is nothing new in parts of northern Europe or Japan, but gen-erally forestry has been considered to be the business of forestry pro-fessionals, while the farmers have been growing food and fodder.

There are as many varieties of village forestry as societies where it is practised, and crops differ as well as the relationship between the local community, the individuals and the government via its Forest Service. Village forestry may also be a part of a greater, government coordinated rural development programme. For the production of locally consumed firewood, village forestry is attractive: individual farmers or the community as a whole establish and maintain plantations, on marginal land within reach of the village. Produce may either be utilized solely by the village or alternatively shared with the Forest Service. In some cases, both timber and firewood as well as other forest products like honey, gum, tanning bark or fruits are produced. Some systems are totally devoted to firewood production and a part of the product may even be marketed.

Village forestry projects are now being started all over the world, and the basic principles are presented in two FAO publications: "Forestry for Local Community Development" and "Forestry and Rural Development". Methods and social strategies are discussed in much detail with numerous examples of successful projects. On the other hand, village or community forestry has not been successful in many places where it has been tried, depending on e.g. land tenure problems, distribution of produce etc.

In South Korea, the Government initiated a nationwide programme in 1973 to create village fuelwood plantations. In every village, a Village Forestry Association was organized, which together with the Government Forestry Office, established the leguminous shrub, Lespedeza on marginal lands. In 1975, 4000 ha were planted, and the yearly area is now even larger. Lespedeza is managed in a coppice system (see below), and is harvested every year. As a result of this and other plantation programmes, the barren hills are now forested, and rural firewood is available.

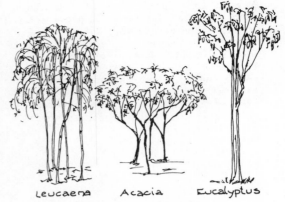

Some common fuelwood species.

In the very densely-populated island of Java, almost deprived of its original lush forest cover, the Central American bush, Calliandra has provided a solution to a serious shortage of firewood. Introduced 25 years ago, villages with only a little government support started planting this leguminous "wonder plant", which now covers 30 - 40 thousand ha, and is gaining ever greater popularity. In many villages it is becoming something of a cash crop, being sold at city markets, but the whole production is run by, and is totally in the hands of, villages and individual farmers.

The above examples all rely on conventional forestry techniques although they have widely different social implications. The interesting question is, what is possible with both a different technical and social configuration? The recent advances in agroforestry mark such an interesting departure from previous practice.

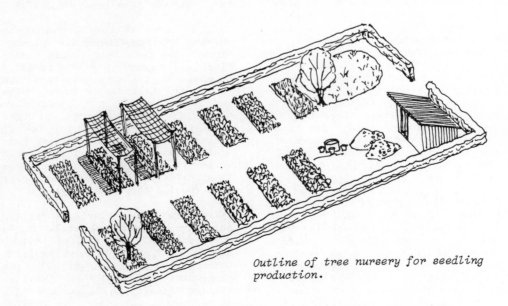

Outline of tree nursery for seedling production.

Agroforestry – Definition and Scope

"Agroforestry" as a term and concept has become firmly established in international development and rural science terminology in a surprisingly short time. It was not until the late 1960s and early 1970s that the word started to appear in writing, then mainly in forestry circles, as a wider, collective name for the various forms of "taungya" afforestation systems practised for a long time in many tropical countries.

Taungya farming.

Since 1977, at least a dozen international meetings have been held which specifically focused on agroforestry, and major UN conferences held during the last five years invariably mention the value of agroforestry in their resolutions and recommendations. All principal donor agencies and development banks have recently adopted agroforestry in their lending and spending programmes, and international and national institutions, journals, consultants and publications featuring agroforestry are mushrooming all over the world.

There are probably many interrelated reasons for this explosive increase of interest – one no doubt being the built-in dynamics of fashion. But there is more to it than that. Agroforestry is the first serious concept that builds on a synthesis of much of the practical experience and scientific knowledge acquired over the past decades in tropical agriculture, forestry, ecology, soil science and rural socio-economics. Our increased understanding of both social and ecological dimensions in tropical environments and the frequent disappointments and failures encountered when trying to implement "modern" agricultural and forestry technologies in ecologically sensitive and socioeconomically complex situations, have led to a realization that alternative approaches to land development must be given a higher priority. It is, of course, not by chance that the increased international concern about deforestation, desertification, food and fuelwood shortages and other social and ecological problems which originate from man´s use and misuse of land resource coincides with the increased interest in agroforestry.

So, what is agroforestry? Many definitions have been proposed, good and bad ones, wide and narrow. Many, unfortunately, make subjective and presumptuous claims that agroforestry, by definition, is a superior and more successful approach to land development than anything else. It would not serve any purpose here to list a large number of definitions but a good general definition is as follows:

"Agroforestry is a collective name for land use systems and technologies where woody perennials (trees, shrubs,

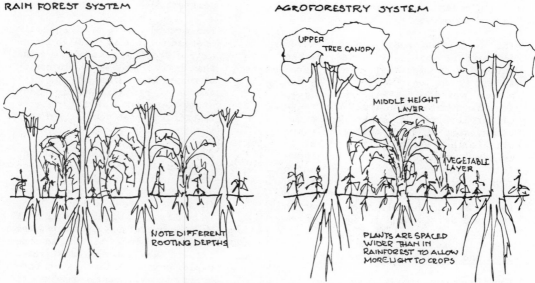

RAIN FOREST SYSTEM

AGROFORESTRY SYSTEM

UPPER TREE CANOPY

MIDDLE HEIGHT LAYER

VEGETABLE LAYER

NOTE DIFFERENT ROOTING DEPTHS

PLANTS ARE SPACED WIDER THAN IN RAINFOREST TO ALLOW MORE LIGHT TO CROPS

Agroforestry systems are often modelled on natural ecosystems, such as rainforests.

palms, bamboos, etc.) are deliberately used on the same land management unit as agricultural crops and/or animals, either in some form of spatial arrangement or temporal sequence. In agroforestry systems there are both ecological and economic interactions between the different components."

Trees may often provide many end-uses.

This definition outlines the broad boundaries of agroforestry and the typical characteristics of such systems:

1   Agroforestry normally involves two or more species of plants (or plants and animals), at least one of which is a woody perennial.

2   An agroforestry system always has two or more outputs.

3   The cycle of an agroforestry system is always more than one year.

4   Even the simplest agroforestry system is more complex, ecologically and economically, than a monocropping system.

The definition and its implication in terms of what is included, brings out one very important point — agroforestry is only a new word, not a new practice. The novelty lies in the realization that all these different land use systems and practices, some of which have traditionally fallen into the field of horticulture, some into agriculture, some into forestry, and many of which have not attracted any systemic attention at all, possess certain common features which hold great promise for land development in the tropics.

Aims and Place of Agroforestry

The aim and rationale of agroforestry systems and technologies are to optimize positive interactions between components (trees/shrubs and crops/animals) and between these and the physical environment in order to obtain a higher total, more diversified and/or more sustainable production from available resources than is possible with other forms of land use under prevailaing ecological and socio-economic conditions.

In order to understand the theoretical potential of agroforestry to fulfil these aims, it is useful to identify the distinct physical and social features of most tropical land and its users. Although these features are quite well known, often being featured in technical literature, their implications in relation to land management are, distressingly, often ignored.

Soil condition and climate show enormous variations within the tropical regions - soil age, origin and its physical and chemical properties vary as much in the tropics as they do in temperate regions. Likewise, annual rainfall spans from practically nil to well over 10 000 mm.

For much of the tropical world one permissible generalization can, however, be made: there is an inherently low land productivity potential. The reason for this is to be found in three environmental conditions which are typical for the tropics and which must be clearly understood when using tropical land:

1   The first is the continuously high rate of photosynthesis, due to the high level of insolation which also promotes high rates of negative biological, physical and chemical processes, such as decomposition and oxidation of organic matter, weathering of minerals, production of pests and diseases and evaporation of moisture.

2   The second and most frequently ignored feature is the high intensity of tropical rainfall, meaning that daily intensity, the intensity of individual storms and the total amount of rain falling in heavy storms are considerably higher in tropical than they are in temperate regions.

3   The third feature is the strong, sometimes extreme, dependence on organic matter for fertility and the maintenance of structure in most tropical soils. This, in combination with the fact that organic matter decomposition rates are five times higher in the tropics than in temperate regions, explains the rapid loss of fertility and structure of tropical soils when organic matter is not continuously added.

An understanding of the dynamic inter-action between these factors, and between them and land management practices, helps in the interpretation of most of the physical land use problems in the tropical developing world. It also helps in identifying the two main principles for successful management of most tropical lands, regardless of the level of capital input available, regardless of whether it is small or large scale enterprises and regardless of whether one produces crops, animals or trees. These are:

1   Protect the soil against direct rain impact.

2   Maintain soil organic matter levels.

Understanding these two seemingly simple principles and realizing that land must, whenever possible be managed to yield more in order to cope with a growing need for basic necessities, are the technical and social points of departure for the agroforestry approach to land use. Trees and shrubs possess many features that can help the land user to adhere to these principles. The most apparent ecological potential for an agroforestry approach to land use exists in areas where soil fertility is low and depends mainly on soil organic matter, where erosion potential and the incidence of surface soil desiccation are high. On such marginal lands, which cover by far the largest amount of land in the tropics, the deliberate use of woody perennials may, if properly integrated in the land use systems, enhance both land productivity and sustaina-bility. The less capital and tech-nology inputs there are available, the more appropriate is the use of trees and shrubs to enhance organic matter production, to maintain soil fertility, to reduce erosion and to create a more even micro- climate.

The potential for agroforestry is however, by no means restricted to marginal lands. Some of the most successful small farmers´ agroforestry practices in the tropics are in fact found on high potential, fertile soils. A further role of agrofor-estry, equally applicable on marginal and rich land, exists in socio-eco-nomic situations where land tenure and/or lack of rural infrastructure (commmunications, markets etc.) and cash incomes make it vital for people to produce most of their requirements for food, fodder, fuel, shelter and cash from a limited geographical area.

In conclusion, there is little doubt that an agroforestry approach to land development, i.e. a deliberate integration and use of trees and shrubs in agricultural and pastoral landscapes to achieve certain protec-tion and production benefits, is theoretically sound. The easily observed negative effects of devege-tation and absence of trees and shrubs underline this. It must be remembered, however, that an agroforestry approach is not always economically and so-cially feasible when actually applied to practical situations.

Potentials of Agroforestry

The potentials of agroforestry can be assessed from three different angles:-

1) General and theoretical

Erosive rains, organic matter de-pendent soil fertility, an increasing fuelwood scarcity and a lack of cash and infrastructure among the vast majority of tropical land users, are some of the most relevant ecolo-gical and socio-economic arguments for tree integration into farming and pastoral areas.

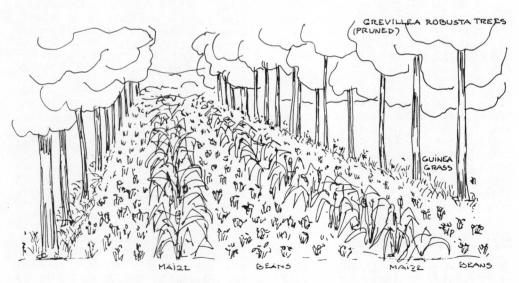

One example of an agroforestry scheme, combining trees and crops.

Another general factor that positively affects the potential impact of agroforestry is the interest and enthusiasm that the approach has aroused in both national and, particularly international, development and technical organizations. Although enthusiasm still far outweighs knowledge and experience on how to realize the potential of agroforestry, there is no doubt that the psychological foundation, including readiness to allocate substantial resources for a major development of agroforestry is there. Many development banks and most major bilateral aid agencies, give high priority to rural tree planting including agroforestry, in their policies and programmes.

2) Technology potential

The most striking potential of agroforestry lies in the practically unexplored field of technology development, e.g. genetic improvement of multipurpose woody species. In agriculture, horticulture and forestry, systematic and determined efforts to improve desired characteristics in crops through selection and breeding have achieved remarkable results. The "green revolution", a major part of which has been the development of improved varieties of rice, maize and wheat, is probably the most widely known example. No less spectacular results have been obtained in forestry and horticulture. There is no scientific reason why selection and breeding for improving features desirable in agroforestry such as fodder and fuel quantity and quality, rooting characteristics and phenology favourable to interplanting with annual crops, nitrogen-fixation, pest resistance, drought resistance, etc., could not result in similar successes.

The potential gains of systematic breeding of agroforestry species has been emphasized by the World Bank which has declared a priority interest in supporting this work. (Thorough discussions on the potentials and problems of genetic improvement of tree species for, among other purposes, agroforestry is found in the references).

Another potential of agroforestry lies in the almost unlimited scope that exists for innovative and imaginative development of new technology packages. The wide scope that exists for addressing particular land productivity problems, e.g. soil erosion, organic matter and fertility, drought, seasonal fodder shortages, fuel and building pole needs, etc., by combining trees and shrubs with desired characteristics in suitable spatial or temporal arrangements with annual crops and/or animals, is a great

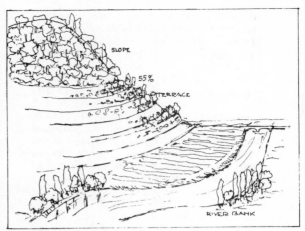

Agroforestry and soil conservation are often complementary.

challenge to research and development organizations.

3) Adoptability into farming systems

It is difficult to assess the potential for agroforestry systems or technologies to be adopted on a large scale by farmers and pastoralists. If, however, the relevant land use problems have been identified in dialogue with the concerned land users, and an ecologically, economically and socially well adapted agroforestry "solution" is demonstrated to be feasible, there is no reason to believe that agroforestry will not be adopted. The crucial thing is to find locally acceptable solutions to problems which are perceived and realized by the farmers. A "diagnostic and design" methodology to enable research and development agencies to do this is being developed at (ICRAF), the International Council for Research in Agroforestry, Nairobi. One factor which will always facilitate the adoption of new technologies by farmers is the demonstration of early cash returns. Today´s situation of a rapidly spreading fuelwood shortage and, as a result an increasing commercialization of firewood and charcoal, may well turn out to be a blessing in disguise from the point of view of eventual adoption of better land use. If fast-growing species, which also have, for example, soil-improving qualities, are marketed as a cash-crop, the adoption process may be very simple. It will, of course, still be up to development and extension services to demonstrate the best location and management of the trees in the farming system in order to obtain the optimum benefit from them.

Constraints of Agroforestry

The many constraints and problems encountered in the development and adoption of agroforestry systems and technologies are certainly not always easily solved. They are summarized and briefly discussed below:

General Constraints

Among the most obvious general constraints to agroforestry contributing significantly to increasing productivity of tropical lands, are the magnitude and number of problems involved. Hundreds of millions of farmers and landless people spread over vast expanses of tropical lands, mean that physical as well as socioeconomic limitations to rational land use are legion. Rapid population growth, insecure land tenure, erosion, droughts, floods, declining soil fertility, lack of infrastructure, political instability, illiteracy and other development problems often characterize those vast areas where agroforestry approaches to land use have a potential role to play. It is obvious that agroforestry development can never be seen in isolation from general social and physical development problems.

Development Constraints

The development of agroforestry systems and techniques requires an integrated and multidisciplinary approach at many levels. Education at technical and professional levels is predominant only along traditional disciplinary lines, e.g. forestry, agriculture, animal husbandry. Similarly, research institutions are normally mandated to work on strictly discipline-oriented problems. At government and administrative levels, rigid boundaries are normally maintained between departments dealing with different aspects of land use, boundaries often being strengthened by increasing competition for scarce development resources.

An example of Agroforestry, involving tree planting and cattle-raising.

Land legislation and supervision of laws often reflect the administrative division of separate laws governing "forest" land and "agricultural" land. Some countries have laws making all trees government property, including planted ones. It goes without saying that such laws efficiently prevent the success of any attempt to convince farmers to plant trees.

With regard to international agencies, e.g. UN bodies, development banks, bilateral aid agencies, etc., most of which strongly promote the ideas of agroforestry and of an integrated approach to land development, traditional disciplinary boundaries still impose severe constraints. Agroforestry is often, as in the case of national institutions, considered as a branch of forestry or even "environmental conservation", which means that resources for its promotion and development are allocated via forest divisions or departments. This, unfortunately means that international research and development funds to agroforestry are, considered relatively, very limited, i.e. limited in relation to the substantial funds channelled through more prestigious agricultural divisions.

Implementation Constraints

Implementing agroforestry on a large scale in rural areas will run into many problems, such as how to reach farmers with information, advice and services, how to convince farmers to adopt new land use practices and abandon old ones, and for the farmers, how to overcome management constraints. Reaching farmers is a problem of extension. This is relatively easy in well established and relatively prosperous agricultural areas where physical and administrative infrastructure is well developed. It is considerably more difficult in the vast expanses of marginal, converted lands where the need for agroforestry solutions to land problems is greatest. Lack of roads, lack of funds, and extension services not competent to deal with integrated land development problems, may impose serious restrictions in this respect.

Even if farmers are reached they need to be convinced about the benefits of new technologies. It may be relatively easy to introduce new and "better" species, or to make farmers adopt marginal improvements in management practices, in areas where trees are already an integrated part of the land management systems, but it may be more difficult where managed trees and shrubs are novelties. The time between planting a tree and achieving measurable benefits from it involves risks that farmers with limited resources may not be prepared to take. Similarly, it will be difficult to convince land users to make long-term investments where land tenure is uncertain. Social and cultural attitudes and beliefs may pose serious constraints at the local level.

Finally, there are many down-to-earth management problems to overcome before functioning agroforestry systems can be implemented. Raising plants as well as establishing, protecting and managing them, requires skills that farmers might not possess. Water availability for nurseries, protection of young plants against domestic animals, increased time needed in managing more than one production component, and so forth, will, in many cases, require additional resources (both labour and capital) which may be outside the limitations of many poor farmers.

Agroforestry and Fuelwood

Looking at the developing world in toto the FAO estimates that presently 112 million people live in areas with an acute scarcity of fuelwood. 1179 million experience a deficit, while still another 323 million are likely to do so in the very near future. Extrapolating 20 years into the future, the FAO estimates that close to 3 billion will suffer a moderate to severe shortage of fuelwood unless drastic measures are taken. "Drastic measures" expressed in conventional terms of area of closed fuelwood plantation, mean that 50 million hectares of such plantations will have to be established between now and the end of the century to ensure a reasonable degree of self sufficiency. This obviously cannot be achieved by relying on donor assisted projects (the cost of establishing such plantations conventionally approximates at US $ 50 billion) and/or existing forest authorities in developing countries (who, at most could only achieve 5-10% of the need). Only through large-scale voluntary adoption of tree planting by farmers and rural populations is there any theoretical hope of meeting the needs. In addition, as pointed out below, vast energy plantation may not be the answer to fuelwood shortages, since most fuelwood is taken from trees outside the forests, from shrubs etc.

There is a much greater hope of satisfying fuelwood requirements if well conceived, locally adapted agroforestry technologies are developed and promoted, rather than concentrating all national and international efforts on trying to persuade farmers to grow single-purpose, fuelwood species. Understanding the whole range of problems facing the farmer in his work so that all his basic needs i.e. not only fuelwood, may be satisfied from his land, is the key to successful agroforestry intervention. For example, it will probably be much more attractive to a farmer to accept the introduction into his farming system of tree or shrub species which, at the same time as it produces fuelwood, also helps in alleviating other problems perceived by the farmer, e.g. production of dry season fodder for his animals.

Examples and Experience of Fuelwood Production in Agroforestry

The examples of fuelwood production in farm situations discussed below are not necessarily the most optimal agroforestry solutions to fuelwood and other problems, but they do serve to show the great variety of practices already in use and, in some cases, also the production potential.

Densely populated agricultural landscapes with and without tree cover.

There are three general possibilities for spatial arrangements of trees and shrubs on farms:

1   <u>Closed tree stands</u>
    woodlot, farm forest
    home garden
    woody fallows (rotated with crops)

2   <u>Tree rows</u>
    wind break, shelterbelt
    living fences, hedge
    "alley" rows in farm fields,
    road and river planting

3   <u>Open tree stands</u>
    shade trees over agricultural crops, shade trees on grazing land.

In many traditional agricultural systems, local farmers have been pragmatic in their choice of how they integrate trees in their farms. Within one area, several different combinations may be found. An example of this diversity in two Latin American countries is given in the Table.

Table	The presence of trees on small farms in costa Rica (10 ha) and Nicaragua (20 ha). Average presence (% of all farms). Sources: Lemckert & Campos (1981) and Jones & Otarola (1981).	
	Costa Rica	Nicaragua
Parts natural	24	No data
Shrubs	27	No data
Trees, natural regeneration	40	42
Trees, planted	24	
Fruit-trees	98	78
Shade trees (for coffee and cocoa)	45	11
grazing land	40	4
Living fences	85	50

Generally speaking, the more or less closed forms of agroforestry are found in the humid tropics, while hedges and scattered trees are found in semi-arid and arid regions. An example of such open systems are the farmed parklands in West Africa. In these areas, 14-18 different tree species can be found scattered on agricultural land, covering normally between 5-25% of the ground with their crowns. Also on traditional agricultural lands in the highlands of Kenya, trees cover 7-17% of the ground. Windbreaks are mostly found in areas where strong, dry winds are dominant. Tree rows are used in irrigation areas. For example, in different parts of South East Asia, farmers plant <u>Sesbania grandiflora</u> or <u>Combertum quadrangulare</u> around their rice fields for woodfuel production.

Despite the traditional character of these systems they are not static. Often exotic trees, originally introduced by the forest service or plantation holders, have been integrated in the local farming systems. The original function of the trees, i.e. the function for which the Forest Department used them, may have become less important, but the trees do play an important role in the local wood production. Examples of trees which were originally used for shade are <u>Grevillea robusta</u>, <u>Cliricida sepium</u> and <u>Erythrina spp.</u>, which now can be found in different areas of South East Asia and East Africa on farmland. The success of these spontaneous introductions is determined by their quicker growth compared with local species, their multifunctional properties and their compatibility for integration into the agricultural systems (small and open crown, soil improvement properties, etc).

Living fences, can also provide e.g. fodder and fuelwood.

Data available at present indicate the importance of wood production on agricultural land. In Pakistan, more than 50% of all local construction wood and woodfuel comes from farmland. In Thailand, 57% of all woodfuel originates from trees outside the forests, while in Tunisia about four-fifths of all woodfuel is derived from shrubs and treecrops. Especially in the production of woodfuel, trees outside the forest play an important role. In most tropical countries, woodfuel consumption lies between 0.5 and 2.0 m³ per person per year. This consumption is often ten times as high as the use of wood for other purposes. For example, the yearly consumption of local construction wood and woodfuel is respectively 0.05 and 0.7 m³ per person in Indonesia and 0.07 and 1.0 m³ in Tanzania. For all tropical countries together, the woodfuel consumption is 84% of all wood consumed.

Most woodfuel is collected and used for home consumption. In Costa Rica and Nicaragua, only 5% and 24% respectively of the small farmers buy their wood on the market, while in rural areas on Java 70-98% of the households collect their own woodfuel and in the Philippines the estimate is 67-80%. Most of the woodfuel is collected within walking distance from the homestead, and, as the majority of the rural population does not have a forest within a distance of 5-10 km, it is understandable that by far the majority of the woodfuel is obtained from trees outside the forest. The relative importance of agroforestry for the production of woodfuel is illustrated with some examples in the Table.

Table   Importance of agroforestry as a woodfuel source in different areas.
         Sources: Bajracharya, 1980, Simon, 1981, Wiersum, 1982a, van Gelder (unpubl.)

			Percentage produced of total used
Indonesia	West Java	Home garden, agricultural land	63
		Plantations and forests	37
	Central Java	Home gardens	55
		Agricultural land	33
		Others	12
	East Java	Home gardens and agricultural land	60
		Village forests	23
		Others	17
	Bali	Home gardens and agricultural land	60
		Village forests	23
		Others	17
Nepal	Pangma village	Farm land	14
		Farm forests	22
		Government forests	64
Kenya	Whole country	Agricultural land	47
		Rangelands	25
		Forests and others	28

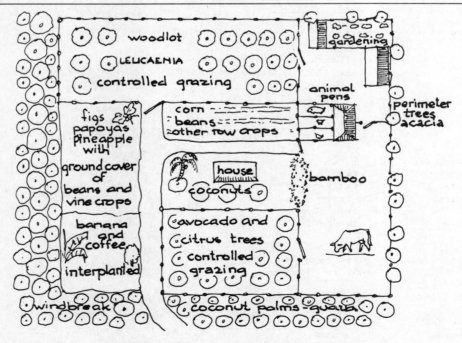

Example of agroforestry on a small farm.

What is the production capacity within agroforestry? Until recently very little attention has been given to this aspect and only incidental figures are known. Within these production figures, a distinction should be made between figures on actual yield and figures on potential yield. Most production figures given are on actual yield, because this is relatively easy to measure. The potential yield is much more difficult to determine, as it depends on a number of factors, such as which species are used, at what age trees are cut, and in what spatial arrangement they are grown. Another important problem in assessing yield is that existing forestry mensuration methods concentrate on production of large sized stemwood, whilst for the production of woodfuel, the total biomass production is of interest – particularly that of branches and small poles. In most tropical countries, the majority of woodfuel used in household consumption consists of branches and small poles of 2 to 5 cm in diameter. With an insufficient supply even smaller sizes are used. This woodfuel is often obtained through pollarding or coppicing. Only recently has forestry research started to pay attention to the methodological problems of production by these methods.

Present data on wood-production within agroforestry are only rough estimates. In the Table, a number of such estimates are presented. From the Indonesian data it can be concluded that about 200 to 300 trees are sufficient to produce the required quantities for one household. Also, for the Usambara mountains in Tanzania, it has been calculated that about 300 trees (Acacia mearnsii) have a sufficient production for one household.

Not just conventional wood-producing tree species are used for fuel production. In many areas branches or dead wood of fruit-trees and other commercial tree crops are also used. In Kenya and Cost Rica, for example, coffee shrubs produce 110-1300 kg/ha/yr of pruned branches, while in Costa Rica the Cordia alliodora shade trees above the coffee produce between 8.2 and 12.3 m^3/ha/yr of wood. In many countries, the wood of certain fruit trees is also in high demand as construction wood, for example Artocarpus indica and the coconut palm. The latter also produces woodfuel from its large leaves and its nuts. In the Philippines, it is estimated that the household requirements for one family can be met by 160 palm trees.

The few figures available at present concerning productivity are, at a first glance, surprisingly high in comparison with wood production figures from forestry. However, the data are not unrealistic since production conditions in agroforestry situations differ considerably from those in a forest. Firstly, the agroforestry data normally include small wood sizes (branches), which generally are not included in wood production figures from forests. Second, forest production figures are often from more marginal soil than the agricultural lands. Trees on agricultural land often profit from agricultural management practices such as manuring, fertilizing and irrigation. When planted with wider spacing, as is often the case on agricultural land, individual trees receive more sunlight than those in a closed forest, resulting in a relatively higher production per tree. The combined effect of these factors is illustrated by data from Rwanda, where it was found that

Table   Estimated wood production within agroforestry practices.
         Sources: (1) Simon, 1981, (2) Singh & Swarup, 1980, (3) Ben Salem & van Nao, 1981,
         (4) van Gelder & Poulsen, 1982.

		Standing stock (trees/ha)	Production capacity (m^3/ha/yr)	Yield (m^3/ha/yr)
Indonesia (1):				
Gunung Kidul	All farmland	33-142	3.7-7.6	
Mangelang	All farmland	191	4.5	
Surakarta	Home gardens	322	7.2-8.9	7.3
	Dry lands	122-150	2.6-3.7	5.6
West Java	All farmland		4	
India (2)				
Himachal Pradesh	Farm and rangeland (excl. fruit trees)	20.5		2.0
South Iran (3)	Hedges and fallowland	200		1.8
Kenya (4)				
Highlands	All farmland	10.6-30.5 (m^3/ha)	0.7-2.0	1.2

four-year old trees planted on agricultural land produced 3.8 times more wood and 2.9 times more leaves than the same tree species planted in a dense stand on forest soil. Finally, it is often more advantageous in agroforestry to use fast-growing species cut on shorter rotations than it is in forestry. Coppicing or pollarding techniques are frequently used. As the root system is not removed, the trees do not have to reproduce a new rooting system and the absorption capacity of the roots to take up water and minerals stays intact. Therefore, greater above ground production is possible.

Such high productivity is confirmed by production figures with yield data from small farm woodlots managed under coppicing systems and which often consist of those species used on agricultural land. In such small woodlots, production is often higher than the 15 to 30 m^3/ha/yr used as a standard of "fast growth" by most tropical foresters. In Java, for example, yields of 27 to 30 m^3/ha/yr for the coppicing tree species <u>Calliandra callothyrsus</u> have been measured in farm woodlots, while in the Philippines woodlots of coppicing <u>Gliricidia sepium</u> and <u>Leucaena leucocephala</u> show yields of 23 to 40 and 28 to 35 m^3/ha/yr, respectively.

Socio-Economic Characteristics of Agroforestry

Discussion is only just beginning on the socio-economic characteristics of agroforestry. The observations and calculations here should be taken as indicative rather than formal formulations. The indicative evidence does, however, suggest the need for careful project planning.

Trees are present within current farming systems. Tree species differ greatly within very short distances and often local arguments and reasoning can be found for the reasons for those differences. It is difficult to generalize the arguments to account for these differences, but some hypotheses deserve further analysis:

On fertile land the population pressure is generally higher than elsewhere. On such land it has been observed that, if the population increases, the tree vegetation also increases. The explanation for this apparent contradiction is that because of land division a family is not able to live from the land and off-farm income becomes important. Less attention may be given to farm crops and instead trees may be grown which need less attention and still have a high cash value. Trees producing poles for house construction or props and sticks for granaries are usually in high demand.

Although the number of trees might increase, the availability of fuelwood might increase only slightly. Because cash has to be produced on the farm in order to pay school fees, medicines, etc., the male in the family looks for all kinds of cash earning alternatives, including the production of trees. His purpose is to sell the wood and only the leftovers are obtainable for the women who are responsible for the fuelwood collection.

The person who is responsible for the production of wood on the farm is not necessarily the one who is responsible for the consumption. The division of labour, already touched on above, is obvious in most cases in terms of tree planting. Often only men plant and cut trees, nearly always the women are responsible for collecting and using the fuelwood.

The preference of species to be used for certain purposes on the farm is often very subjective and difficult to describe. Certain types of tree are planted abundantly by some farmers, while the same species is prohibited by local custom within very short distances.

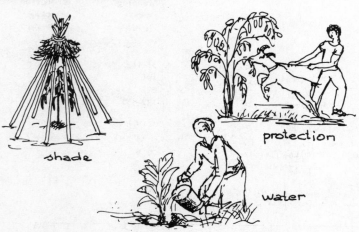

shade

protection

water

Caring for planted trees.

Within the high potential lands the need for fuelwood is much greater than on the so-called marginal lands. Land resources are really scarce in high-potential land where the population is greatest and every effort is undertaken to produce wood. People moving from the highpotential land to the more marginal sites in order to find agricultural land do not have such a problem with woodfuel because much of the original vegetation still exists.

The urbanization process results in a high demand from the cities for wood, particularly charcoal, which is used because of convenience and cheapness. Charcoal can only be successfully produced if wood is available either within existing forests or within the recently settled marginal lands which still have their original vegetation cover. Usually within both areas charcoal production takes place but often by very low standards of production technology (see Chapter 6).

Often production is undertaken illegally and is therefore difficult to control or improve. Production is very erratic (particularly within marginal land). If the crops fail or animal diseases are decimating the herds, charcoal production becomes the only source of income and large areas of trees will be felled without the provision of replacement. Climatic conditions are often severe for natural regeneration if the lands are overgrazed and the erosion process will destroy large areas in these circumstances.

It is the demand from cities and the farmer's need for income, not rural woodfuel demand, that are the main reasons for the destruction of land. Unfortunately improved technologies have hardly any chance to be introduced under such circumstances due to their erratic production character. Only the development of better cash crops for marginal

lands or alternative charcoal production sites will stop this process.

Conclusions

Fuelwood, in addition to water and shelter, is a basic need for the vast majority of people in developing countries. It is unrealistic to believe that a long-term, large-scale solution to the fuelwood crisis can be achieved among subsistence and semi-subsistence rural people if it is not predominantly based on a strategy of individual land users catering for their own needs.

The integration of trees and other woody perennials into farming systems, using species, management techniques and spatial/temporal arrangements which are ecologically and culturally compatible with local practices, holds a great potential. A well conceived agroforestry intervention could achieve an increased production of such wood and at the same time address many of the problems related to land productivity and sustainability.

There is an almost unlimited potential for development of "new" and improved agroforestry technologies and systems, both through the identification and scientific improvement of appropriate "multipurpose" tree and shrub species, and through innovative and imaginative thinking on how to integrate these spatially and management-wise with crops and/or animals in farming systems.

Although it is quite clear that the fuelwood crisis and the resulting substantial resources made available internationally and nationally to alleviate it, will provide the main drive in developing agroforestry technologies, it is important to approach this development in an interdisciplinary fashion rather than as a purely forestry problem.

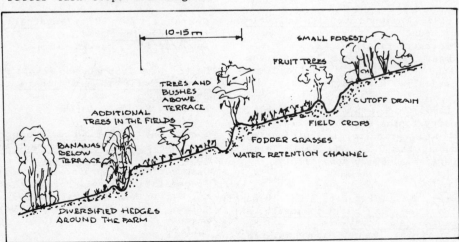

Agroforestry and terrassing for soil conservation.

Bibliography

Agroforestry Systems (1982) "What is agroforestry?"
Editorial Agroforestry systems 1 (1): 7-12.

Andersson, D. & Fishwick, R. (1984) Fuelwood Consumption and Deforestation in African Countries.
World Bank Staff Working Papers No. 704.
The World Bank, 1818 H. Street, N.W., Washington, D.C. 20433, U.S.A.

Arnold, J.E.M. (1979) Wood energy and rural communities.
Natural Resources Forum 3(3): 229-252.

Bajracharya, D. (1980) Fuelwood and food needs versus deforestation: an energy study of a hill village panchayat in Eastern Nepal.
Energy for Rural Development Programme Report PR-80-2. E-W Res. Syst. Inst.

Burley, J. (1979) Choice of tree species and possibility for genetic improvement for small-holder and community forests.
Commonwealth Forestry Review 59(3):311-26.

Earl, D.E. (1975) Forest Energy and Economic Development.
Clarendon Press, Oxford, U.K.

FAO (1981), Agriculture: toward 2000.
FAO Rome. 134 pp.

FAO (1981) Map of the Fuelwood Situation in the Developing Countries at a scale of 1:25 000 000.
Food and Agricultural Organisation (FAO) of the United Nations, Via delle Terme di Caracalla, 00100 Rome, Italy.

FAO (1981) Wood Energy.
Special edition 1 and 2. Unisylva, Vol. 33. No. 131 and No. 133. A journal of forestry and forest industry. Food and Agricultural Organisation (FAO) of the U.N., Via delle Terme di Caracalla, 00100 Rome, Italy.

FAO (1983) Fuelwood Supplies in Developing Countries.
FAO Forestry Paper No. 42.
Food and Agricultural Organisation (FAO) of the U.N., Via delle Terme di Caracalla, 00100 Rome, Italy.

FAO (1985) Forests, Trees and People.
Forestry topics report No. 2.
Food and Agricultural Organisation (FAO) of the U.N., Via delle Terme di Caracalla, 00100 Rome, Italy.

Fleuret, P.C. & Fleuret, A.K. (1978) Fuelwood use in a peasant community: a Tanzanian case study.
Journal of Developing Areas 12:315-322.

Gelder, B. van & Poulsen, G. (1982) The woodfuel supply from trees outside the forests in the highlands of Kenya.
Beijer Institute, Box 50005, 104 05 Stockholm

Hall, D.O. & Barnard, G.W. & Moss, P.A. (1982) Biomass for Energy in the Developing Countries.
Pergamon Press, Headington Hill Hall, Oxford OX3 0BW, U.K.

Jones, J.R. & Otarola, A. (1981) Diagnóstico socioeconómico sobre el consumo y producción de lena en fincas pequencas de Nicaragua.
Proyecto IRENA-CATIE-ROCAP No. 596-0089. Serie Téchnica, informe téchnico No. 21, CATIE, Turrialba.

King, K.F.S. (1968) Agri-silviculture-the Taungya system.
Bulletin No. 1. Department of Forestry, University of Ibadan. 109 pp.

Little, E.L. (1980) Common Fuelwood Crops.
A Handbook for Their Identification.
Communi-Tech Associates, P.O. Box 3170, Morgantown, West Virginia 26503, U.S.A.

Lundgren, B. & Raintree, J.B. (1983) Agroforestry.
In: B Nestel (ed): Agricultural Research for Development: pp. 37-49 Potentials and challenges in Asia. ISNAR. The Hague.

Nair, P.K.R. (1980) Agroforestry Species - A Crop Sheet Manual.
(ICRAF), P.O. Box 30677, Nairobi, Kenya.

National Academy of Sciences (1980) Firewood Crops.
Shrub and Tree Species for Energy Production.
Commission on International Relations, National Academy of Sciences-National Research Council, 2101, Constitution Avenue, Washington, D.C. 20418, U.S.A.

Openshaw, K. (1974) Woodfuels for the Developing World.
New Scientist, January 1974.

Salem, B. & Tran van Nao (1981) Fuelwood Production in Traditional Farming Systems.
Unasylva 33(131): 13-18

Simon, H. (1981) Wood Production and Consumption from Nonforest Areas in Java. In: K F Wiersum(ed).
Observations on agroforestry on Java, Indonesia; pp. 56-61. Forestry Faculty, Gadjan Mada University, Yoyakarta + Dep. Forest Management, LH Wageningen.

Singh, R.V. & Svarup, R. (1980) Estimating forest based needs of the community.
In, community forestry management for rural development. Natraj Publ., Dehra Dun.

Sirén, G. et al. (1984) Energy Forestry, Information on Research and Experiments at the Swedish University of Agricultural Sciences, Uppsala.
Swedish Energy Forest Project, Swedish University of Agricultural Sciences, Box 7072, 750 07 Uppsala, Sweden.

Spears, J.S. (1980) Can Farming and Forestry Co-exist in the Tropics?
Unasylva 32(128): 2-12.

World Bank 1981a, Mobilizing renewable energy technology in developing countries: strengthening local capabilities and research.
World Bank, Washington DC, 52 pp.

World Bank/FAO (1981) Forestry Research Needs in Developing Countries - Time for a reappraisal?
Paper presented to "17 IUFRO Congress", Kyoto, Japan, September 6-17, 1981. Stencil. 56 pp.

Additional Selected Bibliography on Agroforestry.

CATIE (1979) Proceedings of the Workshop on Agro-Forestry Systems in Latin America.
Centro Agronómico Tropical de Investigation Ensenanza (CATIE), Turrialba, Costa Rica.

Hoekstra, D.A. (1985) The Use of Economics in Diagnosis and Design of Agroforestry Systems.
Working paper No. 29.
International Council for Research in Agroforestry (KRAF), P.O. Box 30677, Nairobi, Kenya.

Huxley, P.A. (1983) Plant Research and Agroforestry.
International Council for Research in Agroforestry (ICRAF), P.O. Box 30677, Nairobi, Kenya.

ICRAF Working Paper No. 7 (1983) Draft Resources for Agroforestry Diagnosis and Design.
International Council for Research in Agroforestry (ICRAF), P.O. Box 30677, Nairobi, Kenya.

Lundgren, B. (1982) The Use of Agroforestry to Improve the Productivity of Converted Tropical Land.
Prepared for the U.S. Congressional Office of Technology Assessment.
International Council for Research in Agroforestry (ICRAF), P.O. Box 30677, Nairobi, Kenya.

MacDonald, L.H. (1982) Agro-Forestry in the African Humid Tropics.
United Nations University (UNV), Toho Seimei Building 15-1, Shibuya, 2-chome, Shibuya-kv, Tokyo 150, Japan.

Nair, P.K.R. (1984) Soil Productivity Aspects of Agroforestry.
Science and Practice of Agroforestry No. 1.
International Council for Research in Agroforestry (ICRAF), P.O. Box 30677, Nairobi, Kenya.

NAS (1983) Agroforestry in the West African Sahel.
National Academy of Sciences (NAS), 2101 Constitution Avenue, Washington, P.C. 24018, U.S.A.

Raintree, J.B. (1983) Strategies for Enchanting the Adoptability of Agroforestry Innovations.
Agroforestry Systems 1(3): 173-188.
Martinus Nijhoff/Dr. W. Junk, POB 566, 2501 CN, The Hague, The Netherlands.

Rocheleau, D. (1984) Criteria for Re-Appraisal and Re-Design: Intra-Household and Between-Household aspects of FSRE in three Kenyan Agroforestry projects.
Paper presented at the Annual Farming Systems Research and Extension Symposium, Kansas, October 1984.
Kansas State University, Manhattan, Kansas, U.S.A.

Smith, J.R. (1950) Tree Crops. A Permanent Agriculture.
Devin-Adair. New York.

2 ENERGY CROPS

Introduction

Theoretically any plant could be used for energy purposes. Some are already being cultivated specifically for producing energy, due to their high photosynthetic efficiencies, or ease of conversion to useful fuels. There also exist many other plants which potentially could provide large amounts of energy. The most promising energy crops - trees - have already been discussed in Chapter 1. The growing of plants for conversion into alcohols and vegetable oils, will be looked at individually in Chapter 11. This chapter will limit itself to some of the discussion of the remaining energy crops, such as herbaceous and hydrocarbon crops as well as aquacultures.

Herbaceous Crops

A number of tropical grasses have high growth rates and could be grown and harvested for energy purposes. In some places harvesting of naturally growing grasses is actually being practised, but there have been very few attempts to create such plantations and to survey are fertilizer, labour and capital requirements of an equivalent sugarcane or corn plantation grown for energy purposes.

Napier, or elephant grass (Imperata cylindrica) is a fast growing grass variety which could be a candidate. In Eastern Africa a particularly promising source of energy could be the wildly growing Papyrus which covers thousands of hectares of marshes and lake sides. Harvestable yields as high as 32 t/ha/yr have been reported from Kenya, and if the air-dried papyrus is densified into fuel pellets, the result is a relatively clean and convenient fuel. The environmental impacts of harvesting large areas of papyrus, however, are still unknown and will still have to be studied, along with the overall costs of the harvesting and the pelletizing processes.

Sugar cane is another crop which is known to give very high yields under good conditions. The major energy use of cane - alcohol production - is discussed in Chapter 11. However, recent tests with high-yielding varieties maximizing biomass production, have given experimental yields of such "energy-cane" of as high as over 250 t/ha/yr. A few other crops are also known to have very high biomass yields, and tests are under way in several countries.

Hydrocarbon Plants

There exist a wide range of plant species whose biomass can yield hydrocarbons or hydrocarbon-like substances upon extraction. These can be potential fuels, or feedstocks for the chemical industry, depending on their molecular weights. Generally, lower molecular weights are more suitable for fuel purposes. At present hydro-carbon plantations are only at experimental stages, and it is too early to judge their potential for success and economic viability.

The natural rubber tree in Brazil, Hevea brasiliensis, for example, produces a latex consisting of very high molecular weight hydrocarbons - not usable for fuel. Others, such as the genus Euphorbia and particularly Euphorbia lathyrus seem promising for the production of hydrocarbon fuels. Extraction yields liquids resembling petroleum. Although early hopes of very high yields of Euphorbia under semi-arid conditions (i.e. in Arizona) did not materialize, under irrigation it has been possible to achieve an oil production of 1.5 t/ha/yr.

There are a number of algae which can also produce hydrocarbons, Botryoccus braunii, for example, has been shown to yield 70% of its extract as a hydrocarbon liquid, closely resembling crude oil. The green alga Dunaliella which lives in very salty environments, produces glycerol, beta-carotene and also protein.

Sugarcane can be extremely high-yielding.

22

Aquacultures

In many countries there are long traditions of growing seaweeds or algae, either in fresh or in marine waters. Present production is primarily for food, since algae have a high protein, mineral and vitamin content. In Japan, for example, an area of 200 000 ha is cultivated in the Pacific Ocean, for the production of edible seaweeds. During the last few decades an increasing amount of red and brown algae, has also been used as feedstock for the chemical industry. The total seaweed trade in the world comprises about 700 million t/yr, whose energy content, if it were digested anaerobically in a biogas digestor would be approximately 2400 million GJ, or 2.5% of the total energy consumption of the developing countries.

Theoretically, there is a tremendous potential for increased production of seaweeds and other aquatic cultures to fuel biogas plants for the production of energy, at small, intermediate or industrialized levels.

One of the most promising aquatic plants for developing countries is the water hyacinth (<u>Eichornia crassipes</u>). This is a rapidly growing fresh water weed, and is notorious for its ability to clog up lakes and waterways. It is especially fast growing on nutrient rich, polluted waters. It has high protein, mineral and vitamin content, which makes it suitable as food, animal fodder and fertilizer to recycle into the aquaculture system to keep up nutrient levels. Because of its high moisture content it can be easily digested in biogas plants to yield a methane rich gas plus a residue which retains all the mineral and nutrient contents of the original harvested weed, and which can consequently still be used as fodder or fertilizer.

Aquacultural Techniques

There are a number of techniques available for aquaculture systems, depending on the scale of operation required and the species chosen. Free-floating cultures are used in shallow bays as well as in lakes for the cultivation of many species of green algae, as well as for weeds like the water hyacinth. The algae can, however, grow as attached cultures to ropes and to long lines. Such techniques are widely used in Japan and China and they often tend to be large-scale operations. Spray culture techniques also look very promising. Here the algae are grown on trays, which are continuously sprayed with water from above. In windy areas windmills can be used to pump water.

Depending on which technique, species and site are chosen, production levels of 10-100 t/ha/yr of dry matter have been reported. If used to fuel biogas plants, the production of 50 t/ha/yr would be equivalent to an energy production of 150-200 GJ. One of the main problems with aquacultures is the method of harvesting. For small scale, free-floating or attached cultures, manual cutting and collection can be appropriate. For intermediate and large-scale operations, however, it becomes economic to invest in mechanized harvesters. These are commercially available in Japan, North America and in some European countries. Aquatic plants have to

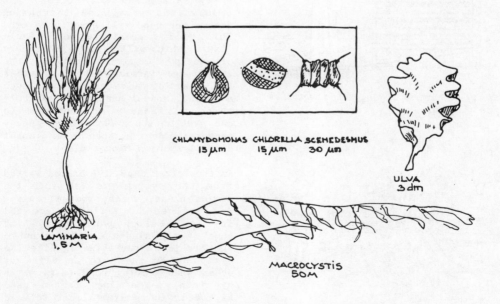

CHLAMYDOMONAS CHLORELLA SCENEDESMUS
15 µm   15 µm   30 µm

ULVA
3 dm

LAMINARIA
1.5 M

MACROCYSTIS
50 M

Algaes used for aquaculture.

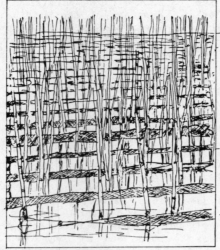

Attached aquacultural system.

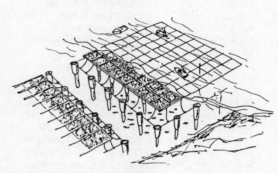

Floating aquaculture system.

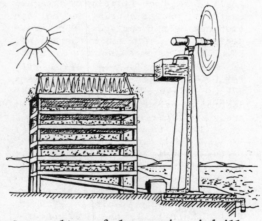

Spray culture of algae, using windmill for pumping.

Harvester for aquatic crops.

be crushed or blended in some manner, before being loaded into a biogas plant in order to shorten digestion times. It may be advantageous, especially in larger scale operations, to invest in a chopper or to combine one with a harvester. Another problem is that because of the large amounts of biomass produced, relatively large digestors are needed.

The Economy of Aquaculture

Apart from a few isolated examples, aquacultures for energy purposes are still in the experimental stage. Their economics depend very much on local labour and material costs, as well as, of course, on the climatic conditions. There are a number of other questions, in addition to the economics which will have to be studied. One of these is the rate at which water hyacinth and other weeds would accumulate heavy metals from polluted waters.

Since weeds, like water hyacinth, grow so well on polluted waters, such aquatic systems make much more economic sense if they are combined with the treatment of sewage, the purification effluents from alcohol distilleries, or with other functions, such as animal husbandry and fish ponds in the rural areas. The technique of combining a number of different components is called Integrated Farming, and will be discussed below.

Integrated Farming

In areas, where there is high pressure on land and other resources, it is essential to yield the highest possible return from all the food and energy-producing activities. As populations grow, the increasing demand for food will have to be met. This will necessitate more fertilizers, which will cost large amounts of energy. In such situations, the Integrated Farming method provides a number of advantages. In such a system, animal husbandry, agriculture and fishery are combined in one system, whose core is an energy conversion device, such as an alcohol distillery or a biogas plant.

In the latter case, the animal wastes from the system are fed into the biogas plant, which produces energy and effluents. Effluents are fed into a settling pond, from which some of the sludge can be taken to the fields as fertilizer, while the rest can be fed into an algae or other weed-growing pond (e.g. water hyacinth). The algae can be used as fodder and as fertilizer to feed the animals (so that they can supply more dung to the biogas plant) and

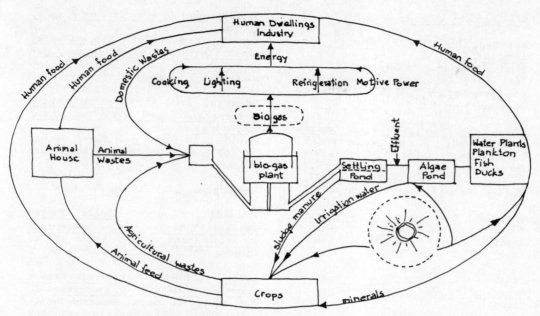

Diagram illustrating principles of one integrated farming system.

to increase agricultural yields, respectively. The system can be complex or relatively simple with just two components.

An interesting example of a complex integrated farm is the Maya Farms in the Philippines. This is an integrated agro-industrial complex with over 22 000 pigs, 6000 ducks and also some cattle. The animals are slaughtered and their meat canned for commercial sale. The animals provide over 50 digestors, each having a capacity of 22 m^3, with dung. The sludge is dried in settling ponds, and is then mixed with pig feed. The savings in pig feed repaid the cost of the biogas installations in just 2 years. The liquid effluents from the digestors, together with the pigsty washings, are sent to another pond, where they supply nutrients for the growth of algae, which are used as food for the ducks. The duck droppings, on the other hand, are used as food for the tilapia that are grown in the pond. The fish are sold, while the ducks are slaughtered and canned, providing the farm with additional income.

Another example of this method, an alcohol distillery integrated with aquaculture, animal husbandry and agriculture has been proposed in Brazil. The core of the system consists of a 20 000 litre per day capacity alcohol distillery, which uses cassava roots grown on its own associated land and generates heat by burning half of the aerial part of the cassava crop together with methane obtained from biodigestion of slop and water hyacinth which are grown in ponds on nutrients contained in the effluent from the biodigestion of slop. Use of the slop

for the production of biogas not only reduces pollution but also provides process heat in addition to biofertilizer and reusable water purified in the water hyacinth ponds. The biofertilizer is used to increase cassava and other fodder yields for the cattle which are also part of the system. Approximately a quarter of the aerial part of the cassava is used to extract protein for animal feed.

Conclusions

The growing of energy crops can, in some circumstances, supplement existing energy supplies. The production of herbaceous and hydrocarbon crops specifically for energy purposes is still in the experimental stage. It is not known how the price of fuels derived from these processes will compare with other energy sources. In both cases the selection of the species is of great importance and requires site-specific decisions. Given their bulk, these crops will either have to be used for on-site combustion, or converted to a higher quality fuel.

In some cultures aquatic plants have traditionally been farmed for food, so their production requirements are known. What is new is their conversion to energy; however, enough experiments have been done to show their economic and ecological viability. What is required is recognition by governments, ministries of agriculture, as well as farmers (both large and small scale) of the tremendous potentials of such techniques, particularly when used in Integrated Farming systems.

Bibliography on Energy Crops

Bergman, K-G. (1982) Energy from Agriculture.
Swedish University of Agricultural Sciences, Department of Economics and Statistics, S-750 07 Uppsala, Sweden.

Biomass - A cash crop for the future.
A conference on the production of biomass from grains, crop residues, forages and grasses for conversion to fuels and chemicals, March 1977.
The Institute, Kansas City, MO, U.S.A.

Cross, M. (1984) Grow your own Energy.
A new scientist guide.
Basil Blackwell Limited, 108 Cowley Road, Oxford OX4 1JF, U.K.

Felker, P. (1984) Short rotation mesquite (Prosopis) energy farms.
Biomass Vol. 5 No. 1.

Hall, D.O. (1983) Biomass for Energy-Fuels Now and in the Future.
University of London, King's College, 68 Half Moon Lane, London SE24 9JF, U.K.

Hall, D.O. & Barnard G.W. & Moss, P.A. (1982) Biomass for Energy in the Developing Countries.
Pergamon Press Ltd., Headington Hill Hall, Oxford OX3 OBW, U.K.

Harris, G.S. (1979) Energy Farming in New Zealand.
Chemistry in New Zealand 43(3), 103-105, 1979.

Johnson, J.D. & Hinman C.W. (1980) Oils and Rubber from Arid Land Plants.
Science Vol. 208. No. 4443.

Margaris N.S. & Vokov, D. (1985) Latex Producing Plants in Greece.
Biomass, Vol. 7 No. 3, Elsevier Applied Science Publishers Ltd., Crown House, Lintin Road, Barking, Essex IG11 8JU, U.K.

Mellado et al. (1984) Euphorbia lathyrus as an Energy Crop.
Biomass Vol. 5 No. 1.

Seiler, G.J. & Adams, R.P. (1984) Whole-plant Utilization of Sunflowers.
Biomass Vol. 4, No. 1.

Smil, V. (1983) Biomass Energies.
Resources, Links, Constraints.
Plenum Press. New York & London.

White, L.P. Plaske, H.G. (1982) Biomass as a Fuel.
Academic Press, London.

Vlitos, A.J. (1979) Creative Botany Opportunities for the Future.
Sugar y Azúcar 74,(7)28-29, 1979.

Wünche, U. et al. (1985) Jerusalem Artichoke for Biogas Production.
Biomass Vol. 7, No. 2.

Bibliography on Aqua-Culture

Becker, E.W. (1984) Biotechnology and Exploitation of the Green Algae in India.
Biomass Vol. 4 No. 1.

Edwards, P. (1985) Aquaculture: A Component of Low Cost Sanitation Technology.
World Bank Technical Paper Number 36, UNDP Project Management Report Number 3.
The World Bank, 1818 H. Street N.W., Washington, D.C. 20433, U.S.A.

Eldvidge, P. (1984) Yields, Photosynthesis Efficiencies and Composition of Microalgae.
Biomass Vol. 5 No. 3.

Lydén, A. (1984) Pilot-scale Growth Experiments with Water Hyacinths.
Swedish Environmental Research Institute (IVL) Hälsingegatan 43, Box 21060, 100 31 Stockholm, Sweden.

National Academy of Sciences (1976) Making Aquatic Weeds Useful.
some perspectives for Developing Countries.
National Academy of Sciences (NAS), 2101 Constitution Avenue, Washington, D.C. 20418, U.S.A.

Wolverton, B.C. & McDonald, R.C. (1979) The Water Hyacinth: from profilic pest to potential provider.
Ambio Vol. 8, No. 1, Royal Swedish Academy of Sciences, Box 50005, 104 05 Stockholm, Sweden.

3 AGRICULTURAL RESIDUES AND ORGANIC WASTES

Agricultural Residues

As woodfuel has become more scarce in many parts of the developing world, an increasing number of people have been forced to turn to straw, crop stalks, animal dung and other agricultural residues as an alternative cooking and heating fuel. Though often ignored in energy statistics, for many millions of families agricultural residues have become the single most important domestic fuel.

The term "agricultural residues" is used here to describe the full spectrum of biomass materials that are produced as byproducts from agriculture. It includes woody residues, crop straw, crop processing residues, green crop residues and dung.

The greatest concentration of agricultural residue burning today is the densely populated plains of Northern India, Bangladesh and China. Here, dung and crop residues are the major domestic fuel, providing as much as 90% of household energy in many villages, and a considerable proportion in urban areas too.

Both rich and poor in these and many other areas have come to depend on agricultural residues. For many of them, the "woodfuel crisis" is essentially over. What they have entered is a new phase in the evolution of fuel scarcity where the struggle is not to find wood, but to obtain enough dung, straw and crop stalks to cook their food and heat their homes.

Numerous examples of agricultural residue burning can be found elsewhere in the world. Dung is the principal domestic fuel in some of the highland regions of Ethiopia, and in the high altiplano region of Peru. It is also extensively used in parts of Egypt, Turkey, and Lesotho and Transkei in southern Africa; and there are isolated reports of dung burning in Tanzania, Sudan, Iraq, Yemen and other countries.

Since the term "agricultural residues" covers such a large variety of materials, with widely differing properties as fuels, it is necessary to treat the different types separately in order to get useful results.

1   Woody residues such as corn cobs, millet stalks, jute sticks, coconut shells, pigeon pea, cotton stalks etc. burn well and are often popular fuels. These are slow to decompose in the soil and sometimes harbour pests and diseases. To use these as a fuel usually has little detrimental effect on the soil.

2   Crop straw, like rice, wheat etc., makes a poorer fuel; it tends to flare up and burn too fast. Straw can be made into bundles, bales, or briquettes to become a more convenient fuel. Straw has also many competing end-uses e.g. as fodder for animals, building materials, thatching and a range of other important functions. Provided that ashes are returned to the fields the burning of straw makes little difference to the nutrient balance in the soil.

Much agricultural residue is wastefully burnt in the field.

27

3   Crop processing residues such as rice husks, coffee husks, coir dust, peanut shells etc., are generated in large quantities at the crop processing facilities. These are rarely recycled to the fields and often mount up as wastes outside the factory. These residues generally make a difficult fuel and without briquetting they need special stoves if they are to be used at all. But if made into briquettes they often make an acceptable fuel, both for domestic and industrial uses (See Chapter 5).

4   Green crop residues such as ground nut tops etc. make good organic fertilizer but they are also valuable as animal feed. They make a poor fuel, and have to be dried before they can be burnt. An attractive way of using them as energy is to use them in a biogas-plant, then the decomposed residues can be used as a fertilizer to maintain and improve the properties of the soil (see Chapter 9).

5   Animal dung. Dried cow, buffalo, and horse dung makes a relatively good cooking fuel. Its main problem is that it tends to produce a lot of smoke, that can be irritating for eyes and lungs. Dung makes an excellent fertilizer and the removal of the dung from the fields to be used as a fuel will have a direct impact on the soil. Biogas digesters can be an answer to this dilemma as these produce both energy and fertilizer from dung (see Chapter 9).

By using agricultural residues, the supply of burnable biomass materials in an area is greatly increased. Using a variety of technologies, agricultural residues can also be upgraded to make them into more versatile and convenient fuels.

In most farming systems, the amount of agricultural residues produced each year is extremely large. With most cereal crops, for example, for every ton of grain at least a ton and a half of straw is produced. A healthy cow or buffalo produces as much as five times its own weight in dung each year. Though rarely measured, dung and crop residues are in fact a major resource, often constituting the largest single element in biomass production at the village level.

Rough estimates of agricultural residue production at a national level can be derived using livestock and crop production statistics by multiplying crop-to-residue ratios with average cereal production figures. Crop yields depend on many factors, such as type of crop, farming methods, soil and climatic conditions, etc. But for most developing countries, production of crop residues and dung, together, works out at more than a ton per person in rural areas. In some it is considerably higher.

Based on such data it is possible to estimate the total amount of cereal residues produced. This ranges between 1-10 GJ/cap/yr in the developing countries, with an average of 5.6 GJ. This is considerable, when compared to typical rural energy consumption levels of 5-20 GJ/cap/yr.

Table   Crop-to-Residue Ratios for Selected Crops

Crop	Residue	Residue Production (tonnes per tonne of crop)
Rice	straw	1.1 - 2.9
Deep water rice	straw	14.3
Wheat	straw	1.0 - 1.8
Maize	stalk + cob	1.2 - 2.5
Grain sorghum	stalk	0.9 - 4.9
Millet	stalk	2.0
Barley	straw	1.5 - 1.8
Rye	straw	1.8 - 2.0
Oats	straw	1.8
Groundnuts	shell	0.5
	straw	2.3
Pigeon Pea	stalk	5.0
Cotton	stalk	3.5 - 5.0
Jute	sticks	2.0
Coconut (copra)	shell	0.7 - 1.1
	husk	1.6 - 4.5

Source: Barnard & Kristoferson, Agricultural Residues as Fuel in The Third World.

But availability for energy is also heavily dependent on the many other competing uses of agricultural residues. Besides being employed as organic fertilizers, many have important roles as animal fodder, thatching, building materials, animal bedding, and suchlike. Very little is a waste product, as such. The notion of agricultural residues being low-value wastes is very much an imported one, originating in Europe and North America. For traditional farmers, crops and animals are not simply food producers; they are seen as having multiple purposes. Often their by-products are as esential to the continued viability of the local farming system as their principal outputs.

Equally important is the question of access to agricultural residues at the village level. The availability of residues to particular families is often heavily affected by their individual circumstances. A prosperous farmer, for example, with several hectares of farmland and a sizeable herd of cattle, may produce a large surplus of agricultural residues. In contrast, a landless farm labourer may have automatic access to hardly any. What his family is able to use may in practice be determined by the generosity, or otherwise, of his employer.

The use of agricultural residues as fuel is not new. What is new is the scale of agricultural residue burning. As many as 800 million people now rely on agricultural residues as their principal cooking fuel, according to one estimate. This is little more than an educated guess, however. Until recently, agricultural residue burning has been almost totally ignored by energy planners. It still rarely features in official statistics, and the figures that do exist are often highly suspect.

Three other important limitations have to be mentioned here. Crop residues are usually bulky and difficult to transport. Consequently it may not be practical to transport them away from the place of production. Some fraction of the total residue will thus always be wasted. Secondly, the removal of large quantities of biomass from agricultural lands may result in unwanted removal of nutrients from the soil. This can be a problem particularly in areas where soil fertility is already low. The result is low yields and eventual exhaustion of the soil, unless appropriate fertilization takes place. Finally, a further factor pertaining to the use of crop residues for energy is the seasonality of supply. Residues are only available for a limited time during the year. In order to be able to supply them year around, sizeable storage capacities will be required. The storage problem can be reduced significantly, as shown by recent experiments, by compacting the residues into pellets or briquettes and by employing appropriate drying methods (see Chapter 5). In general, storage is expensive and quite difficult, especially in wet climates.

It can thus be seen that assessing the availability and the potential of using crop residues for energy purposes is a complex and site-specific task. If the present patterns of crop residue consumption are to be improved, careful consideration will have to be given to all their different and often competing usages.

Crop Residues as Fuels

Most crop residues, when they are oven dried, have a fairly similar heat content. When tested in a laboratory using a bomb calorimeter, gross calorific values usually fall within the range 15-20 MJ/kg. This compares with a typical value for oven dried wood of about 20 MJ/kg.

Sample figures are shown in the Table. Data for the ash content of different residues are also given. One notable trend is that the residues with the highest calorific values tend to be those with the lowest ash contents. Materials such as rice straw and rice husks, which contain a larger amount of non-combustible silica, produce more ash and less heat when burnt, so tend to be at the lower end of the range.

The energy available in practice also depends on the moisture content of the fuel. Air dried crop residues typically contain between 10% and 25% moisture, depending on how hot and dry the weather is. Under these conditions, net calorific values work out at approximately 10-16 MJ/kg, compared to wood at 13-18 MJ/kg.

Making dung-sticks for fuel.

From the point of view of a Third World cook, however, calorific value figures are of little relevance. What is more important in determining which residues make the best fuels is how convenient they are to collect and store, how easily they dry out, and how well they burn in a village kitchen.

The Case of Cotton Stalks

One interesting example of a systems approach to crop waste utilization for energy purposes can be found in Nicaragua. A joint project between the Instituto Nicaraguense de Energia (INE), SAREC and the Beijer Institute is looking at the different ways of using cotton stalks for fuel. Cotton is one of the most important products in Nicaragua, and the heat content of the cotton stalks produced every year is equivalent to 30% of total petroleum consumption in the country. At present the stalks are either burnt on the field or are ploughed into the soil. Since cotton stalks are quite woody, they do not contain much nutritional value. In fact, the ploughed in stalks actually reduce the erosion resistance of the soil.

Experiments are being run to see which method of collection would be best suited to the local situation. The collected stalks are chipped to be used as fuel in small and large industrial boilers. The stalk chips can either be improved by compacting into briquettes for combustion, or converted further into producer gas. Preliminary investigation indicates that the price of energy from cotton stalks will be approxi-

mately competitive with petroleum, but that it will not be able to compete with local firewood.

Food Processing Wastes

There exists a large category of waste products in the food processing sector which could potentially be used as energy sources. These are

Cotton stalk puller.

Cotton harvesting machinery can also be used for transport of residues.

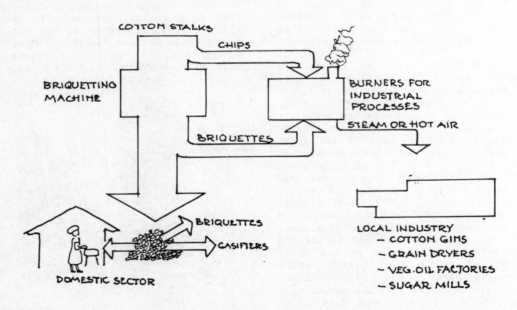

Utilization of cotton stalks for energy purposes.

sometimes difficult to distinguish from crop wastes, since often they concern the same plant, although usually different parts of it. They include bagasse, from the production of sugar, husks and shells of various crops and fruits, such as rice, coconuts, etc, as well as wastes from canneries and slaughterhouses. Usually the quantities produced are rather small and would not make significant impacts on the national energy balances. Their utilization for energy purposes, locally, to fuel the agricultural and food processing activities which had produced the wastes in the first place could, however, be very beneficial. Very often these wastes represent a major pollution problem, thus using them for energy purposes would provide double benefits.

An approximate calculation shows that the main wastes from food processing activities represent slightly more than 2% of the total energy consumed in the developing countries, as shown in the Table.

Sugarcane bagasse can be used in direct combustion to fuel activities in the sugar industry and to produce electricity for the grid. This is already done in many places. Husks and shells are also suitable for burning. They could be made into more convenient fuels by some kind of densification, such as briquetting which also makes it possible to store the fuel for use outside the harvest season. Some of the wastes with higher water content, like fruit residues could be used to feed biogas plants.

Animal Wastes

Animal manure when dried has an energy content similar to that of wood. This energy can be liberated by direct combustion, or by conversion into biogas through anaerobic digestion. In many areas, animal manure, particularly cowdung, and horse manure are widely used as cooking fuels. They are often mixed with, for example, crop residues in India or grasses in Peru, in order to bind the manure together, so that the dung cakes do not fall apart.

In the Peruvian highlands where the climate is cold and dry, biomass production is very low. As a result, people have traditionally depended on cow and llama dung. In other areas, such as much of India, people turn to cow dung because of the disappearance of trees from the countryside, to such an extent that it has been estimated that approximately one-third of all cow dung produced in India is used as fuel, representing roughly one third of all biomass consumption for energy. In the hills of Lesotho, people burn cow dung in the winter, when it is difficult to find firewood. The cow dung bricks which are made in this area burn slower than wood and are therefore better for space heating, giving an additional reason for being the preferred fuel in the winter.

People will usually turn to animal manure when woodfuels are not available. When crop and animal wastes are burned, most of their fertilizer value is lost, depriving the soil

Estimates of food processing wastes in developing countries

Product	Approx. energy content (million GJ/yr)	Present level of use for energy purposes
Sugarcane bagasse	1060	High
Rice hulls	790	Low
Coconut husks	185	Low
Cotton husks	110	High
Groundnut shells	100	High
Coffee husks	35	Low
Oil-palm husk	35	High
Oil-palm fibres	20	High
Total	2,330 million GJ (=2.5% of total energy use in developing countries)	

Source: Stovit (1979)

of nutrients needed to sustain plant life. Worldwide, the use of livestock droppings as fuel is estimated to lower grain production by some 20 million tons, enough food to minimally nourish 100 million people. On the other hand, about 500 million people depend on dung or crop residues as fuel resources, and in countries like India they account for 10% of the country's total energy supply and 50% of rural household energy. The biogas digester opens one way for developing countries to increase the energy value of dung and agricultural residues without incurring the heavy environmental costs (see Chapter 9).

It is no easier to estimate the production of animal manure than that of crop residues. The former depends obviously on the number of livestock, but also on the state of health of the animals and on the ease of manure collection which varies according to the husbandry method employed. Typical values for manure production by different animals in developing countries are shown in the Table.

Knowing the total number of livestock in each country, it is possible to arrive at an approximate figure of dung production, expressed in energy terms. There is, of course, rather high variation from country to country, depending on the amount of animal husbandry taking place. Countries like Botswana and Argentina, with high animal populations, produce 55 and 40 GJ/cap/yr, respectively, India 6.3 and others, such as the Congo as low as 1.0 GJ. The average for all the developing countries is 5.7 GJ, very similar to the figure for cereal residues. Animal residues, therefore represent a potentially large source of energy, on average, being comparable to cereal residues.

One of the main limitations for increased reliance on animal wastes has to do with the problem of collection. Partially this is a technical problem insofar as the different husbandry practices do not always lend themselves to easy dung collection. Freely grazing animals leave their manure on the fields while on the other hand, the manure of stall-fed animals is easily collected. However, collection is also a social and cultural problem. In some cultures, tradition or religion forbids people to touch animal residues. In others, some animals are associated with dirtiness, such as pigs in Indonesia, making the collection of dung almost an impossible task.

From an agricultural point of view, direct burning of animal manure is not recommended as it is more ecologically sound to use it as a fertilizer. It is, however, an unfortunate fact that cow dung and horse manure are very important fuels in those developing countries where other fuels are not presently available. Using manure in biogas plants would be much more advantageous since not only would the energy be liberated from the manure, but its original fertilizer content would be retained - in fact even improved after the anaerobic digestion process (see Chapter 9).

Forestry and Wood-processing Wastes

A great deal of wood is presently wasted in forest and wood-processing operations. In the former, whole trees, as well as stumps are often left in the forests after large-scale logging operations. In the latter, even more wood is wasted, especially in saw mills and in the paper and pulp industries.

Forest residues could be utilized for the production of charcoal. It has been estimated, for example, that in Kenya, about 20% of the wood felled during commercial logging operations is left in the forests. An additional 50-60% is wasted at the saw mills. Often, if these wastes were properly utilized, the energy needs of a paper and pulp factory or a large saw mill could be entirely satisfied from the wood wastes produced. It has to be noted, however, that some of the wastes resulting from the wood-processing operations are actually recovered through the "non cash" economy by the poorer people. Where this is the case, care has to be taken that these people will still be able to satisify their fuel needs.

Table	Animal Dung Production from Animals	
Animal	Annual Dung Production tonnes (air dry)	
Buffaloes	1.5 - 2.0	
Cows	0.8 - 1.6	
Camels	0.6 - 1.0	
Horses & Donkeys	0.3 - 0.7	
Pigs	0.2 - 0.3	
Sheep & Goats	0.1 - 0.2	

Source: Barnard & Kristoferson, Agricultural Residues as Fuel in The Third World.

Household Wastes

Disposal of household wastes, particularly of night soil, is a difficult task in most developing countries. In rural areas this does not represent a major problem, because there is sufficient space in which these organic wastes can be degraded. In semi-urban and urban areas, however, which generally do not have sufficient infrastructures for sewage disposal, the existence of large quantities of household waste represents a major health hazard.

The anaerobic digestion of these wastes in biogas systems would not only solve the pollution and health problems, but would at the same time provide biogas, which could be used for cooking or by industries, depending on the scale of the operation. In the semi-urban areas the digestion may take place in small individual or community size plants, while in big cities large plants are envisaged (e.g in Singapore).

Conclusions

This discussion shows that there is a potential in using wastes and residues for energy production. The main limitations of increased reliance on these include: competing demands, existing energy uses, seasonal supply, etc. Particularly large potentials exist in the fuelling of certain agricultural and food-processing activities, which themselves produce these wastes.

The economic viability of using wastes and residues for energy purposes depends very much on the specific waste and the purpose for which it will be used. In many situations the wastes and the residues either have to be disposed of at great cost, or are simply dumped, resulting in pollution. In the case of alcohol production from cassava, for example, the final alcohol cost per litre is halved if the various wastes and effluents are also utilized in the production process.

Very little information exists on the use of organic residues, either for energy or for other purposes. This reflects the lack of interest in this area of government and other agencies. A great deal of work needs to be done in the assessment of available resource potentials, existing demands, as well as of the economics and the environmental impacts of increased reliance on residues for energy purposes.

Bibliography

Barnard, G. & Kristoferson, L. (1985) Agricultural Residues as Fuel in the Third World. Earthscan, 10 Percy Street, London WC1H ODD, U.K.

FAO Agricultural Services Bulletin No. 47 (1982) Bibliography 1975-81 and quantitative Survey of Agriculture Residues. Food and Agricultural Organization (FAO), Via delle Terme di Caracalla, 00100 Rome, Italy.

Hall, D.O.& Barnard, G.W. & Moss, P.A. (1982) Biomass for Energy in the Developing Countries. Pergamon Press Ltd., Headington Hill Hall, Oxford OX 3 0BW, U.K.

Ho, G.E. Crop Residues – How Much can be Safely Harvested. Biomass Vol. 7 No. 1.

Massaquoi, J.G.M. (1985) Assessment of Sierra Leone´s Energy Potential from Agricultural Wastes. Paper presented to the "International Conference on Research & Development of Renewable Energy Technologies in Africa". Mauritius March-April 1985. Dep. of Mechanical Engineering, Fourah Bay College, University of Sierra Leone, Freetown, Sierra Leone.

Ministère de la Cooperation, Republique Francaise (1979) Biomasse: Comparaison des valorisations des sous-produits agricoles. Technologies et Development NO. 5. Group de recherche et d´échanges technologiques (GRET) 34, rue Dumont-d´Urville, 75116 Paris, France.

Shacklady, C.A. (1979) The Use of Organic Residues in Rural Communities. Proceedings of a Workshop in Denpasar, Bali, Indonesia, Dec. 1979. United Nations University (UNU) Toho Seimei Building, 15-1, Shibuya 2-chome, Shibuya-kv, Tokyo 150, Japan.

Stovit, B.A. (1979) Energy for World Agriculture. Food and Agricultural Organization (FAO) of the United Nations. Via delle Terme di Cavacalla, 00100 Rome, Italy.

Svenningson, P.J. (1985) Cotton Stalks as an energy source for Nicaragua. Ambio No. 4/5 1985. The Royal Swedish Academy of Sciences, Box 50005, S-104 05 Stockholm, Sweden.

World Bank (1979) Prospects for Traditional and Non-Conventional Energy Sources in Developing Countries. World Bank, 1818 H. Street, N.W., Washington, D.C., 20433 U.S.A.

4 PEAT

Introduction

Strictly speaking, peat is neither a renewable nor a new energy source. Yet, as a result of the high price of alternative commercial fuels, peat is again gaining importance in the industrialized countries. In those developing countries, where sizeable peat deposits exist, peat could contribute not only to the substitution of oil, but also to domestic fuel supplies.

Peat can be regarded as the lowest grade of coal. It is a heterogeneous mixture of partially decomposed organic matter (plant material) and inorganic minerals that have accumulated in a water-saturated environment. Such an environment inhibits active biological decomposition of the plant material and promotes the retention of carbon which would normally be released as gaseous products of the biological activity. Consequently, there is an accumulation of organic material which is known as peat. It is formed in swamps, bogs and salt waters, some of which are still under water; others are now on dry land. The rate of peat formation is very slow. It is estimated that in Europe 20-80 cm are formed every 1,000 years, which effectively makes peat a non-renewable energy source. Most peat has been formed in the last million years, and under further compression it would turn into brown coal, and eventually into coal of a higher quality.

There exist approximately 420 million hectares (some 100 billion tons) of peat deposits in the world. Since many areas in the developing countries have not yet been adequately surveyed, the potential global deposits could run as high as 500 x 106 ha which is approximately equivalent to half of the natural gas deposits of the world. Since peat is usually found near the earth's surface, it is relatively easily available compared to other fossil fuels, although not necessarily at lower economic cost.

The largest peat deposits in the world are found in Northern Europe, North America, China and Indonesia. Today, the largest producers, the so-called "Peat countries" are the USSR, Ireland and Finland, while a number of other countries, such as Sweden, have made sizeable commitments to research and development in peat production and utilization systems. There are, however, a number of other areas, including the developing countries, where large peat deposits exist, or are thought to exist, such as Malaysia and some eastern African countries. In Burundi, for example there is a national programme producing peat on large and small scales for industrial and domestic uses respectively. Many countries are, however, not even aware of the fact that they have considerable peat deposits, and as a result, know nothing about the costs and benefits of exploitation.

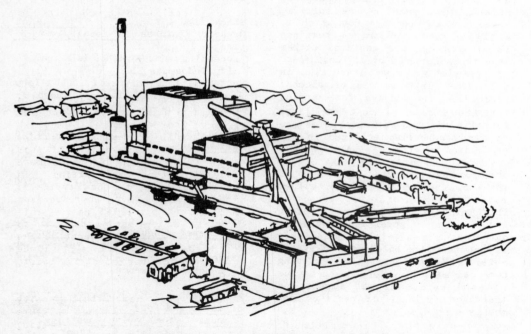

Large-scale peat power plant in Finland.

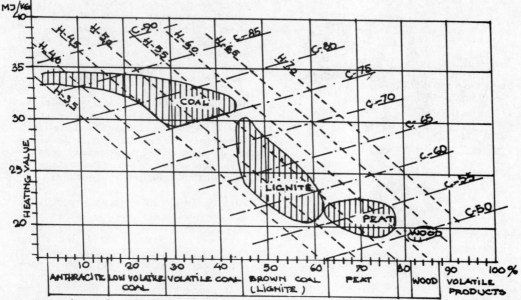

*Comparison of solid fuels with respect to heating value and content of volatile
matter. The dotted lines indicate percentage of carbon (C) and hydrogen
(H) in the composition of the different fuels.*

Peat, in its natural state, has a
moisture content of over 90%. Even
after the drainage of the bog it
comprises 70-80% moisture. In general
peat has similar combustion properties
to wet wood, both needing relatively
higher temperatures for ignition.
Once burning, it does so with a smoky
flame, giving off an acrid smoke.
It has an energy density of 8-18
MJ/kg, depending on its moisture
content and bulk density. The relatively high moisture content is the
major obstacle to using peat as a
fuel. De-watering, therefore is
the main problem, the moisture content
having to be reduced to 55% or less.
Although peat is particularly suitable
for rather large scale mechanized
exploitation, and for fuelling medium
to large sized boilers, it can also
be used as a domestic fuel to ease
pressure on local wood resources.
Reducing the moisture content to
acceptable levels of 30-50% is one
of the major problems. Furthermore,
due to its bulk, peat cannot always
be economically transported long
distances. In spite of this and its
relatively difficult combustion,
peat is a promising energy source
in those developing countries where
suitable deposits exist.

Site Preparation

Peat deposits are located on the
surface and it is therefore easy
to find them and analyse their quality. Drying a peat bog for production normally takes from 2 to 6 years
depending to some extent, on the
quality of the peat. Besides drainage, treeclearing, road building,
stump-extraction and levelling of
the bog surface must be carried out
before the area is ready for produc-

tion. This will require substantial
capital investments for machinery.
Special machinery for bog preparation
operations has been developed in
some countries.

Peat Production

At present there are three main peat-
producing methods: hand-cut peat,
milled-peat, and sod-peat, resulting
in three different kinds of fuel.
Milled peat can be compressed into
briquettes, resulting in a fourth
type of peat fuel. Characteristics
of the four main peat fuels are listed
below:

In small-scale production both the
cutting of the peat and the digging
of the ditches for its drainage can
be done by hand.

Hand-cut peat is traditionally used
as domestic fuel in a number of areas,
where easily accessible peat deposits
are found, such as Ireland or the
outskirts of Calcutta. Hand-cut
peat is usually in the shape of bricks
or cubes. The hand-cutting of peat
is usually lighter work and more
productive than wood cutting. In
eight hours one man can typically
cut 10 m^3 of raw peat, corresponding
to 1.5 t of air-dried peat. This
quantity is approximately equal to
5-6 m^3 of wood. The peat bricks
have to be stacked up in a dry place
for a long time for air drying before
use.

Hand-cut peat has a bulk density
of 200 - 400kg/m^3, average moisture
content of 25-40%, and a heating
value as received of 11-15 MJ/kg.
It can be used in individual homes
for cooking and heating. In local

Handcutting of peat.

small-scale use, transportation is not usually a problem. Organized in a proper way, small-scale production of peat can be very economic, especially for producing cheap fuel for local domestic use. The investment costs for hand-cutting are very low. In fact only a shovel is needed, and of course the necessary labour.

The second stage of development is when farm tractors become available. They can be used on drained fields under dry weather conditions for all operations using special light equipment developed for sod and milled peat production. The tractor should be equipped with special wide or extra wheels to prevent it from sinking.

Sod peat is a mechanically-cut and compressed fuel product. It can be cylindrical in form, with a diameter of 5-10 cm and a length of 10-30 cm. It is mainly used for domestic purposes and in moderate/small commercial projects. It has a bulk density of 300-400 kg/m^3, an average moisture content of 30-40%, and a heating value of 11-14 MJ/kg.

The traditional and still widely used method in mechanized sod-peat production is to excavate the peat to a depth of 2-7 m with a bagger, which macerates the peat and spreads it in sod form on the bog surface. Owing to the high moisture content, by the time the sods are spread they are usually badly shaped and the drying time is rather long. Sod-peat technology is now veering towards the milled-peat technology. Special light equipment, attachable to tractors is available for the harvesting of sod-peat. A small sod-peat cutting machine with a capacity of 1-2 tonnes per hour, suitable for a standard horse power farm tractor is available for US$ 3000. The fields are drained by open drains 20-40 m apart. Peat

is taken from the upper layer, which means relatively low 80-83% moisture sods. This results in well-shaped sods and a short drying time, often of two to four weeks. The two main types of sod-peat-cutting machines are the screw and disc cutters.

Sod-peat cutters usually compress the sods slightly during cutting resulting in a briquette-like fuel containing somewhat less moisture and consequently higher energy density than raw sods. Upon air drying the sods shrink and harden further.

Turning is often unnecessary because sod peat dries quite well. In sod-peat production it is possible to begin with only a simple small and cheap cutting machine. Because the peat machinery designed to fit farm tractors is simple and cheap, smallscale peat production can be a good sideline to normal farming.

Sod peat cutter, attached to a farm-tractor.

Milled-peat production is the most widely practised method in the major peat producing countries, as it allows large scale, mechanized harvesting.

The production cycle includes milling, harrowing, harvesting and stock-piling.

Milled peat production.

In good weather conditions (with mechanical harvesting) the length of a cycle is from two to three days. The harvesting season varies considerably with climate. Rainy conditions make the mechanical harvesting of milled peat very uncertain. In hot climates it is often very difficult to prevent peat from drying below 40% moisture, at which point milled peat begins to be very dusty.

Characteristics and Utilization of Peat.

Fuel type	Heating value (MJ/kg)	Moisture content %	Utilization
Hand-cut	11-15	25-40	Domestic cooking and heating
Sod-peat	11-14	30-40	Domestic and intermediate commercial use
Milled-peat	8-11	40-55	Large boilers, power plants
Briquettes	17-18	15	Domestic and intermediate commercial use

Milled peat is a heterogeneous mixture of loose, small particles that have been cut mechanically from the surface of the bog. The average particle size is 3-8 mm. First the upper 0.5-1.5 cm layer of the field is loosened and pulverized by mechanized millers. If the weather conditions are right, the thin upper layer dries in a few hours. This acts as a good insulator, so the milled peat layer has to be turned once or twice a day, with spoon harrows, until the whole pulverized layer reaches the 40-50% moisture level. At this stage the peat can be collected (harvested) into ridges before it is loaded on to bog trailers, or harvested directly with self-loading trailers. It is mainly used in large boilers, power plants and heating plants. It has a bulk density of 300 - 400 kg/m3, average moisture content of 40-55%, and a heating value as received of 8-11 MJ/kg.

In order to make peat a higher quality, more transportable fuel, it can be compressed into briquettes (see Chapter 5).

Peat briquettes.

Peat briquettes are normally made from milled peat, which is thermally dried to a moisture content of 10-20% and then compressed into briquettes. In size and shape a briquette is similar to a brick. It is mainly used in moderate commercial applications and in individual fire-places for heating and cooking. It has a bulk density of 700-800 kg/m3, average moisture content of 15%, and a heating value as received of 17-18 MJ/kg.

The Vyr-method, developed in Sweden, is an interesting attempt to use the naturally existing methane-producing bacteria in the peat bog to produce methane gas for collection. Water is circulated in pipes laid down in the bog. The methane gas produced by the bacteria dissolves in the water, which can be removed at a "de-gassing" station on site.

Environmental Impacts

Experience in Northern European countries shows that there are a number of important environmental impacts associated with peat production.

1   Exploitation of peat land, especially on a large scale, can affect ecology, climate, hydrology and people. The various interest groups must agree on the ultimate usage of the area and what are the main impact results from the conversion of wetlands to peat production. These areas are often unique wildlife sanctuaries, with complex hydrological and ecological cycles.

2   A problem that frequently occurs is one of conflict with agriculture. In many parts of the world bogs are cultivated with the result that agricultural land is coming under increasing pressures.

3 Another conflict can be one of water usage. Traditional exploitation of bogs respects the delicate balance of hydrology in the bogs. Too strong an intervention can have serious and sometimes irreversible consequences such as subsidence or the drying-up of the ground.

4 One major negative impact is dust formation during large-scale production, especially with milled peat. Health studies have indicated that peat dust can cause itching and smarting eyes and symptoms of rhinitis. Proper protective measures are recommended.

5 Peat contains little sulphur in comparison with fossil fuels, but the sulphur amounts vary a lot between different reserves. In general, sulphur-dioxide emissions from peat-fired plants are less than one-fifth of those from oil-fired plants. Emissions from peat-fired plants are very similar to those from wood-fired plants. Solid dust emissions can be controlled by similar devices used in coal-fired units.

Some peat, however, contains small amounts of heavy metals and other environmentally dangerous substances, which will have to be accounted for.

In many cases peat-land utilization has a positive effect; after providing energy and jobs for people during peat production, it also creates permanent employment when alternative uses of the exploited land area (e.g. forestry, agriculture) have been established. In many cases the area for fuel-peat harvesting is a very small proportion of the total wetland area and there should not be any major conflict of interest.

There is, however, a new awareness of the value of bogs and wetlands for wildlife. With appropriate planning and good management, however, these impacts can be reduced to acceptable levels. Once the peat production has finished, the land can be converted to agriculture or forestry. It is, however, usually impossible to return it to its pre-production state.

To date there is very little experience with peat projects in developing countries, with which to evaluate the major environmental problems. Given the limited resources available to governments or energy ministries in developing countries, it is perhaps unreasonable to expect that the peat production will be "appropriately managed" and that land will be reclaimed after production has ceased. It has been reported, that large-scale cutting of papyrus swamps (in order to get to the peat below) could be environmentally disastrous in the long term. This is especially so if the cutting is done in an uncontrolled fashion.

Economics

It is difficult to estimate the economics of peat used in small scale, hand-cut systems. It can be stated, however, that where peat is available, people tend to use it as domestic fuel, implying that for some people it makes sense.

The investment for small-scale production is very dependent on the preparation of the bog. Sod-peat production on a small scale can be carried out with farm tractors and relatively cheap equipment. The investment required for the smallest sod-peat-cutting machine with a capacity of 1-2 tons per hour and suitable for a 40 horse power farm tractor is US$ 3000. The economic viability

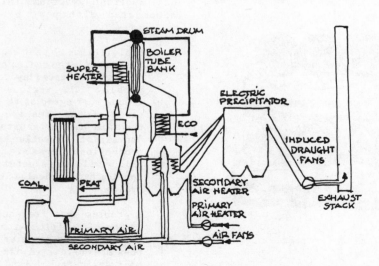

Outline of large-scale peat combustion plant.

of peat as a fuel depends on such local conditions as available fuel alternatives, labour and material costs, transportation distances, climatic conditions and operational scale. In commercial applications, the cost structure of peat is rather similar to that of coal or wood use. The cost of combustion equipment is usually higher than for oil or gas use but the fuel cost may be considerably lower. Since peat technology is already well-established costs can be reliably estimated and there are some figures from Northern European, particularly Finnish and Swedish experience, which show that peat can be an economic fuel in many instances.

Sod-peat and milled-peat production require roughly the same investments for bog preparation, peat production and peat transport equipment. Investment costs for the production of one million of milled peat per annum (100 000 tons of oil equivalent) are about US\$ 12.5 million (bog preparation US\$ 7 million, production machinery US\$ 1.5 million). The total investment cost for 100 000 m^3 of sod peat (12 000 tons of oil equivalent) is about US\$ 1.5 million. These are Finnish figures where the production season is only about three months long, so under better climatic conditions production will be increased for the equivalent investment.

In Finland the price of milled-peat is about US\$ 6/MWhr at the bog site. When transported 100 km the price rises to US\$ 9/MWhr. Sod-peat costs about US\$ 3 more per megawatt hour than milled-peat and the price of peat briquettes to the consumer is about US\$ 16/MWhr (1981 prices). The equivalent costs of light fuel oil and coal at this time were US\$ 30 and 11 respectively.

There are presently some large-scale peat projects going on in developing countries, but it is still too early to draw conclusions about their economic viability. For a developing country to develop its peat reserves often requires financial resources for an extensive programme of supporting actions and pre-investment activities. As an example, the Burundi programme had a total budget of US\$ 11 million for a five-year period (1978-1982).

In large-scale use, the peat transportation distance is often quite far. Because of its low density, peat is regarded as a local fuel, with a maximum economical transport distance of 150-300 kilometres. The transport can be done by rail or trucks.

Because of these transport limitations peat is generally utilized within the country of origin. Peat utilization, especially on a large scale, offers a good opportunity to replace imported fuels and will have a positive effect on the balance of trade.

Peat Combustion

Reduction of the moisture content is the most demanding task in fuel-peat production. In addition to the moisture content, other important characteristics of peat that affect the feasibility of its use in energy conversion are: bulk density, ash content, chemical composition and calorific value of dry substance.

In general, peat can be burned using the same equipment as wood or coal (see Chapter 8). The conversion of existing oil-fired boilers for peat burning is, however, difficult and in many cases almost as expensive as buying a new peat-fired boiler. There are many boilers on the market designed to burn various solid fuels, which are also suitable for peat.

In large systems the peat is transported to the unloading station and dumped into a silo. The stumps and clods are crushed in a hammer crusher, then the peat is transported to the boiler house by conveyor belts. In pulverized peat firing, the peat is crushed and dried in two to four pulverizers, each feeding a set of combination peat and oil burners. On a smaller scale peat may also be burned in a grate. In this method, water-cooled inclined grates and after-burning grates may be utilized. Because control of the combustion and air flow, especially with partial load, may be difficult, it is often recommended that milled-peat should be fired in conjunction with some other fuel, e.g. bark. Sod-peat, however, is well suited to grate firing.

Dried milled-peat, or gasified peat, can be used as an energy source in several industrial furnaces and ovens. The use of peat in small heating boilers, from 20 kW to 1000 kW is increasing. Such boilers normally use sod-peat or briquettes. Peat burning in small boilers is especially competitive when compared with light fuel oil. Competitiveness will, however, depend on the personnel costs, because peat-firing normally requires daily charging and ash-removal operations, whereas oil-fired plants can operate for several weeks without service.

Today peat is a competitive fuel in many countries. In those countries, where there is no tradition of using peat for domestic purposes, the introduction of peat may be difficult. Transport and storage problems should be anticipated for targeted retailers since the present domestic cooking fuel market deals in lightweight, high calorific, and non-bulky fuels (charcoal). Peat is heavy, bulky, and readily absorbs water. A peat marketing network has also to be established. People have to be accustomed to peat as a fuel and understand that it takes a long time to dry and that it has to be stored under cover so that it does not get wet again.

The other problem is to get appropriate cooking stoves for peat which are acceptable to their users (see Chapter 7). This includes all the problems of stove efficiency besides which peat smokes profusely and is more difficult to ignite. Peat has a high ash content so the stoves have to be cleaned regularly.

Very high temperatures are needed both to get a peat fire started and to keep it going. This means that stove materials must be durable enough to withstand the high temperatures and, as peat gives a much hotter flame than charcoal, cooking methods might have to be adapted to the new fuel. As a fuel, dry peat resembles wet wood. Both require sustained high-heat for ignition, emit dense smoke as they approach combustion temperatures, eventually produce long flames, and finally settle into low-flame beds of coals.

Burundi peat stove.

The use of peat briquettes is preferred because of their cleanliness, but sod-peat can be used in outdoor stoves. The factor that prevents general domestic consumer acceptance of raw peat is principally the smoke nuisance. If used locally in stoves with bad combustion, problems with smoke and sulphur emissions can arise in villages and towns. The control of smoke emissions from stoves is especially important in crowded urban areas. Experience from Burundi shows

some of the problems, but work is being carried out to develop domestic peat-fuelled cooking stoves, and experience will show which way to go in the utilization of peat as a household energy.

Implementation and Constraints

The implementation of selected peat programmes in a number of developing countries might be very beneficial for their energy sectors, as well as the national economies. Small-scale, hand-cutting of peat is labour intensive, and where labour is available, these programmes could provide continuous employment for many people. For the large-scale production of peat, a great deal of capital investment, as well as specialists and other skilled labour, will be required. The costs of these, however, have to be looked at in the light of the fact that large-scale peat production may substitute for imported oil in, for example, the production of electricity.

There are many developing countries with known peat deposits and many others are expected to have peat. Of these countries, China is the only one which has been using peat as an energy source, mainly for rural, domestic fuel supplies. Finland has taken practical steps by assisting financially and technically in the peat development programmes in Burundi, Kenya, Jamaica, and Senegal. The Burundi programme is currently the most advanced.

The most accurate statistics on peat-land areas are from Europe. Only fragmentary data are available on tropical and sub-tropical peat-land. Recent surveys made in Indonesia show surprisingly large bog areas with thick peat deposits, indicating the possibility of vast peat deposits in some tropical areas of the world.

In spite of the lack of data on peat deposits, World Bank/UNDP have compiled a list concerning the prospects for peat use in developing countries. Eight countries are listed which are very likely to utilize their peat resources within the next decade. These are Bangladesh, Burundi, China, Cuba, Jamaica, Rwanda, Uruguay and Zimbabwe. For these countries, an extensive programme of geological assessment, training of technical staff and feasibility studies (if not included in the national plans) should be formulated as soon as possible. A small programme of assessment is recommended for Angola, Argentina, Brazil, Ecuador, Guayana, Indonesia, Ivory Coast, Malaysia, Mozambique, Paraguay, Peru, Senegal, Sri Lanka, Thailand and Yugoslavia.

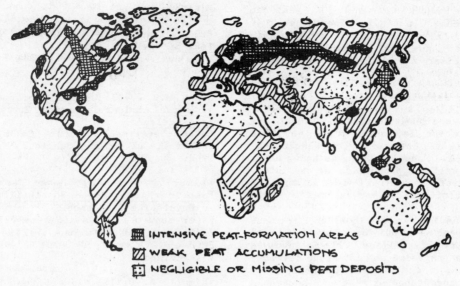

41

- INTENSIVE PEAT-FORMATION AREAS
- WEAK PEAT ACCUMULATIONS
- NEGLIGIBLE OR MISSING PEAT DEPOSITS

World peat resources.   Source: *Ekono Consulting Engineers 1983.*

Although peat seems to be a viable alternative energy source in many developing countries, there are many constraints on the use of peat for energy. The most important are:

- Lack of information on peat technologies as, at present, it is only used as fuel in a few countries;

- Adequate data on indigenous peat resources in most developing countries are not available;

- There is no comprehensive international information system on peat;

- Peat has not been included in the energy development programme of the international organizations;

- National energy planning is still at an early stage in many developing countries.

In all the main peat-producing countries, the national government plays a major role in promoting the use of peat energy and helping overcome constraints to its production which include: planning and programming, research development and demonstration, transfer of technology, dissemination of information, education and training and financing. A system of international collaboration would also further this aim.

It can be concluded that peat utilization is very limited but that the potential is high. If 2-10% of the total energy consumption of those developing countries which do have peat reserves could be covered by peat-fuel, then this would amount to the equivalent of 20 to 80 million tons of oil.

Burundi Case Study

Burundi has a manual peat production of about 5000-10 000 tons per year. It is being used on a small scale in households and the brick industry. Burundi now has two major peat programmes for industrial and small-scale usage. One aim is to develop peat production methods for the purpose of fuelling the Musongati nickel development project. The project would need 200 000-800 000 tons peat per year depending on the final technical solution and extent of nickel production.

Burundi´s Ministry of Energy and Mines is actively pursuing a national peat programme, in collaboration with the Finnish government and the Finnish peat industry. The country has 13 000 ha of peat deposits, mostly in the "Grand Marais", along the Akanyaru River. The deposits can be as thick as 30 m. The programme includes an assessment of the actual quantity of peat in the country, and research and development into the best suited methods of peat production and utilization in Burundi, both for large, industrial and small-scale domestic purposes.

Since the water level of the Akanyaru river is much higher than the bottom of the bog, conventional draining techniques cannot be used. It was decided that first the bog has to be cleared of papyrus vegetation and roots. Following this the peat will be pumped out as a slurry with a water content of 94-97%.

The peat will be dried in two different ways, the first of which is by solar energy: the peat is pumped as a slurry onto levelled fields. Once it dries to an acceptable moisture content of about 30%, peat can be cut. It has been envisaged that peat harvesting will be done in various manual and mechanical stages, to offer a lot of employment opportunities.

Peat deposits are also found at higher altitudes and can be drained using conventional techniques. Peat extraction is taking place at these deposits, but at a lower technological level. Manual cutting is being used, as well as production at an intermediate technical level, such as tractor mounted harvesters.

Burundi is also looking into uses of peat as a fuel for other industries and as a fertilizer. VITA is involved in a programme of peat stove development, which should contribute to the acceptance of peat as a household cooking fuel.

Conclusions

Peat is not a panacea for the solution of the energy problems of developing countries with peat resources. It may, however, represent sizeable contributions to national energy matrixes in those countries where sufficient deposits exist. As the example of Burundi shows, peat can supply a whole industry with fuel. Limited experience seems to show that peat production, except for the most capital intensive projects (which will probably be carried out largely by foreign companies and consequently will have to be paid in foreign exchange), will often be cheaper than imported oil which they will replace.

Peat is, however, not a convenient fuel because of its bulk on the one hand, and its relatively difficult ignition and combustion on the other. Its acceptance in areas where no peat-using tradition exists may be difficult to achieve.

Bibliography

Ekono. Consulting Engineers (1983) Peat Energy.
P.O. Box 27, SF 00131 Helsinki, Finland.

Jones, M.B. (1983) Biofuel Development in Central Africa: Technology Transfer from Ireland.
Botany Department, Trinity Collage, University of Dublin, Dublin 2, Ireland.

Kalmari, A. (1982) Energy Use of Peat in the World and Possibilities in Developing Countries.
Paper submitted at seminar on Peat, in Bandung Indonesia June 1982.
Ekono, P.O. Box 27, SF-00131 Helsinki 13, Finland.

Kalmari, A. Utilization of Peat Resources in Developing Countries.
Ekono, P.O. Box 27, SF-00131, Helsinki 13, Finland.

Mark, K. & Bogart, A. (1982) Oil Shale, Tar Sands and Peat.
Paper No. 7, Panel of High Level Experts.
World Bank - UNDP Joint Study.
Financial Requirements of Supporting Actions and Pre-Invest Activities in Developing Countries.

U.N. (1981) Report on the Use of Peat for Energy.
U.N. Conference on New and Renewable Sources of Energy, Nairobi 1981, A/CONF.100/PC/32.

VITA (1982) Peat Stove Projects in Burundi: A Review.
Volunteers in Technical Assistance (VITA), 1815 North Lynn Street, Arlington, Virginia 22209, U.S.A.

5 BRIQUETTING OF BIOMASS

Introduction

Briquetting is one of several techniques which are broadly characterized as densification technologies. Agglomerating residues and making them more dense can expand their use in energy production since such measures improve the calorific value of fuel, reduce the cost of transport to urban areas and may also improve the fuel situation in rural areas.

Raw materials for briquetting are commonly available and include residues and waste products from the wood industries, loose biomass, rubbish and other combustible waste products (see Chapter 3). Many of these waste products create problems and using them as an energy resource can solve some disposal difficulties. When using agricultural residues the ecological consequences must be considered. In many areas of the Third World, soil quality and the productivity of agriculture are being undermined as more people turn to organic wastes for fuel.

However, certain amounts of selected agricultural and processing plant residues, such as groundnut, coffee or rice husk, are available in large quantities and can be used for fuel without major implications for soil productivity. If technologies are used for agglomerating the residue or making it more dense, its handling can be considerably facilitated. At the same time the energy content is raised to a level approximately equivalent to that of fuel wood.

Another example is the conversion of charcoal fines (small fragments, powder etc.), into charcoal briquettes so that the fines are used and not wasted.

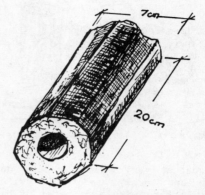

Biomass briquette from screw extruder.

The briquetting technologies can be divided by scale and compacting method. The division according to scale can be taken as:

1   large scale, done by motor on an industrial or regional basis

2   medium scale, done by bullock or motor on a village basis

3   small scale, done manually, on a family basis.

The compacting methods can be divided into:

a)   high pressure compaction

b)   medium pressure compaction with a heating device

c)   low pressure compaction with a binder or string.

Some raw materials, like straw or charcoal, do not remain together with high pressure alone so a binder has to be added.

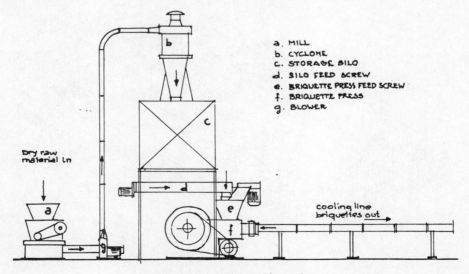

a. MILL
b. CYCLONE
c. STORAGE SILO
d. SILO FEED SCREW
e. BRIQUETTE PRESS FEED SCREW
f. BRIQUETTE PRESS
g. BLOWER

Diagram of briquetting plant.

Biomass briquettes from piston extruder.

Briquettes vary a lot in size and form, but usually they are of a cylindrical shape with a diameter of between 25 and 100 mm and lengths ranging from 10 to 400 mm. Square or rectangular briquettes also exist. Pellets are smaller and more suitable for large-scale, high-technology production and are therefore, not so well suited to developing countries.

Large-scale Briquetting

Where large amounts of residues are concentrated as in the wood industry (at sawmills) or in cash crop agriculture (at harvest time) another approach to briquettes has to be taken. The industrial presses for briquettes currently available are capable of exerting a pressure of about 1000 kg/cm^2. The combination of high pressure and the heat gene-

rated during the compression process breaks down the elasticity of the materials used and enables hard, solid briquettes to be produced without the need for a binder which would only hinder the functioning of the machine. The capacity of industrial presses ranges from 250-3000 kg/hr. The press is only one element in a complete production chain which will include: collection of waste, storage, drying, grinding, pressing, cooling, storing and transporting to user. The more stages that can be avoided, the more economical will be the briquetting. But the investment cost is high (about US$ 100 000). So in order to make it more economical when used for seasonal agri-wastes, it could be made movable so that it could be used at different times and places, according to the availability of raw materials. For a briquetting press with a capacity of 1000 kg/hr the power consumption is approximately 20-30 kW. To this must be added the power needed for the grinder and other auxiliary equipment.

At present there are three different briquetting technologies used: rind dye or roller head pelletizing machines, piston extruder and screw extruder devices.

Traditionally, briquettes were produced in rolling presses consisting of two drums with half impressions which line up at the moment of compaction. Much work is going on to find better compacting methods and other methods will soon be available.

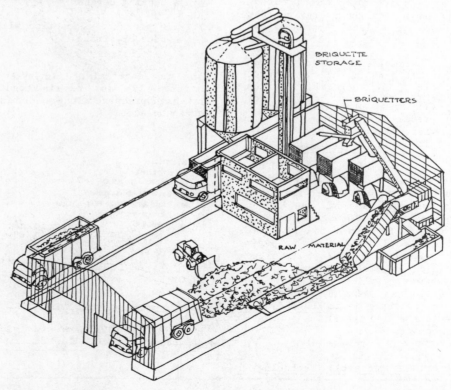

Biomass briquetting plant in Sweden, with a capacity of 120 ton/day.

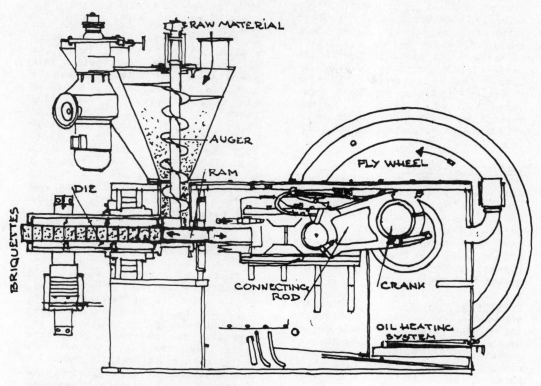

Principle of piston extruder briquetting machine.

The pelletizing machines consist of a circular die with holes where the pellets are formed (the die can be stationary or rotating) and edge runners (wheels) which press the material through the die while rotating inside it. This kind of equipment is expensive and complicated and economical only on a fairly large scale. It will usually not be suitable in a developing country.

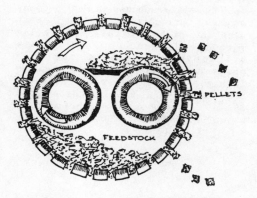

Principle of rotary pelletizing machine.

The piston extruder is the most frequently used briquetting equipment and is available from many producers around the world. It consists of a fly-wheel that operates a piston, which presses the material through a die where the briquette is formed like a long log and is later broken into convenient lengths.

The screw extruder consists of an extrusion screw and a die; the screw compresses the material and forces it through the die where the briquette is formed. Sometimes the material is pre-heated to enble the power to be reduced on the screw, and often the die is water cooled.

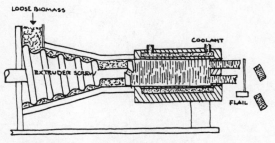

Principle of screw extruder briquetting machine.

Briquetting is being used more and more in developed countries and some projects are also operating in developing countries.

Intermediate Briquetting Technology

A small inexpensive machine and simple press that could convert small quantities of residues to briquettes economically would be very useful (especially if portable) for places where available waste is not sufficient to warrant the installation of the large-capacity machines now available. The capacity of this kind of village equipment should be somewhere between 25 and 250 kg/hr. Several different methods are available but each seems to have its problems.

Bullock-operated machines. Briquetting machines, with dies and punches, driven by a single bullock, have been developed by the School of Applied Research in Maharashtra, India. They cost about US$ 2400 each. The machine is very sturdy but the problem is the limited maximum production 25 kg/hr and the price of the equipment.

Power-operated machines. The same school has also developed a briquetting machine with two plungers driven by a 3 horse powermotor. The maximum capacity is 100 kg/hr and the price about US$ 4000. However, the pressure on the briquettes is not very high and it is necessary either to use a binder or to handle the briquettes with great care.

GAKO-Spezialmaschenen in West Germany produces briquetting equipment that uses the piston extruder compacting method and produces good quality briquettes because of the high pressure – although this results in higher prices and power consumption. A 150 kg/hr machine costs about US$ 12 900 and a 60 kg/hr machine about US$ 8800 and requires a power load of 8.5 kW.

T & P Intertrade Corporation Ltd in Thailand markets a press-screw system briquetter that heats the agro-waste before compression. This means that good briquettes can be produced without needing a binder and at lower pressure, resulting in cheaper equipment. Their "Eco-fumac" has a capacity of about 150 kg/hr, needs a 15 hp motor and three 2000 watt heaters and costs about US$ 5850. The grinder needs a 5 hp motor. Unfortunately a lot of energy is used by the heaters and there have also been some problems with other components.

It can be seen, therefore, that even if equipment does exist, the problems are not totally solved. Either equipment is too expensive with little capacity and too high an energy use, or poor quality briquettes result. There is still a need for a medium-size briquetting machine that is inexpensive, easy to operate, repairable using local tools and commonsense, energy efficient, reliable and which can handle different types of raw material. The advantage of medium-sized equipment is that capital investment is low and mechanized drying and special storage space is not required. In addition it would be practical for use in villages and in places with small wood industries or small agro-industries like groundnut oil mills, sugar mills, saw mills and paper mills. The briquettes could be used locally in bakeries, brickworks, potteries, curing houses, breweries, drieries or simply for cooking.

Bullock operated compaction machine.

Hand-made Briquettes

In many rural areas there are vast amounts of agricultural residues available, at least seasonally, together with biomass materials such as shrubs, twigs, bark, straw, hay, weeds, dry leaves, etc. However, in many such places there are great difficulties in obtaining firewood for cooking needs. The major drawback with the above-mentioned materials is that they all burn too quickly to be used for cooking purposes but this difficulty can be overcome if they are made into bundles or small faggots. Such a process diminishes the access of air and thereby slows combustion. Efficiency can be further enhanced if each bundle or faggot contains a piece of wood in the centre. Bundles may be fashioned, very simply, by hand, but their heat value increases if they are pressed using simple devices, eg presses made from rope, wood or metal. Bundles and faggots should not be looked upon merely as a poor wood substitute. Indeed, for many uses they are even more suitable than larger pieces of wood. Bakers, for example, traditionally use faggots composed of branches of between 3 and 5 cm in diameter to heat bread ovens.

Materials such as dry weeds, husks, cotton waste, coconut fibre, olive residue, fish waste, wood sawdust and municipal rubbish - can be converted into briquettes. This kind of material can be compressed quite adequately with the aid of simple hand operated presses. The pressure exerted varies from 10-1000 kg/cm^2 depending on the design of the press, and with such low pressures a binder is needed in order to prevent the briquette from falling apart. The higher the pressure that can be exerted by the press, the higher the density and heat value of the resulting briquettes. Thus, wherever possible, it is desirable to use equipment capable of producing high pressure.

The first operation in producing briquettes is the chopping of the chosen material for which a machete, broad axe or a hand-operated straw chopper may be employed. The next step is to blend it with a suitable binding material which can be done in a cement mixer or a specially made cylinder drum mixer. Next it is necessary to compress the material in some kind of press. Manual briquette presses have been designed by The Bellerive Foundation and VITA and prototypes have been made and tried. They consist of a mould and a piston. The piston can be operated through the hitting power of a hammer or by the pressing power of a lever.

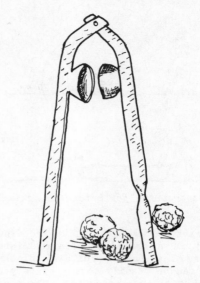

Simple hand operated briquetting press.

Tying the bundle.

A "baker" press for larger bundles.

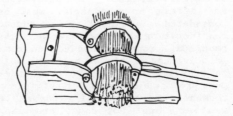

A small metal press.

Lever press for briquette making.

Mould for briquette making use in the lever press.

An earth block press (like the Cinva Ram) is a manually-operated piston press for making building blocks, but can be readily converted to produce excellent briquettes. The press can also be adapted to produce several briquettes in one movement. This design is called the TERSTRAM and is produced in Belgium by Fernand Platbrood.

The limitation of a hand-operated briquetting press is the low production of about 5 kg/hr or 50 kg in a 10-hour day. The advantage is the relatively low cost (an Indian hand press costs US$ 360). Storage is no problem because of the small amount of briquettes, but drying might be necessary depending on the moisture content of the raw material, and can be done outdoors in the sun.

Binders

Binders are needed when the pressure produced by the compacting equipment is too low for "self-bonding", or when materials are compacted that do not self-bond such as straw, rice husk and charcoal. In simple and cheap briquetting equipment, binders could be a solution to producing good quality briquettes. Some raw materials have internal binders and can be compacted with low pressure like bitumen in soft coal, gums in southern pines and tars in partially carbonized wood.

If satisfactory briquettes are to be produced economically, binders must meet a number of stringent requirements. Overall cost is the primary consideration in which cost of material, cost of application and effect on production must be considered. Availability is a second consideration where relative quantity required, transport needed and competitive use of binder material must be considered. Thirdly, the binder must produce a briquette of sufficient toughness to withstand exposure to weather, must not cause crumbling or excessive softening and during combustion exposure to heat must not cause disintegration and conse-

quent loss of fine pieces through the grate. Finally, the briquette should not produce much smoke, gummy deposits, an objectionable odour or dust during burning, storage or handling. Added binders should be combustible and preferably have a heat value at least as high as wood.

The majority of binders most suitable from a physical standpoint are too expensive to use in the proportions necessary for good briquetting. Inorganic materials such as cement, clay and silicate of soda are sometimes used but are objectionable because of increased ash, decreased combustibility, and disintegration during combustion. Organic binders usually increase the heat value, do not add to the ash content and do not disintegrate during combustion. Consequently, these are the ones most commonly used.

Commonly used binding agents include starches from corn, wheat or cassava, sugar cane molasses, tars, pitch, resins, glues, fibre, fish waste and certain plants like algae. Dung is also widely used, but this is unsatisfactory as its combustion is a major cause of lung and eye disease and it has other important uses as a fertilizer.

Of non-combustible binders, ash, clay or mud are the most widely used. Adding used motor oil increases heat value, but actually acts as an anti-binder and makes the briquette crumble.

Another way to make good briquettes is to use mixtures of several raw materials, e.g. hay, dry leaves, wood shavings, charcoal dust, sawdust and some binder.

One of the most interesting opportunities for utilizing waste products like sawdust, bark, agricultural residues of fine structure and grasses, is afforded by carbonizing and briquetting. Many kinds of neglected biomass e.g. lalang grass, water hyacinth, reed, lantan, etc. can be converted into char and char briquettes. Charcoal briquettes may be produced by preparing the charcoal first and then pressing it, by carbonizing wood briquettes after formation, or by heating the material under pressure so that semi-charcoal briquettes are formed.

Economy

Economy of briquetting is very site-specific. It depends on the cost of collecting the residues, the scale of production and transport requirements to the end-users. A better idea of the costs involved can be gained by considering some examples.

The UNSO/DANIDA project in Gambia.

The plant capacity is 5.52 t/hr. The annual production period is 5.5 months. The raw material is groundnut shells which have no other value. The work is carried out by one shift of 8 hours for two months, and two shifts the rest of the time. This means 1360 effective hours of operation and a total production of 7510 tons of briquettes (160 000 bags of 47 kg each).

The economic data assumed for this plant are the following (in US$ 1981 prices), estimated production is 7.507 tonnes per year. See Table below.

The price of fuelwood varied between 57 $/ton-62 $/ton.

It should be pointed out that in this case transport costs are unusually high. In spite of that, the cost of briquettes on an energy basis (i.e. per energy value) is, however, somewhat cheaper than the cost of commercial fuelwood with a similar calorific value of 4000 kcal/kg. It is also about one third of the corresponding cost for oil fuels.

In addition to this plant, a similar plant of equal capacity is also now being planned for the Senegal, again sponsored by Denmark. Very briefly, the comparable indicative economic data given for that plant are very similar to the Gambian plant and results in a retail price of 46.75

Raw material costs are also taken as zero in this case. The high labour costs in this case are based on official fixed minimum wages, although actual market wages are about one third of those. Labour costs may therefore be significantly overcalculated. However, in the high wage case, commercial fuelwood is also much more expensive.

Similar cost calculations for a Swedish sponsored plant in Nicaragua based on cotton waste with no competing value, gives a considerable cheaper cost, mainly due to smaller plant (1 ton/hour), more continuous usage pattern, and lower transport costs due to local use in small industry. Plant and interest costs are USD 7/ton, and maintenance, electricity, labour and transport costs amount to USD 2/ton each, adding up to a total of only USD 15/ton.

Going down the scale to still smaller plants of the village level size, calculations concern a plant with an output of 250 tons/year. Again costs do not include any charge for the raw material, but includes only plant etc., interests, energy, maintenance and labour costs, divided as follows (see Table on the next page).

Annual Cost of the Gambia briquettes project.

Plant	399 000 (10 years depreciation)	39 900 $/year	5.31 $/ton
Planthouse	69 825 -"-	6 983	0.93
Storage	83 125 (20 years depreciation)	4 156	0.55
Interest	(10%)	34 865	4.64
Maintenance of plant and generator		25 365	3.38
Labour		28 215	3.76
Energy (diesel)		28 262	3.76
Bags (for distribution)		15 152	2.01
Transport		71 345	9.50
Administration 10% of total cost		25 460	3.39
		279 917 $/year	37.29 $/ton
Wholesale price			37.29
Transport			1.82
Retailers cost and profit margin			9.8
		Grand total	48.91 $/ton

Source: Danida (1981)

The price of fuelwood varied between 57 $/ton - 62 $/ton.

Annual Cost of a Village Scale Plant

Plant machinery	4500 US$
Plant building	1000 US$
Depreciation 5 years	900 US$
Interest 20%	540/year
Energy	350/year
Maintenance	200/year
Labour	2000/year
Total annual cost, abt	4000/year
Total annual output	250 tons
Total cost/ton	16 US$/ton

Source: Dillner et al. (1983).

Briquette Technology Dissemination

The dissemination of the use of briquettes is dependent on a whole series of circumstances. There has to be knowledge that such a possibility does exist, there has to be an adequate raw material that does not have competing end-uses, there has to be a way to get the right equipment, there has to be knowledge of how to run and maintain it and finally, there has to be an accepted end-use for the briquettes.

The main problem experienced in the Danish project in Gambia seems to be that people do not want to use the briquettes for cooking purposes because they give off too much smoke. The same was experienced by the Aeroglide Corporation in Chile, Ecuador and Mexico; people were reluctant to use briquettes. However, recent experiments with improved briquette stoves in the Gambia have given a clear indication that those problems might be overcome in the near future. Although the Danish plant shows that the briquettes can compete economically with firewood and charcoal, the transportation cost is very high when one has to distribute the briquettes to many small end-users. They also had problems with the running of the briquetting equipment, with the feeding of raw material and with the electric motors.

Conclusions

In summary, all this indicates that it would probably be much easier to introduce briquettes as a fuel in projects with somewhat larger end-users. This would solve several problems as it would be easier to get briquettes accepted as a fuel, it would be simpler to get a combustion process that reduces the problems of smoke and it would make it possible to minimize transport and to organize the whole briquetting chain in a more efficient way. In a more industrial application, it would also be easier to run and maintain the process machinery.

This would mean that briquetting could be introduced at sawmills, paper mills, rice mills and in other factories that produce agroindustrial waste. Other bigger end-users could be bakeries, brickworks, potteries, curing houses, breweries and drying processes for agro-industries (e.g. tea, tobacco, coffee, spices, vegetables and cereals).

Another dissemination possibility would be locally on a village scale. Here the whole briquetting chain would be in one place, the transportation would be easier and the obvious advantages of briquettes in areas with fuelwood shortages should be incentive enough for the end-users. But here the machinery still needs to be perfected. The small scale equipment now available is too expensive, unreliable, uses too much energy or produces briquettes of too poor a quality. When these technical problems are solved it should be worthwhile to try production on this smaller scale. Other problems are the need for organizing finance at a village scale and until this happens, larger scale industrially orientated briquetting seems to be the most viable solution. Problems to be solved involve the training of workers to operate and maintain the equipment, and the eventual production of complete machines or spare parts in the country. Even if low-cost densification technologies such as briquetting were perfected, they would add value to previously "free" goods which would have effects on the distribution of income and serious consequences for the poor.

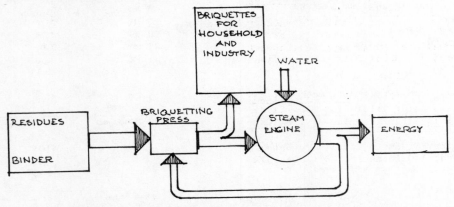

Principle for briquetting plant, producing its own fuel.

Bibliography

Agrowaste Compaction Machine (1982)
School of Applied Research, Vishram-
bag, Sangli, 416 415, Maharashtra,
India.

Danida (1981) The Gambia. Improving
the Production of Fuel Briquettes
in Kaur Plant and the Use of Agricultural Waste Products for Fuel.
Project no. UNSO/DES/GAM/81/006.
United Nations Sudano-Sahelian Office.

Dillner, P. & Saask, A. (1983) Bri-
quetting of Agro-waste in Villages.
SCARAB, Gärdesvägen 11, S-183 30
Täby, Sweden.

Hall, D.O. & Barnard, G.W. & Moss,
P.A. (1982) Biomass for Energy in
the Developing Countries.
Pergamon Press. Ltd., Headington
Hill Hall, Oxford, OX3 OBW, U.K.

Hausmann, F. Briquetting Wood Waste
by the Hausmann Method.
Fred Hausmann Ltd., Basel, Switzerland.

Janczak, J. (1980) Compendium of
simple Technologies for Agglomerating
and/or Densifying Wood, Crop and
Animal Residues.
FAO Forestry Dept., Rome, Italy.

Joseph, S. & Hislop, D. (1984) Residue
Briquetting in Developing Countries.
Intermediate Technology Development
Group, 9, King Street, London WC2E
8HW, U.K.

Journey, T. (1981) Charcoal Briquett-
ing Experiment at Kilifi Plantations
Ltd. Kilifi, Kenya.
A.T. International, 1709 N. Street,
N.W., Washington, D.C., U.S.A 20036.

Lichtman, R. Briquetting Agricultural, Animal and Forest Residues.
Energy Division, The World Bank.

Mandley, E. Briquettes the Alternative
Fuel.
Bogma Maskin A.B., Ulricehamn, Sweden.

McChesney, I. (1985) Briquetting
Wastes and Residues for Fuel.
Article in Appropriate Technology
Vol. 12, No. 2, September 1985.
Intermediate Technology Development
Group, 9. King Street, London,
WC2E 8HW, U.K.

Pinson, G.S. (1983) Report on Initial
Commissioning Trials for V.S. Extru-
der/Briquetting Machine at TPI, Cul-
ham.
Tropical Products Institute, Culham,
Abingdon, Oxfordshire OX14 JDA, U.K.

Reed, T. & Bryant, B. (1978) "Densified Biomass: A new form of solid
fuel".
SERI, 1536 Cole Boulevard, Golden,
Colorado 8401, U.S.A.

Reinecke, L.H. (1964) Briquettes
from Wood Residue.
Forest product laboratory, forest
service. U.S. Department of Agriculture.

UNSO (1981) Senegal. Development
of New and Renewable Energy Sources
and Strengthening of Energy Conservation Activities.
UNSO/DES/81/001.
United Nations Sudano-/Sahelian
Office.

6 CHARCOAL PRODUCTION

Introduction

For over two billion people in the developing world, fuelwood, charcoal and agricultural wastes are used as the predominant fuels, with fuelwood being the preferred fuel in rural areas since it has historically been readily available and requires no complex or expensive equipment to gather and use. Those in urban areas of the developing world, however, rely more on charcoal since it is a more convenient, energy dense and virtually smoke-free fuel and can be transported to urban areas relatively easily compared with fuelwood. Fuelwood, because of its bulk is more difficult and expensive to transport from distant plantations and open forests to urban areas than the more energy compact charcoal and hence charcoal has tended to dominate as an urban fuel. The energy content of charcoal is about twice that of air-dried wood.

However, charcoal is a wasteful fuel compared with fuelwood since conversion efficiencies in terms of energy content rarely exceed 25% when fuel wood is converted to charcoal in primitive earthen kilns. Also the burning efficiency of charcoal in traditional stoves is in the order of 20% leading to a total fuelwood to fuel use burning efficiency of only 5% for charcoal as against about 15% for fuelwood burned in its natural form on the traditional three-stone fire.

Taking into account the increase of urban populations at the expense of rural populations, as rural inhabitants drift towards the cities, the fuelwood use per capita may thus triple, as more and more people become dependent on charcoal. In addition, rural fuelwood use tends to come from branches, twigs etc, cut from tress outside the forests, whereas charcoal is usually produced from whole trees by clear-felling areas of existing forests. The increase in charcoal use has resulted, in many cases, in the complete removal of woody vegetation around urban areas and, as areas close to towns and cities become denuded of potential fuelwood and charcoal material, the pressure is transferred to areas parallel to main transport routes between urban areas. The pressure on the country´s forest resources is thus spread over the whole country. This pressure is not eased by the growing dependence of industry on charcoal as it seeks to move away from reliance on increasingly expensive oil fuels.

One way of reducing the pressure on trees is by introducing more efficient stoves and kilns. Existing energy-efficient woodstoves can achieve energy efficiencies of about 27%, existing improved earth kilns or brick kilns can achieve energy efficiencies of about 50% and existing improved charcoal stoves can achieve a figure of about 36%. This would improve the ratio of wood use between fuelwood and charcoal from about 1:3 to about 2:3.

More sophisticated and well-managed kilns can achieve energy efficiencies of 60-70%. Whether this can be achieved in practical and continuous use in a developing country is however uncertain.

Other means of reducing charcoal usage is to use more wood (as overall efficiencies for fuelwood are bigger than for charcoal), or to switch to other energy sources, if that is economically possible.

The growing of greater areas of forest for energy uses is also of major importance in alleviating the pressure on what remains of the developing world´s natural forest areas, and a great number of energy plantation

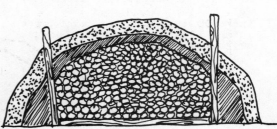

Simple earthen kilns dominate charcoal making in developing countries.

studies and trials are now under way. A rapid solution to the problem is urgently needed since it was estimated at the United Nations Conference, Nairobi 1981, that of the 3050 million of wood removed from the world´s forest cover in 1978, 59% was used either as fuelwood or was converted to charcoal.

The Carbonization Process

Charcoal is produced as a result of the chemical reduction of organic materials under controlled conditions i.e. the heating of wood with restricted air flow.

The carbonization process takes place in four main stages determined by the temperature reached in each stage. The first stage is endothermic (a reaction requiring heat input) and results in the initial drying of the wood to be carbonized. The stage is divided into three sub-stages in which water in wood pores is driven off through heating up to 110°C; water in cell walls is driven off with the wood heated up to 150°C; and finally, when chemically bound water in the wood is driven off by applying heat up to 200°C.

The second stage in the process is the pre-carbonization stage which is also endothermic. During this stage, which occurs in the 170-300°C temperature range, some pyroligneous liquids (in the form of methanol and acetic acid) and a small amount of non-condensible gases (mainly CO and CO_2) are obtained. This is followed by the third stage in the 250-300°C temperature range which is exothermic (a reaction producing heat). During this stage, the bulk of the light tars and pyroligneous acids produced in the pyrolysis process are released from the wood and the stage continues until only the carbonized residue of the wood - charcoal - remains. This is followed by the fourth stage in temperatures exceeding 300°C during which the carbon content of the charcoal increases as the remaining volatile content is driven off. Maximum operating temperature for this stage is about 500°C.

Once these stages have been allowed to occur, the final charcoal produced is allowed to cool - a process which can take from a few hours to many days depending on which of many different charcoal production kilns is being used. Once cool, the charcoal is exposed to air for at least 24 hours during which time it stabilizes and the risk of spontaneous combustion is reduced making it safe for packing, storing and transportation.

Ideally, charcoal should be produced in kilns capable of reaching high temperatures since higher temperatures drive off greater proportions of the acids, tars and volatile gases making the resulting charcoal purer with a higher carbon content. Good quality charcoal requires temperatures up to 500°C which produce charcoal with a fixed carbon content of 75%. However, high purity charcoal is more friable than charcoals produced in lower temperature kilns and so a maximum kiln temperature of 450-500°C is often considered the optimum.

A soft burned charcoal (carbonized at lower temperatures) has a high volatile matter content and a low fixed carbon content, is corrosive to metals, paper, fibre and packaging material (but not to plastic) and it also tends to smoke during burning.

Hard burned charcoal, on the other hand, is high in fixed carbon content, low in volatiles and is much more friable. Hard burned charcoal burns

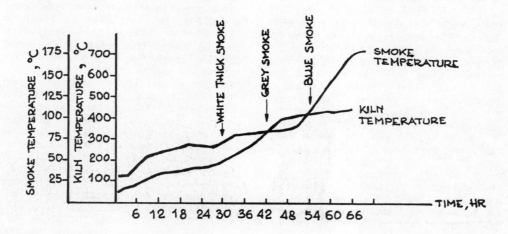

Relationship between smoke temperature, kiln temperature and time during a typical carbonization cycle for making charcoal.

Carbon contents of charcoal made
at different kiln temperatures.

Carbonization temperature degrees C	Proportion of carbon as % of dry weight of charcoal
200	52.3
250	70.6
300	73.2
500	89.2
700	92.8
800	95.7
900	96.1
1000	96.6

Source: Quoted in Foley, G., 1985.

cleanly but may be difficult to ignite. The ideal compromise product, when dry, has a fixed carbon content of about 75%; a volatile content of about 20%; an ash content of about 5% and a bulk density of around 250-300 kg/m^3. It is moderately friable.

In terms of net calorific value, charcoal has a value of about 30 MJ/kg compared with about 15-16 MJ/kg for air-dry wood meaning that double the energy can be obtained from a unit of charcoal than from the equivalent unit of wood. In terms of volume, charcoal produced represents about half the original volume of wood used in the carbonization process.

The Charcoal Chain

In order to fully appreciate the real energy content of charcoal compared with that contained in wood, it is necessary to take account of the complete chain of events in the charcoal production process. First, wood for the process has to be cut. If not cut with a hand axe and saw, an amount of fuel is required to power a chainsaw which acts as a negative factor in the final energy yield calculation. On the average approximately 35 m^3/ha of wood can be obtained although the production yield can vary from 20-100 m^3/ha.

Once cut and stripped of bark for the charcoal production process, approximately 20% of the area's wood is left as residue in the forest. Debarking of the trees is necessary in charcoal production to reduce the final charcoal's ash content. Wood stripped of bark produces charcoal with 1-5% ash content whereas charcoal made from unstripped wood

has an ash content of up to 29%.

Once cut, the wood must be allowed to dry for two reasons. First, dry wood is lighter than wet wood and is therefore easier to transport and second, if wet wood is put in a kiln a lot of energy is used to evaporate the water so the energy efficiency of the charring process drops. Moisture content in wood can fall from 100% to 40% in two months in countries with high ambient temperatures, if the wood is cut in half, thereby reducing the wood's weight by about 25%. When cross-cut and stacked, wood can be dried to a 30% moisture content in 6 months, 20% in 12 months and 16% in 18 months (Australian eucalyptus).

Since the weight of the wood against the weight of the final charcoal product has a great bearing on the ease of transportability (and hence the cost of transport and the amount of fuel used in transport), the distance from wood source to the kiln should be kept as short as possible since the charcoal's weight will be far lower than that of the wood to be converted. If the distance from wood source to the kiln is short, draught animal power can be employed whereas trucks requiring fuel are necessary for longer distances, thereby further reducing the energy efficiency of the charcoal production process. In this context the choice of earthen kilns, brick kilns or moveable metal kilns is of importance, as well as the grouping of kilns for better management and use of skilled manpower. In Brazil brick kilns are often built in groups of 7 which has been found to be the ideal number, for a cyclical management system (loading, firing, cooling, unloading etc.).

Charcoal manufacture can be made in different kinds of kilns with different yields and energy efficiencies. Charcoal yields depend on type and size of equipment and vary from 10-35% (dry weight basis) or 20-70% (thermal energy basis). Traditional earth kilns have yields according to weight of about 10-20%. These figures vary with tree species and with charring temperature from case to case. Improved kilns have higher yields of about 20-35% and will be described in detail later. However, the conversion efficiency is affected by a number of factors such as the skill of managing the production of charcoal in the kilns, how the kilns are packed with wood to be converted and the extent to which the carbonization process is controlled during the conversion stages.

A large proportion of the energy content of the wood lost during conversion to charcoal is in the form of by-products such as wood gas, wood tar and pyroligneous acids. Some kilns are capable of recovering these products either for return to the firing process to enhance heating or for other processes. By and large, however, the by-products are not recovered. The Casamanche kiln with a metal chimney can recover wood tar (about 25-35 kg/t of wood). In addition to the energy losses in the form of by-products, 20-40% of the final charcoal product can be lost in the form of "fines" - chips and powder. In a kiln about 10-25% of the charcoal is produced as fines, whilst during transport and handling a further 10-20% of fines are produced. These fines are difficult to utilize. Although briquetting is possible, it needs binders and presses and this makes briquetting of charcoal fines complicated and expensive.

The transport and distribution of charcoal (in bags) is expensive and therefore short transport distances with a minimum of reloading are preferred if possible.

Earth kilns: These have evolved from the earliest known methods of charcoal production in which wood was burned under turf or in a hole in the ground. Present day earth kilns have progressed relatively little since those early times and today they are divided into mound kilns and pit kilns.

Mound kilns, which are found throughout the developing world, were also the traditional form of charcoal manufacture in Europe. They take the form of a circular or rectangular stack of wood with a volume of 30-100 m^3 around a central post inserted for stability and covered with earth and leaf material with draught holes at the base.

A typical village type mound kiln is about 4 m in diameter at the base and about 1-1.5 m high. Fire is first kindled at the base and kept burning for about two days. When the centre of the stack is well lit, the top chimney hole (formed when the central support post is removed) is blocked and vent holes are made in the mound's shoulders. As the carbonization process continues, smoke from the mound changes from white to blue showing the rate of carbonization progress in the mound and the vents are blocked when the blue colour is seen. At this stage, other vents are opened up further down the mound. Once the burning is complete, all vents are blocked and the kiln is allowed to cool. The complete process takes 14-30 days depending on the mound size, and average charcoal yield from the wood is 10-20% although up to 25% can be achieved with a skilled kiln operator overseeing the process.

The mound kiln's main advantage is that it can be constructed in areas of rocky soil or high water table where pit kilns cannot be constructed. Such kilns need no capital investment other than for the tools needed for cutting the wood and clearing the land, and can accommodate large logs so reducing the time and effort expended in cutting the wood. However, controlling the through-draught, on which productivity of the kiln depends, requires a considerable element of skill and, unless an experienced operator controls the process, the extent of charring of the wood is rarely uniform.

The fuelwood to be carbonized can be gathered slowly over a period of months, stacked in position and allowed to dry out well before covering and burning. This fits in well with the life-style of the small farmer who may gather scrap wood, branches and logs and stack them carefully in the mound. In practice, poor yields of charcoal are common because it is difficult with large logs rolled into place, to make a well-packed stack, gas circulation is erratic and large amounts of uncarbonized wood result.

Swedish chimney kiln.

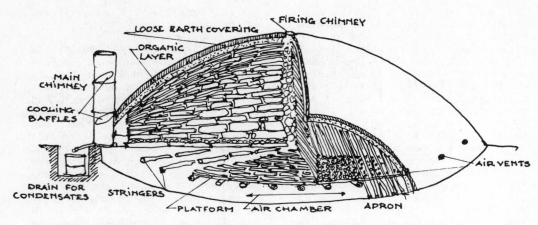

The Casamance kiln, an improved version of the earthern kiln using a chimney.

Research work in Sweden has improved the mound kiln by optimizing the flue system and the use of an external chimney to improve gas circulation. The chimney is connected to a flue construction beneath the pile. In Senegal, Ghana, Cameroon, and a number of other countries, this type of kiln has been adopted and modified under the name of the Casamanche kiln. This kiln gives yields of about 25%, or up to 30% with good management. The external chimney can be made of steel drums, and in some kilns the chimney contains baffles which encourage condensation of volatiles. The kiln is cheap, a drum chimney might cost about US$ 200, and it has all the advantages of the mound kiln. Therefore this is a very interesting technology for more efficient charcoal making, with little cash investment.

Pit kilns consist of pits dug in the ground and their size can vary considerably. In Sri Lanka pits are circular, about 2-3 m in diameter and up to 3 m deep. Timber of a wide size range is stacked until the pit is full. The burn is started from the base and the pit is left uncovered until the level of the stacked timber has fallen by 25-30%. The pit is then filled with fresh timber and the process is repeated three times. Once the third filling has been completed, the pit is covered with flattened oil drums and sealed with earth and sand. The pit is then left for about four days to allow

cooling to take place and the complete process takes about seven days. Charcoal yields from such pits are extremely low, in the region of 10-15%.

Larger pits, with volumes up to 30 m^3, are also used in some parts of the world. These consist of pits with a shallow end and a deep end both with vents. The timber to be carbonized rests on a crib so that the air entering at the shallow end can flow down the floor's slope and out through the vent at the deep end. The burn is started at the shallow end and carbonization continues towards the deep end with the timber stack sinking as carbonization continues. The complete cycle takes in the region of 60 days and the yield of charcoal from the timber is far better than from the smaller pit kiln types. Charcoal yields from big pit kilns are about 20-25%.

Capital investment for this kind of kiln is minimal, nothing more than a shovel, an axe and a box of matches are required.

Pit kilns can be improved by lining them with bricks or iron sheets to prevent contamination of the charcoal with earth and by installing chimneys and air vents to improve air flow through the kilns. If chimneys are installed, they should ideally be made of metal with diameters of 25-30 cm and should be installed on both sides of the pit, extending to the bottom. If the pit is covered with iron sheeting, a gap of approximately 15 cm should be left at either end, and the sides sealed with sand and clay. Kindling of the timber in the pit is carried out through one of the gaps which should then be closed when the fire has taken a firm hold, leaving a single gap open to regulate flow, its opening being gradually closed as carbonization progresses. When carbonization is complete, this gap and the chimneys should be sealed, and the pit allowed to cool.

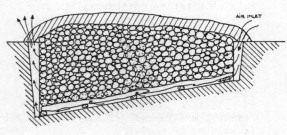

Pit kiln.

 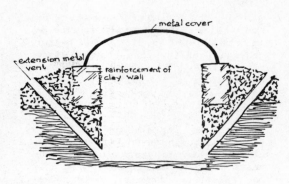

Improved pit kiln with metal chimneys and cover.

Charcoal yields from this type of improved pit kiln can reach 25% and, in kilns with brick linings, 30% yields have been reported. Moreover, in addition to the improved yields from pit kilns over mound kilns, the kiln can be used up to six times as against the mound kiln's single use, and less skill is required during operation. However, they require more labour input during construction, they can prove dangerous in that unwary inhabitants of the area could stumble into the pit by mistake and the proportion of "fines" produced can be as high as 30% of the charcoal's potential volume. In addition, the pit's nature makes uniform carbonization difficult and there is great variation in the amount of volatile matter released. Nevertheless, skilled operators using pit kilns can make excellent quality charcoal.

Metal kilns: Cylindrical metal transportable kilns for charcoal production were developed and widely used throughout Europe in the 1930s and the technology was refined in the UK during the second World War and later transferred to the developing world in the late 1960s. The design was taken up in particular by the Uganda forestry department which developed the Uganda Mark V kiln.

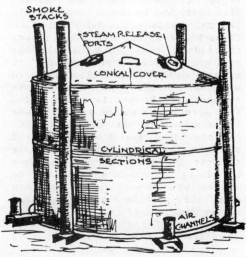

The Mark V kiln, a transportable metal kiln.

A version of the Uganda Mark V kiln was produced by the Tropical Products Institute (TPI) of Britain's Overseas Development Administration (ODA) with particular attention being paid to economy of construction, robustness, durability, ease of operation and maximum efficiency and productivity under conditions prevailing in the developing countries. There are lots of other designs and many manufacturers in different countries.

The TPI kiln's main features include a 3 mm thick steel cylindrical bottom section, a 2 mm thick cylindrical steel top section, a 2 mm thick conical steel cover section, eight inlet/outlet channels positioned under the kiln's bottom section and four chimneys which can be moved between the eight channels. Inside the kiln, 50 mm thick angle iron shelves, supporting the top section and cover, are welded to the inside of the top rims of the two main cylindrical sections. The rims can be filled with sand to provide a firm fitting for the joints. Four steam release ports are fitted in the cover section.

To fire the kiln, the lower section is first loaded with wood of specific sizes and shapes (450-600 mm long and up to 200 mm in diameter) and the top section is then placed onto the lower section. After the top section has been filled with wood, the cover is placed on the top section and all joints are filled with sand. The burn is started through the top cover's ports or through the firing points at the base of the kiln. The various air inlets are closed as carbonization continues, its progress being monitored by the change in smoke colour, or by observing the fire through the kiln's ports. The wood stack may be shifted between the air inlet and outlet channels as necessary to ensure even carbonization of the wood. Once carbonization is complete, all ports, vents and chimneys are closed off and the kiln is allowed to cool.

The main advantage of metal kilns over earth kilns is the extent to which their air flow can be controlled, which leads to yields in the 20-25% region, and better charcoal quality than charcoal from earth kilns. In addition, they can operate in all weathers, whereas earth kilns are prone to flooding in wet weather.

Metal kilns can also be used many times, although their life span can be under three years (up to five years in good conditions), and can be transported from place to place relatively easily. The production cycle is shorter, sometimes as low as three days, which means that under suitable organization the productivity can be very high. Production per cycle is 300-400 kgs for the TPI kiln. The main disadvantage is the cost which works out at about US$ 2000 - 5000.

The most vulnerable part of the kiln is the bottom section, the manufacture of air channels with more robust steel can increase lifespan but at increased cost. Besides the corrosive nature of the materials used in metal kilns, they also have the disadvantage of having to be filled with relatively small pieces of wood which increases the preparation time. From an environmental point of view the transportability of the kiln can be considered a disadvantage in that it can lead to increased deforestation if used in an irresponsible fashion over large areas of wooded landscape.

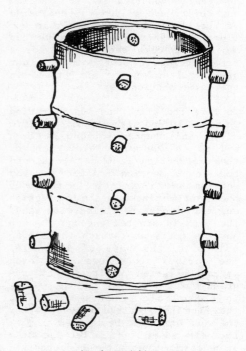

Oil drum kiln.

In addition to the portable steel kiln described, there are large numbers of smaller steel kilns, both portable and non-portable, normally constructed from used oil drums. They have been used for many years in south-east Asia and the Pacific for domestic and small-scale charcoal production.

In its crudest form, as used in the Philippines, for the production of charcoal from coconut shell, the drum is charged, partially burnt, and recharged 2-4 times, with water being added at intervals to prevent combustion proceeding too far at any stage. This can give 12-15 kg of charcoal per firing per drum, on a 4-8 hour cycle.

A 45 gallon oil drum can be used to char fast burning materials such as coconut palm timber, coconut shells and scrub wood. However, when operating with dense hardwoods, complete carbonization is difficult to achieve and the resulting charcoal is likely to have a high volatile content, although this is not a major problem with charcoal for a local domestic fuel. One man can operate a group of up to 10 oil drum kilns. The process involves a carbonization period of about 2-3 hours followed by a cooling period of about 3 hours. An experienced operator can cycle ten drums twice each day to produce a total output of up to 30 kg of charcoal from each drum. This means that a one man operation, using ten kilns, can produce 1.5 tons of charcoal per 5-day week, if supplied with adequately prepared wood.

Improved versions of the oil drum kiln have been developed to include chimneys to regulate the release of gases and to conserve heat within the kiln, and the addition of air vents to regulate air flow.

Compared with traditional methods of production, the conversion efficiency obtained in oil drum kilns is comparatively high with yields of about 23%-28%. The main disadvantage is that the raw material must be less than 30 cm long, with a maximum diameter of 5 cm, to achieve a satisfactory result. This means that a considerable amount of labour is required in the preparation of the raw material. Although capital costs are usually low, used oil drums are sometimes difficult and expensive to obtain. The drums tend to burn out rather quickly due to the thin metal used and have to be replaced fairly frequently.

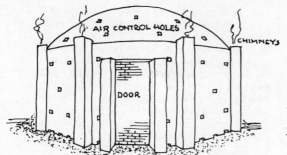

Brazilian brick kiln, the "bee-hive" kiln.

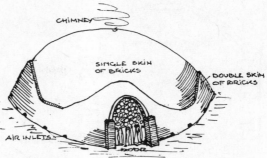

Argentine brick kiln, the "half orange" kiln.

Brick kilns: Properly constructed and operated brick kilns are without doubt one of the most effective methods of producing charcoal. They have proved themselves over many decades to be low in capital cost, moderate in terms of labour input required and capable of giving good yields (25-35%) of high quality charcoal suitable for both industrial and domestic uses.

The best known brick kilns are the Brazilian "beehive kiln" and the Argentinian "half orange kiln". The beehive kiln has reached a high level of development due to its extensive use in large-scale charcoal production for Brazil's steel industry.

In the plantation forests there are large-sized beehive kilns often operating in groups of 7-9, making up more or less centralized charcoal plants of up to 10-11 batteries (70 to 100 kilns).

A 45 m³ beehive kiln is usually 2 m high with a 5 m diameter and is topped with a dome roof. A firing hole is left at the top of the dome and a steel bank reinforces the bottom section where the dome meets the vertical walls. Two doors, one for charging, the other for discharging the produced charcoal, are built into the walls together with a number of air inlets and ports. Each kiln has several chimneys and the entire brick construction is plastered with mud to act as a sealant.

Operating the kiln is carried out by first charging the kiln with logs, the same length as the wall height, in vertical configuration on stringers, and with smaller logs laid horizontally on top of these up to the roof. Firing of the wood is carried out in much the same fashion as in other kiln types.

In the native forests there are smaller beehive kilns called "hot tail kilns" grouped in batteries of 4 located at the cutting site and operated by one man. Native wood charcoal making is, to a large extent, derived from a nomadic prac-

tice, involving the utilization of woody materials generated from the clearing of land for agricultural and cattle-raising operations. This "always on the move" characteristic, leads to the setting up of small charcoal-making systems, at low investment, and limited life-span (1-2 years). Work is totally non-mechanized; felling is done with axes, and the wood is transported in oxen-drawn carts over distances of 100-500 metres.

Another small-sized version of the beehive kiln is the slope-type beehive kiln which is built onto the side of a hill. They require fewer bricks and are less costly. They also require less skilled operation as there is only one air inlet. These kilns are extremely popular among the small independent charcoal producers in Brazil.

The Argentinian half orange kiln is similar to the beehive construction but the design is simpler in that the kiln has fewer vents and no chimneys.

Brick kilns have several advantages over other charcoal kilns. Although they cost US$ 500-1500 more than earth kilns, they are far cheaper than metal kilns, they require few steel parts, can burn far larger wood pieces, have higher yields than both other kiln types, produce good quality charcoal, have a lifespan of 5-8 years and require no manufacturing or maintenance skills which

Brazilian slope kiln.

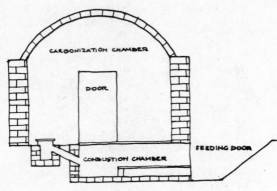

Improved brick kiln.

are not locally available. However, disadvantages include their lack of transportability and their long production cycles of eight days (four for carbonizing and four for cooling) to which should be added the time for loading and unloading.

Improvements continue to be made to brick kilns. Experiments are being carried out in Brazil on kilns with a separate combustion chamber underneath the main carbonization chamber (with yields of up to 36%) and attempts have also been made to develop methods of recovering tar.

Cement or masonry block kilns: Used widely in the USA in particular, these kilns go under the common name of Missouri kilns after their use in the hardwood forests of that State. They are large, up to 350 m^3 in volume, permanent kilns designed to operate in areas with severe winters and where labour costs are high. These kilns are rectangular in shape with a domed roof and are constructed using masonry blocks or poured concrete with steel re-inforcing. Large thick steel doors are fitted to allow mechanical handling equipment to load and unload the wood and charcoal. Burn and gas circulation control is similar to the operation of the metal kilns but thermocouples positioned at various points in the kiln to detect

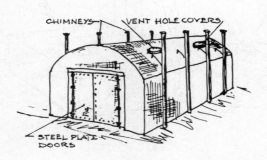

Missouri kiln.

hot and cold points are used to improve carbonization standardization.

The charcoal production cycle is long: in cold climates 25-30 days, in warm climates up to 33 days, and double that period if mechanical handling is not possible. There are very few Missouri kilns in the developing countries, because of their cost, which can be about US\$ 15 000 depending on size (this price is for a 350 m^3 kiln). Life can be about 10 years.

Unlike brick kilns they cannot be demolished and rebuilt. Hence a ten year wood supply must be available within economic haulage distance of any group of kilns. The amount for a group of some 180 m^3 kilns would be 120 000 m^3 approximately. About 4000 hectares of forest, capable of yielding 30 m^3/ha would have to be set aside for ten years to supply this amount of wood. Such an area would give a mean haulage distance of about 2.5 km, which is reasonable.

Charcoal yields from the kiln can be as high as 25-33% but their high cost and large quantities of sophisticated materials and equipment make them generally unsuitable for use in developing countries at the present time.

In Uganda a smaller and simpler kiln, the Katugo kiln, was tried in the 1970s. It had a capacity of 70m^3 of stacked wood and it was labour intensive. The cost was about US\$ 8000 and charcoal yields were about 25-30%. Its life was estimated at 10 years. But this is still double the price of a beehive kiln.

Retort kilns: These are ovens in which wood is heated to produce charcoal. Their advantage over other kiln types is the improved control over the charring process which leads to high quality charcoal. Retorts also produce gases (which, when condensed, are known collectively as pyroligneous acids) which can be fractionated into commercial products such as methyl alcohol, acetic acid and pitch. Retorts are more expensive per unit weight of charcoal than kilns because they are more elaborate and require an external energy source.

The Lambiotte retort was developed in Belgium; it cost about US\$ 2 million but has low labour costs, produces high quality charcoal, and has yields of 30-35%. It can produce some 9 x 10^3/t/yr of charcoal. It is an internally heated continuous system. A portable Lambiotte furnace operates in the Ivory coast with a production of 3 x 10^3/t/yr, and a capital cost of US\$ 500 000.

Most retorts developed are extremely high cost, high quantity systems but a number of smaller prototype systems are being developed for use in Africa.

One such system is the Constantine portable batch retort which consists of a steel cylinder mounted above a furnace designed to burn wood wastes. Hot flue gases pass through a jacket around the cylinder heating the inner cylinder in which wood is placed. Gases produced are routed into the furnace to assist furnace operation and to reduce the quantity of wood burned. Production rate of this retort type is 1-2 tons per day.

Partial retorts (or continuous kilns) also fall into the bracket of retort kilns. They consist of vertical insulated cylinders in which material to be carbonized enters through the top and the finished product is withdrawn from the base. Such systems are said to have short operating cycles, and to make uniform and high quality charcoal. An example is the Cornell retort which is heated by oil; with about 11 litres of oil burned in the process per hour, the Cornell retort can convert 4 tons of wood to one ton of charcoal in 4-9 hours. By-products are returned to heat the retort and overall yields vary from 22-33%.

The Norwegian Defence Research Establishment has tested a Cornell pyrolysis unit with a capacity of 2-3 t of charcoal per day. Most of the oil and gas produced is used to dry and heat the fuel, but some additional source of heat may be necessary. Overall efficiency is 22-33%, but the unit is small, lightweight and not extremely expensive. The price of a fullscale version is about US$ 40 000.

The Kumasi University in Ghana together with Georgia Tech., USA, has since 1980 tested a pyrolytic conversion plant which produces oil, char and combustible gases from sawdust and woodshavings. A similar, but more automated plant, is being tested in Thailand, using rice husk. The gas production is used to power a diesel engine for a rice husk mill. The plant consists of a burner, a dryer, a reactor, a condenser, and a demister. From 2,6 tons of wood it produces 330 kg of charcoal, 200 litres of oil in the condenser and demister in addition to combustible gases. Many problems have, however, occurred. The fixed bed dryer is inefficient and needs modification. The pyrolytic oil becomes very acid, and thus is difficult to store and use. Current plans for modification include a shift from producing oil to the production of char. The char could then be used as a wood preservative replacing dieldrin. Charcoal is produced as a powder, and one of the main problems is therefore how to best convert the powder into a usable fuel. Some work on briquetting is done, with pyrolytic oil as a binder. In addition, the current plant design results in low yield of the derived products. It therefore needs improving to optimize the yields.

At the University of Zambia a charcoal retort has been constructed for production of charcoal at high efficiencies and with recovery of by-products. The results from tests show that the retort developed is technically viable. However, analysis shows that the initially designed prototype with 4.5 m^3 retort chamber capacity does not render the project economically viable in view of the low return of investment obtained. Yields were too low in this small retort. A life-time of ten years was estimated, with a change of bricks

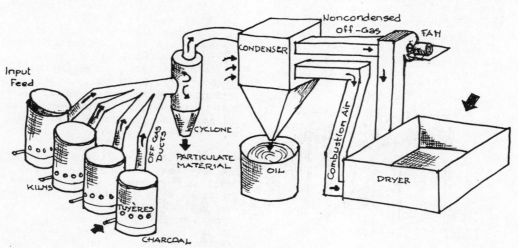

Diagram of a Pyrolytic Conversion Plant installed in Ghana.

every 2 years. A theoretical study was also made on a 10 m³ retort chamber, which seems to give better results. A new test plant is now being discussed.

Transport Of Charcoal

Distribution of charcoal, i.e. the packing, loading and transporting of the charcoal from the kiln to the point of wholesale or large-scale industrial use, can represent up to 25% of the total production cost. Therefore, charcoal should be produced as close to the market as possible. However, there are no longer many localities today where wood resource is close to the final charcoal user. That means that transport costs are significant and are becoming more so as forest resources are used up. Moreover transport requires scarce and costly liquid fuels. One way to get around this is to plant greenbelts round the cities, and fuelwood plantations close to industry in order to utilize the wood for charcoal production.

The cost of transporting charcoal in terms of energy is much lower than transporting fuelwood, because the calorific value of charcoal is about twice that of fuelwood. Hence, charcoal can tolerate a relatively high transport cost and implies that charcoal should be produced close to the wood supply and then transported to the market.

Charcoal easily absorbs water and therefore covers must be used during transport and storage to prevent wetting. Charcoal is fragile and fines are produced in loading and unloading operations. Hence the number of loading and unloading operations should be kept to a minimum. Costs are usually lowest when large unit loads are transported, but bulk handling and transport is not regarded by traditional producers as practical, partly because the lorries get dirty and are then difficult to use for other purposes. For this reason charcoal is usually bagged before transportation.

For short distances, up to 20 km, and for small quantities, transportation by mules and horses should not be overlooked. It has been used for many years in Brazil. Each mule carries two baskets of 60 kg each. Average transportation speed is in the order of 6 km/hr.

Water is the main adulterant found in charcoal. Good quality charcoal will normally have a moisture content of 5-10%. More than 10% is excessive and indicates rain wetting or adulteration to increase weight. That is why purchasing charcoal by volume has much to recommend it. It renders invalid the addition of water and other materials as adulterants and discourages mixing in fines to increase weight or handling the charcoal carelessly, thus producing fines, since this would, at the same time, reduce the bulk volume. Therefore, agreement to buy charcoal on a volume basis is fair to both buyer and seller, providing there is a previously agreed method of measurement.

Charcoal Production and Economics

Charcoal can be produced in a variety of systems. It can be produced in earth kilns by local farmers where the cost of the kilns is zero, but production is low and a lot of cheap labour is needed. It can be produced in a semi-industrial way, which uses animal draught transport and smaller more simple brick kilns. It can be produced more industrially and mechanized with chainsaws, mechanized transport and batteries of brick kilns managed by experienced workers. Or it can be made in huge, expensive but labour-saving high-technology retorts.

Brick kiln battery in Brazil, used in the iron industry.

Production costs for 12 000 kg of charcoal made in a Casamanche kiln in Senegal:

	US$	% of total
100 steres of fuelwood at US$ 1.77/stere*	177	28.0
Labour including packing	249	39.2
Loading trucks at kiln site (US$ 0.81/tonne)	10	1.5
Transport to main depot	34	5.3
Unloading at main depot	7	1.1
Bags	48	7.6
Tax at US$ 0.0048/kg	58	9.2
Contingency costs at 10% of cost pre-tax	52	8.1
Total	635	100%

Source: FAO, 1983, Simple technologies for charcoal making.

*A stere is 1 cubic metre of stacked firewood.

In most areas of the developing world where capital is scarce but labour is plentiful, improved earth kilns such as the Casamanche kiln can be an economic technology. Costs of production from such a kiln can be derived from an example of a 100 m^3 volume kiln in Senegal, the costs shown relating to 1978. Three men are needed for the work - see Table.

It can be seen that the main costs are the labour and wood, with bags and taxes also being of importance. As a result, the cost of the charcoal at the depot is in the region of US$ 0.05/kg to which has to be added the cost of transporting the charcoal to the main markets. This adds US$ 001-0.03/kg to the total price to the consumer plus, of course, the profit margin required by the supplier which can be up to 100% of the cost price.

With traditional clay brick kilns, studies have shown that although kiln construction costs are higher, labour costs decline dramatically.

In many developing countries there is a shortage of capital and a surplus of labour, therefore a serious social problem might arise if many charcoal makers were put out of business. In these circumstances some kind of compromise has to be sought. In Brazil it has been shown that manual methods with simpler brick kilns have the lowest total cost of US$ 8.08/m^3 of charcoal, when wood is taken from native forests and that semi-mechanized methods (with manual loading) give the cheapest total cost of US$ 11.64/m^3 of charcoal with wood taken from Eucalyptus plantations.

The same argument can be applied to the economic comparisons between earth kilns and metal and retort type kilns. Although the quality and production yields from the more advanced systems are far superior to earth kilns, their high costs prohibit their widescale use. For example, depending on size, brick kilns can cost in the region of US$ 500-1500; metal kilns such as the

Illustration of mechanized charcoal production.

Illustration of manual charcoal production.

TPI kiln can cost US$ 2000-5000; cement or masonry block kilns can cost up to US$ 8000 (as in the case of the Katugo kiln from Uganda); and retorts (such as the Belgian Lambiotte retort) with a 9000 ton per year production rate, can cost in the order of US$ 2 million.

The most promising ways to rapidly disseminate more energy-efficient kilns in a developing country therefore seem to be improved earth kilns (Casamanche kilns or lined pit kilns), or brick kilns (beehive kilns, slope kilns, and half orange kilns). To keep costs to a minimum and to maintain high energy efficiency in the operation, mechanization cannot be viewed as a main priority and draught animal power for transport should be used where ever possible to reduce dependence on fuel-consuming mechanized transport.

Environment

The environmental problems of charcoal production can be divided into problems with forests (i.e. deforestation and intensive cultivation of monocultures) and environmental problems in connection with carbonization (i.e. smoke, heat and byproducts).

The environmental problems of deforestation are well known and unregulated charcoal production is often contributary. Pressure arises from urban and industrial charcoal needs and, as charcoal is relatively energy dense and practical to transport, the pressure can, in rural areas, be far from the end use. This can only be prevented by adequate planning, so that the needs are filled within an ecological cycle of planting and cutting so that the stock is preserved and the yearly yields are efficiently used. The environmental problems of intensive forestry, and especially monoculture plantations of eucalyptus, are now more researched and understood. Man-made forests, like other agricultural monocultures, obviously bring about changes in existing ecosystems. Resulting environmental damage can be soil im-

poverishment, water table dry up and wildlife disappearance, but all these problems can be met with adequate planning and adequate forest technologies. Soil impoverishment can be controlled by leaving leaves and branches in the forest and by interplanting other plant species. Thus, with adequate management forestry soil can actually be improved. Mixed planting and animal husbandry between the trees can also give a diversity that allows different end uses (i.e. the needs of the local people) to be met at the same time as fuelwood is produced for energy use.

The total number of kilns at one centre must be limited due to chimney fumes as these irritate eyes and lungs. Charcoal manufacturing centres should therefore be located at least 2 km away from villages. The prevailing wind direction should also be taken into consideration and in some cases protective clothing (mainly against heat) should be used.

If by-products are recovered, care must be taken over their handling. Pyroligneous acid for example is a highly noxious corrosive liquid which must be treated properly to produce by-products for retail, or be disposed of by burning in conjunction with other fuel such as wood or wood gas. Although the economics

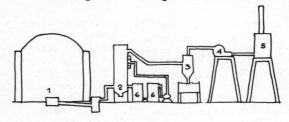

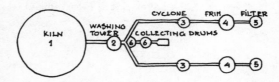

Diagram of equipment for byproduct recovery in charcoal production.

of by-product recovery are poor, recovery of byproducts does reduce atmospheric pollution from wood carbonization. The combined benefit might make it worthwhile to work in this direction. Another side line is the small scale recovery and sale of tars produced during the process. "Stockholm tar" has uses in veterinary medicine, as a caulking agent in boat building, and as a wood preserving paint or mastic. For its recovery a metal chimney is needed. The amount of tar which can be obtained is small: 40 kg/t of dry wood, i.e. about a 4% yield, but, in practice this often falls to 25-35 kg.

Charcoal Briquetting

Substantial amounts of charcoal are lost in the form of fines at the production site or through handling in transit and at the market site. The loss is about 20-30% of charcoal produced.

The more the charcoal is handled and transported, the more fines are produced. Fines cannot be burned by the usual simple charcoal burning methods and hence are more or less unsaleable. The introduction of briquetting equipment into charcoal production could save a vast amount of the resource lost in this form (see Chapter 5).

Charcoal is a material totally lacking in plasticity and hence needs the addition of a sticking or agglomerating material (binder) to enable a briquette to be formed. The binding material should ideally be combustible itself in order that the briquettes have a low ash content. The preferred binder, starch, is a food material and costs about ten times more per unit weight than charcoal. Hence, as a 4-8% starch addition to charcoal fines is needed to make the briquettes, it is a very significant cost item.

Some clays can be used providing they do not comprise greater than 15% of the briquette's volume since a high ash residue is produced with this binding material. Tar and pitch produced as a by-product from coal distillation or charcoal making can also be used as a binder under certain circumstances. Polyvinyl acetate and crude petroleum have also been tried as binding materials but few if any of these materials appear to be as effective as starch.

Four major steps are involved in the production of charcoal briquettes, namely:

1   preparation of charcoal fines; a hammer-mill type crusher and disintegrator are usually used;

2   mixing the charcoal fines with a binder; this requires efficient mechanical mixing and is considered one of the most critical operations; a kneader type, double shaft mixer is normally used;

3   briquetting of the mixture; two opposite-rotating rolls, that contain hollowed half moulds are most commonly used. The briquetting presses must be well designed, strongly built and are usually precision machines capable of high output;

4   drying of the briquettes; artificial drying of the briquettes is required, especially when starches are used as binders.

Briquetting can also be done on a very small scale by hand-mixing charcoal dust with a binder and placing this into moulds where it is left to set for 24 hours. It is, however, a very labour intensive method and of doubtful economic viability. Brick-making presses have been used but there do not appear to be any commercially effective, really low-cost machines for this purpose.

Briquette quality can be modified by the inclusion of a suitable additive. Low value materials such as sawdust and woodchips can be included in the mix to impart a "woody" tang to barbecued meat, and to lower production costs. Wax-coated charcoal briquettes are dust-free and wax mixed in the briquette ensures kindling.

Unfortunately, experience has shown, up to now, that although it is technically possible to briquette charcoal fines, it is too expensive, except where the price of lump charcoal is very high and the fines are available at a very low or zero cost. Most charcoal briquetting processes are currently found in the developed world where the briquettes are in demand as a fuel for barbecues, and where the briquette's special dust and smoke-free properties are a higher priority than the cost of the fuel.

Conclusions

Most charcoal is made by small producers either for their own local requirements or for a restricted market, usually a city. Relatively little international trade in charcoal is carried out and virtually no competition exists between producers in separate regions, the market for the product being largely localized. This being the case, the bulk of the developing world´s charcoal production comes from producers lacking access to sufficient capital with which the production process could be improved. They may not even possess an efficient axe or saw and must choose a method which requires the absolute minimum of capital investment. Improved technologies could therefore make a substantial impact on reducing the amount of wood required to produce the same quantities of charcoal, in places where it is possible to introduce them on a large scale.

Tradition, the embodied wisdom of rural societies, plays an important part in influencing the method of charcoal production and keeping efficiencies low. To use the established method which is known to work successfully in a locality is the logical option for those who cannot afford to take risks because of their precarious economic situation. Where social factors are dominant, it is usually very difficult to introduce a new technology of charcoal making unless the social factors are changed. Frequently one sees attempts to modify the technology of charcoal production by providing aid: inputs such as chainsaws, new kilns and so on. When these inputs are no longer available, economic necessity forces producers to revert to the traditional, successful method with all its obvious technical faults. Therefore carbonizing methods cannot be evaluated just on the basis of technical factors; social factors are of equal importance.

However, not least since charcoal production has a direct impact on deforestation, good technology is important in the long run in improving social and environmental conditions. Therefore, if social factors permit, methods which give higher yields of better quality charcoal at lower cost should be used.

In the short run, the three realistic possibilities found are improved earth kilns, which require the least cost, brick kilns of intermediate cost, and steel kilns which are the most expensive.

Earth kilns and pits even when operated efficiently, are slow burning and slow cooling, contaminate the charcoal with earth and have low yields. However, where capital is limited or non-existent and wood is plentiful, earth kilns are preferable. Improved earth kilns like the Casamanche kiln or the lined pit kiln can reduce contamination, improve quality and increase yields, and should be promoted.

Steel kilns have two advantages: they can be moved easily and they cool quickly, allowing a shorter cycle time. However, portability is not always a good thing; it is more difficult to organize and it can be misused and contribute to deforestation. Steel kilns may find use where mobility is of such over-riding importance that it overcomes the high capital and repair costs. Where oil drums are cheap and readily available, oil-drum kilns may be a good option.

Where some capital is available and a serious effort is to be made to produce quality charcoal efficiently, brick kilns will probably be preferred. Although cycle times are around 6 to 8 days compared to two days for steel kilns, the greater volume and much lower cost of brick kilns make them preferable except where portability is essential. Simpler brick kilns like the "hot tail kiln" and the slope kiln (which does not use so many bricks), and more manual and draught animal work can reduce investment costs when capital is scarce and cheap labour is plentiful.

However, improved methods are usually much less labour-intensive than the current ones, and charcoal production is often the only off-farm cost income available for rural peasant farmers. This means that interventions in the charcoal market, particularly those involving large scale undertakings with modern methods, may lead to serious social disruption, if not properly managed.

Such demonstration and dissemination should also be accompanied by large-scale forestation projects designed to alleviate the pressure on available forest areas, and by ecological studies aimed at protecting land areas from over-cultivation, the effects of mono-culture over long periods and other environmentally- damaging practices. Without these safeguards, there is a danger that more widely disseminated charcoal production systems could ultimately wreak the sort of havoc that the improved technologies are designed to eliminate.

Bibliography

Earl, D.E. (1974) Charcoal - an André
Mayer Fellowship report.
Food and Agricultural Organization
(FAO) of the United Nations.
Via delle Terme di Cavacalla, 00100
Rome, Italy.

FAO Forestry Paper 41 (1983) Simple
Technologies for Charcoal Making.
Food and Agricultural Organization
(FAO) of the United Nations.
Via delle Terme di Caracalla, 00100
Rome, Italy.

Florestal Acesita S.A. Belo Horisonte,
Brazil (1982) State of the Art Report
on Charcoal Production in Brazil.
Bioenergy Systems and Technology
Project No. 936-5709.
U.S. Agency for International Develop-
ment (USAID), Washington, DC 20523,
U.S.A.

Foley, G. (1985) Charcoal Making
in Developing Countries.
Earthscan, 3 Endsleigh Street, London
WC1 ODD, U.K.

Hislop, D. & Barnard, G. (1983) Char-
coal Report - Working Draft.
Earthscan, 10 Endsleigh Street, London
WC1 ODD, U.K.

Karch, G.E. (1980) Carbonization
- Final Technical Report of Forest
Energy Specialist, SEN/78/002, Casa-
manche Kiln.
FAO, Ziguinchor, Sénégal

Karch, G.E. & Boutette, M., Char-
coal: Small-Scale Production and
Use.
German Appropriate Technology Exchange
(GATE), GTZ, Postfach 5180, D-6236
Eschborn 1, Federal Republic of Ger-
many

Mobonga-Mwisaka, I. (1983) Charcoal
Production in Developing Countries.
Swedforest Consulting AB for SIDA.
(Swedish International Development
Authority). Industrial Division.
105 25 Stockholm, Sweden.

Paddon, A.R. & Harker, A.P. (1980)
Charcoal Production Using a Trans-
portable Metal Kiln.
Rural Technology Guide 12.
Tropical Products Institute, 56/62
Gray's Inn Road, London WC1X 8LU,
U.K.

Paddon, A.R. & Harker, A.P. (1985)
The Construction, Installation and
Operation of an Improved Pit Kiln
for Charcoal Production.
Rural Technology Guide No. 15.
Tropical Development and Research
Institute (TDRI), 127 Clerkenwell
Road, London EC1R 5DB, U.K.

Whitehead, W.D.T. (1980) The Con-
struction of a Transportable Charcoal
Kiln.
Rural Techn. Guide 13.
Tropical Products Institute, 56/62
Gray's Inn Road, London WC1X 8LU,
U.K.

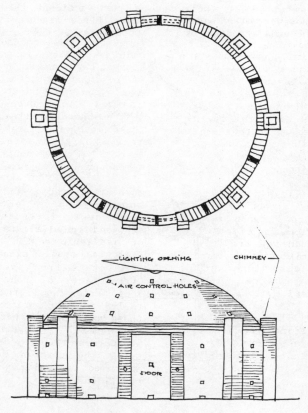

Brazilian brick kiln.

7 WOOD AND CHARCOAL STOVES

Introduction

For most developing countries wood is the economy's most important energy source. This means that about 2 billion people depend on various forms of biomass to meet their primary energy needs – mainly for cooking. In Nepal, fuelwood supplies 97% of the domestic sector's energy; in Upper Volta this figure is 90% and there are a great number of countries whose energy demand is predominantly met by burning wood. Actual consumption of fuelwood varies substantially in relation to climate, culture and availability, but 1 ton per capita/year is a general rate of the wood per capita used in Africa.

For more than a third of the world's people, the real energy crisis is thus a daily scramble to find the wood they need to cook food.

Woodstoves

Most people in developing countries use open, three-stone fireplaces for cooking. These fireplaces are often dirty, dangerous and sometimes inefficient. Dirty because smoke and soot settles on utensils, walls, ceiling and people. Dangerous because the fire is open and because the pots can easily tip over. The smoke irritates and is a known danger to health. The argument about their inefficiency is, however, often based on an incomplete understanding of

Three stove cooking place.

the role of the stove. In addition, the 10-15% energy efficiency often quoted for three stone fire places, is not that bad either. Rather advanced, and costly, designs of new stoves are necessary for a substantial increase of this figure.

Three-stone fireplaces, however, do not cost anything and they are very flexible. They can provide either slow or fast cooking, such as simmering and boiling. They give light, provide heat and a place of social focus. The smoke keeps insects like flies and mosquitoes away and is often used to preserve foods. Tar deposits from the smoke may also preserve thatched roofs. It is not by pure accident then, that three stove fires are as widespread and appreciated as they are!

Diagram of cooking hut with chimney stove.

Early stove projects (in the 1940s and 1950s), like the Magan-chula and Herl-chula in India, were aimed at the four-fold problem of health, housing (i.e. smoke and dirt), fuel economy, and forest economy. Later stove programmes e.g. the Singer-stove in Indonesia in the 1960s and the Lorena-stove in Guatemala (in the 1970s) were focussed on energy efficiency. Recent programmes have continued to look at energy efficiency mainly because of the problem of deforestation, and the increased workload for women to collect wood.

Most ongoing stove programmes are currently in their early stages, having been started in the 1980s. They also tend to be low budget and low profile. It is therefore too early to make evaluations. They need more time, money and not least opportunities to exchange results so that they can learn from each other's mistakes and successes. In particular, the two main approaches to stove development - the laboratory and the "in-the-field" approach - must be given enough time to benefit from each other.

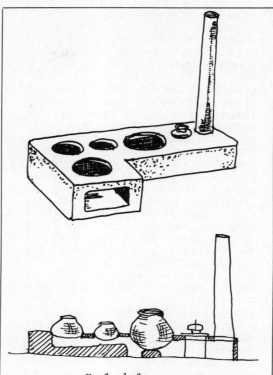

Herl-chula.

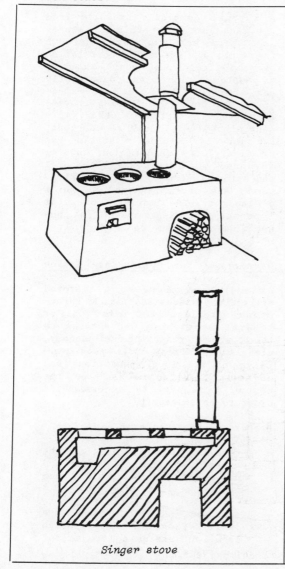

Singer stove

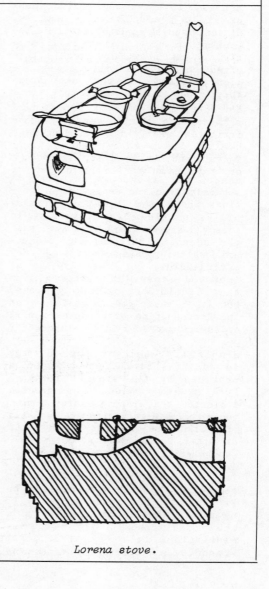

Lorena stove.

There has also been much controversy about the actual role of improved woodstoves in reducing deforestation. Laboratory tests show that a typical more efficient stove (like the Tungku Lowon) has an efficiency of about 20-25% compared to the 10-15% of an open fire. Women using such improved stoves usually claim fuel savings of 20-50%, varying from family to family. Such variation is expected, since stoves are used in many different ways. This is why laboratory tests are not enough. They only have limited validity as they seldom reproduce all the different, real-life, cooking conditions. Many studies also show that rural families seldom use fuelwood, originating from cut whole trees, but rather burn dry branches, dead wood, twigs and leaves, not adding to deforestation, but rather obtaining their fuel from trees outside the forest. In most places, deforestation is also much more often caused by agricultural and forestry practices like land clearing for agriculture and the logging of forests for industrial, and urban needs. For example, commercial charcoal making around cities usually involves the cutting down of whole trees in large numbers, directly adding to deforestation. The link between rural wood use and deforestation is therefore weaker than often assumed. For stove dissemination programmes in the rural areas this weakened link means that their effect on deforestation will be limited as long as they do not achieve almost full penetration.

Nevertheless, as discussed above, improved woodstoves have a significance that is much wider than is apparent from the energy considerations above. The fact that in many places people start investing in improved stoves means that they want the benefits offered: e.g. cleaner and healthier homes, improved fuel efficiencies, etc. The adoption of improved stoves is rather a sign that development is taking place, but it is misguided to believe that the introduction of woodstoves will necessarily result in development.

People will welcome fuel savings in addition to the other benefits provided by the stove. Experience from several woodstove programmes indicates that making energy efficiency the main criterion will usually not work since, in most places, wood or other replacement no-cost fuels can be still found.

The concepts of efficiency and the economics of woodstoves have to be broadened. Since these stoves will be used primarily in areas outside the "cash economy", it is usually meaningless to apply traditional economic criteria to the process

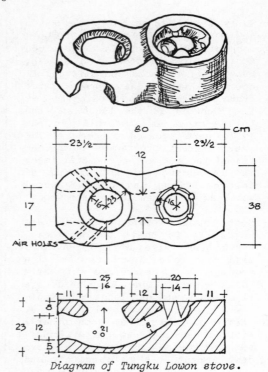

Diagram of Tungku Lowon stove.

of stove dissemination and adoption.

A further complexity results from the fact that more efficient stoves also become more expensive (although there is no automatic correlation between efficiency and cost). However, as people get richer (and consequently can afford to buy the more expensive, and better performing stoves) they tend to switch to other, more convenient fuels, such as charcoal, kerosene or eventually bottled-gas or electricity rather than wanting advanced woodstoves. Agricultural and animal residues, as well as firewood, are often regarded as fuels of the poor. The "market niche" for improved efficient stoves is thus not always easy to identify.

Development of Appropriate Stoves

As indicated above, any improved stove aimed at substituting the three-stone fireplace must offer one or several substantial advantages to the local user regarding fuel economy, cleanliness, safety etc. A good knowledge of the local conditions and contact with the users, i.e. the women, is therefore necessary in order to be successful.

Factors to be considered are:

- food and cooking habits: can the stove handle local dishes and different types of cooking, such as fast boiling, simmering etc.?

- size required: how large are the families; does the stove match local pot design and number of pot-holes required etc.?

- fuel requirements: does the stove work well on the preferred fuels?

- conveniency and safety: is the stove safe and convenient; is it mobile?

- space-heating or lighting requirements: does the stove give off light or excess heat (and smoke, if required for roof preservation)?

- stove price and life-time: is it affordable enough to become attractive?

- efficiency in practice, not only in the test-lab, and time needed to do the cooking?

The trade-offs between all the different factors involved in changing stove practices will also involve social issues like status and gender issues for example, is a husband willing to invest in a new stove rather than marrying another wife. In the final balance, then, the much publicized issue of energy efficiency and fuel economy may well appear far down the list of priorities for a husband or woman in a developing country, when considering all their options and trade-offs.

The complexities can be further illustrated by the case of having a chimney or not on a household stove. Using a chimney takes away the smoke inside the home, which would constitute a very important health improvement. In fact, recent pioneering work by K. Smith at the East-West Center, Hawaii strongly suggests that smoke from indoor cooking fires causes extremely severe health hazards for very large numbers of people in the developing countries. Smoke inhalation may in fact be the major contributor to the increased rates of respiratory diseases, reported by WHO to be the chief cause of death in developing countries. From this point of view, chimneys (or any other improved ventilation system letting out smoke efficiently) would

therefore be of utmost importance, and a change-over to chimney-stoves could mean a dramatically improved health situation in many developing countries.

In practice, however, chimneys may not always be the preferred alternative. They require closed-in stoves, letting out no light. Smoke preservation of thatched roofs against insects is also lost. In addition, if a chimney stove is operated with an open door (e.g. to let out light), energy efficiency is usually lower than a three-stove fire due to the increased draught caused by the chimney. Chimney stoves are also usually heavy, non-mobile and comparatively costly.

All this makes the introduction of woodstoves a very complex task, which is also the common experience of all serious stove dissemination efforts to date. There is no one perfect stove. Different stoves are appropriate for each culture, tradition, area and cooking habit. Sometimes even within one village, different stoves may be appropriate, since family sizes and their incomes are different. Fuels to be used in the stoves also vary a great deal, depending on the area, incomes and the amount of rain. A range of models is needed in each stove programme, or better, a systems approach is necessary where the basic components and concepts of a global design meet individual requirements.

The functions and maintenance methods of a new stove must be understood by the women. The users have to understand the new stove, they have to know how to use it, maintain it, and how to repair it. Especially the introduction of new devices like chimneys or dampers needs careful explanation and training. In order to achieve a long-term success, check-up of the new knowledge will be needed by the users. A technology that is not understood and accepted will soon be abandoned, as shown repeatedly by various aid programmes.

Stove Technologies

The technological principles that govern the behaviour of a stove, such as combustion processes or turbulent heat transfer etc., are well known and it is therefore possible to apply ordinary engineering notions to stove design problems, at least in principle. In practice, however, it is advisable to use also experience and practical advice from prospective users, since stoves are not going to be used in a lab but under very varied circumstances, and subject to view, mistakes and widely differing styles of cooking. In addition, con-

Open cooking fire with ventilating hood.

Table	MATERIALS CONSTRAINTS FOR WOODSTOVE DESIGN[*]	
	Advantages	Disadvantages
CLAY	- abundantly available - building easy - stable 4 safe - user built	- non uniform quality - control difficult - heavy, not portable - not easy to commercialise
CERAMIC	- abundantly available - quality control better - lighter, portable - marketed move easy	- material stringent - special kilns - risk of shattering - uncertain life
METAL	- relatively available - quality control possible - light, portable - excellent marketable	- not easily available - special machinery

[*]Modified from Krishna Prasad 1983.

siderations about cost, available material, local versus industrial manufacturing, quality control etc. are also important (see Table).

There are two basic approaches to making a stove efficient. One is concerned with the design of the combustion chamber, the other with regulating the air flow using a chimney and dampers. Question of heat transfer, insulation and regulation are other important issues.

For example, how big should the combustion chamber (firebox) be? What shape should it have? What should the distances be between the fire and the cooking pot and between the fire and the stove wall? Naturally, available pot sizes and fuel types determine the size of the firebox, which in turn is crucial for energy efficiency. Baffles and bends under the pot holes increase heat transfer to the pots, while the shape and material of pots and pans and available construction materials are also of importance.

Chimneys create draughts, so dampers are needed to regulate combustion. A chimney has to be properly constructed and regularly cleaned, or it might turn into a dangerous fire hazard. Chimney stoves also have to be attended more during operation, but allow for better regulation.

Depending on the skills, economic resources and materials available, different materials can be chosen for making the preferred stove, whether it is in a high mass, stationary design or a small portable metal bucket-type design. These designs represent the two basic stove types presently used in most stove programmes, even though there are many intermediate designs. The high-mass type is most useful when space-heating is also required, or when cooking habits require long preparation times. The lighter design is faster to heat up and seems to be preferred when no space heating is required, and meals are prepared quickly.

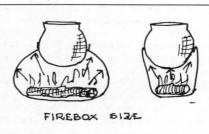

FIREBOX SIZE

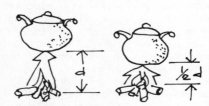

DISTANCE FROM FIRE TO POT

SITING OF POT

Factors of importance in combustion chamber design.

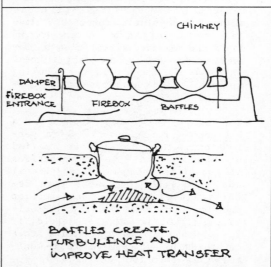

BAFFLES CREATE TURBULENCE AND IMPROVE HEAT TRANSFER

Factors of importance in regulating the air flow through a stove.

Stove Tests

The purpose in measuring the efficiency and other properties of stoves and fires is to obtain a quantitative knowledge of their behaviour. This is necessary when trying to improve a stove, or to compare different stoves. However, there is no "correct" definition of efficiency. A lot of the reports on efficiency that appear in the literature are thus not comparable. The problems involved for example, include moisture content of fuelwood, the density of different wood, the mass in cooking pots and stoves, variation in thermal performance over time, etc. To be taken seriously, efficiency claims of various stove types should therefore be received critically, and the test procedure taken into account.

In December 1983 a meeting was held at VITA (Volunteers in Technical Assistance) where agreed Provisional International Standards for Testing the efficiency of Wood-burning Stoves were established for the first time. Three basic tests were agreed, involving procedures of increasing complexity.

The first test is the laboratory-based water boiling test, a short and simple simulation of a standard cooking procedure. A water boiling test was accepted to measure accurately fuel consumed and time required for simulated cooking tasks. A high power and a low power phase of the test was agreed upon to demonstrate fuel efficiencies for boiling and frying as well as for the simmering and slow cooking of foods.

The second test, labelled a controlled cooking test, was designed to compare cooking on different stoves and to determine whether a stove can cook the range of meals normally prepared in a given area. This test, to be conducted in a laboratory or demonstration centre, provides a measure of standardization while allowing for local differentiations: a real meal is prepared using local foods, cooking methods, and fuel, but under controlled conditions.

The third test agreed upon was a kitchen performance test. With permission of the cook and over a period of time, normal cooking practices are observed using both traditional and improved stoves. The test determines relative fuel consumption and demonstrates efficiency and fuel-saving potential to the household. It allows study of the impact of a stove on household energy use, and it monitors long-term fuel savings.

It must be understood that a lab test efficiency and an actual efficiency in practical use is never the same thing. It must also be understood that people perform differently when a stove is new and when they are monitored.

One of the major difficulties in reviewing the evidence for actual fuel saving with improved stoves is that very few detailed follow-up studies have yet been carried out. This, in turn, depends on the fact that very few successful stove programmes have yet reached the stage where they can be properly evaluated. Even if it is obvious that stove tests can only give a limited answer to the question of actual fuel efficiencies, everyone seems to agree that well designed and well understood stoves can help people save fuel. On the other hand, it is not possible to predict in advance how much they will save and how much this will vary from user to user. Tests make it possible to compare different stoves from an energy point of view. But the figures cannot be used to calculate precisely how much energy would be saved if a certain number of woodstoves with a certain efficiency were disseminated.

A well-managed open fire has an efficiency of about 10-15%, a shielded fire, like the Luoga stove, has an efficiency of about 20%. Ceramic liner stoves like the Tungku Sae have efficiencies of about 25%. Higher efficiencies with woodstoves of 30% or more have only been accomplished with small pieces of dry wood properly cut and well attended.

Charcoal stoves of metal have efficiencies of about 20%, and charcoal stoves of ceramic material have efficiencies of about 25%. More efficient designs with ceramic or vermiculate liners have achieved efficiencies of 30-35%. High efficiencies of close to 40% have only been accomplished with very sophisticated and well-attended charcoal stoves, too expensive to be mass-disseminated.

Recent evaluations of kerosene and other types of stoves provided some information on urban energy problems, where the primary energy source is different and the distribution system is better for the commercial types. Efficiency of the wick stove, pressurized stoves and gas stoves were compared under standard conditions. The wick stove showed an average efficiency of about 42%, the pressure stoves 53% and the gas stoves 57%. The spirit burner has the highest efficiency (61%) among the stoves tested. As a comparison, electric stoves are considered to have efficiencies of about 50%, or more, although special devices like immersion boiling heaters can achieve efficiencies of about 90%.

High-Mass Chimney Stoves

Much of the early work on improved stoves concentrated on heavy stationary stoves with several potholes, equipped with chimney and dampers, made of brick, cement or lorena (a sand-clay mixture). The main problems with this kind of stove are that they are expensive to build and do not save much fuel because much energy is used to heat all the heavy material involved.

The Lorena sand-clay stove takes a long time to build and includes a great deal of material. A heavy mass is advantageous if space heating is required and the stove is burning most of the day. But if it is only used for cooking, especially during limited times, much heat is lost in heating up the stove.

Another main drawback to such stove designs is their short lifetime. They have to be replaced regularly, and often are not properly built. There is usually a considerable wear in the inner parts of the stove in time. Since the inner dimensions of the stove are important for its performance, efficiencies go down as the stove deteriorates. These drawbacks in performance have led stove development to concentrate on somewhat more light-weight designs.

Two other examples of heavy mass stoves are the Nouna brick stove and the Kaya precast concrete stove, both used in Bourkina Fasso. The main problem is that the costs tend to increase for those types. The Nouna stove for example, costs US$ 20 even after the cost has been substantially reduced by government subsidies.

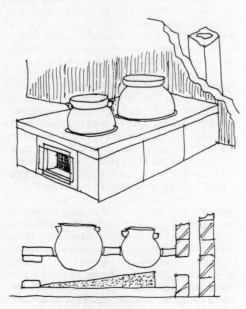

Kaya stove.

One Pot Chimney-less Stoves

A common experience from many of the stove testing programmes has been that one pot-hole stove without chimneys are usually the most energy-efficient ones in practical use. In fact, even if several pots have to be cooked at the same time, it is often most energy efficient to use several one pot-hole stoves, not least since each can be individually regulated.

Most home-built stoves are made of locally available materials like mud or sand-clay mixtures, sometimes mixed with straw, dung or cement to reinforce the material. Usually, the life-length of those constructions is very limited. The Louga stove in Senegal, for example, rarely lasts more than a year. In order to increase the life-time, three stones or ceramic liners could be incorporated in the structure as support for the pot. Training and information about maintenance is also of importance for increasing the life of the stove.

Ban-ak-Suuf or Louga stove.

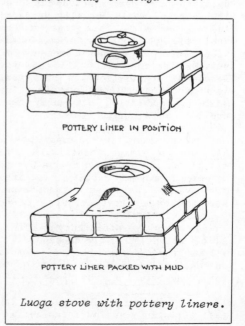

POTTERY LINER IN POSITION

POTTERY LINER PACKED WITH MUD

Luoga stove with pottery liners.

Ready-made stoves can be made of ceramics or metal. Most ready-made stoves are made locally in small workshops and such stoves are by far the most efficient of the stoves now being developed. Unfortunately, though, the use of more sophisticated materials in the construction and the fact that they have to be manufactured by local artisans, adds to their cost which could prove a serious limitation on their widespread dissemination.

There are, however, several stove projects that claim success in constructing and disseminating cheap, ready-made one pot stoves. The Botswana Renewable Energy Technology (BRET) Group's efficient metal stove is one example, the "Thai bucket" another.

The BRET stove involves double metal walls and a metal shield to protect the pot.

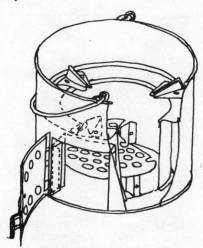

BRET double wall metal stove.

The Thai stove involves a ceramic liner. There are models of ceramics only as well as more expensive models with a metal bucket on the outside of the ceramic stove. The stove has no chimney and the walls are sloped to fit different sizes of pots, thus reducing the gap between the stove wall and the pot. The efficiency of these stoves is about 30-36%.

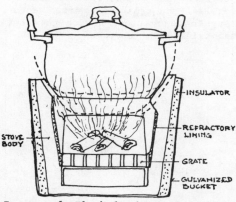

Improved Thai bucket stove for wood.

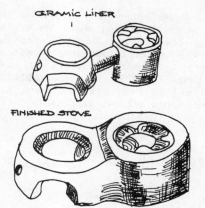

Tungku Lowon stove with ceramic liner.

Two Hole Chimneyless Stoves

In Sri Lanka and Indonesia, two-hole stoves are usually preferred to the one hole model. One typical design, the Tungku Lowon, is made of mud, but has problems with its short life length and difficulties with quality control. In the Tungku Lowon programme, quality control was quite critical, since the efficiency of the stove depended heavily on the correct sizing of the fire chamber.

One way to reduce this problem is to use ceramic liners, produced by trained craftsmen, but incorporated in the structure of the home-built stove. This approach increases the effective life of the stove as well as the efficiency, while still not increasing the cost too much. This type of stove is called the Tungku-Sae and has been introduced with some success in both Sri Lanka and Indonesia.

Light Chimney Stoves

As discussed above, in many places the smoke is a greater and more acute problem than fuel shortage. Despite their increased costs, chimney stoves are therefore preferred. Recent work on home-made stoves has concentrated on designs with smaller mass, and designs that are simpler to build, like the Nada Chula. As with other stove types, an important part is to understand how to build the stove properly and how to use it. Therefore extensive training programmes and follow-ups are needed, as well as good handbooks in the local language for the training of builders.

The Nada Chula consists of sun-dried mud slabs covered with a mixture of clay and chopped straw. The Nada Chula has a back damper in front of the chimney and has two pot holes with baffles under each one. The size and number of potholes and the overall size of the stove are determined in accordance with the user's requirements.

The success of the programme is partly due to the fact that women are training other women on how to build and use the new efficient stoves. Either you build your own stove or have it built by a female stove builder. The metal chimney and dampers are bought from a local blacksmith.

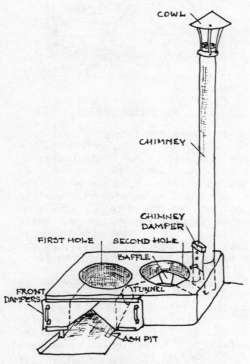

A 2-holed Nada Chulha.

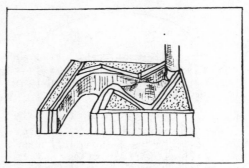

Making a Nada Chulha stove.

The Nepali ceramic liner stove, is a similar chimney stove with ceramic liners. Ceramic liners have a cost factor since they have to be made by local artisans and sold to users. The material for producing the ceramic liners is available in developing countries and the cost of such parts (about US$ 1.5 each) is sufficiently low to allow such systems to come into widespread use, providing the quality of manufacture does not deteriorate. Expected lifetimes, however, will only be attained provided that the stoves are cleaned regularly and training is often necessary to teach users how to go about maintaining their stoves.

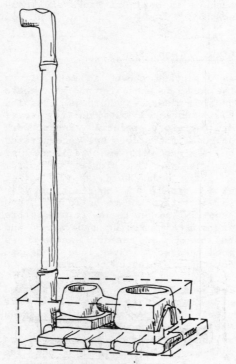

Nepali ceramic liner stove.

Charcoal Stoves

Charcoal is often preferred as a domestic fuel, especially by the urban population. This is mainly due to the fact that charcoal has approximately double the energy content of wood per unit weight, and is therefore easier and cheaper to transport. Charcoal is also virtually smoke-free and consequently the stoves do not need a chimney. Stoves are usually lit outdoors and then carried indoors to do the cooking.

Simple charcoal stoves are usually made out of metal by local metal workers, often out of scrap metal such as used oil drums. The stoves made in Kenya, for example, are circular and are known as jikos while rectangular stoves called malgaches are generally found in West Africa. Most jikos and malgaches measure approximately 25 cm across by 15 cm high and contain a perforated grate.

Pot supports are fixed to the top of the device and charcoal is fed in by lifting the pot. Ash is removed from a small side door which can also be opened and closed to regulate air flow.

These stoves have efficiencies of around 20% and cost Ksh 20. Besides the use of metal, charcoal stoves are also made from pottery as in the Indonesian Anglo which has a fuel-burning efficiency of 25%. In Thailand the metal and clay Thai Bucket was introduced for charcoal use in 1920. The charcoal version being a smaller version of the wood-burning type discussed above, consists of an outer metal skin with an inner lining and grate of fired clay. The space between the metal and the clay was insulated with ash from rice husk. It is now used all over Thailand and has an efficiency of around 30%.

Many tests were made in Kenya on different charcoal stoves. Some of the models of metal jikos were lined with cement or vermiculate. The latter is a light-weight rock that is produced by the heating of mica - a mineral composed of aluminium silicate with other silicates - which expands to about 10 times its original volume. Some jikos were fully lined, others only partially, at the top. The tests showed that this way it was possible to make both efficient and cheap charcoal stoves. The tests also showed the importance of the design and the position of the grate, as well as that it had to have a sufficient number of air holes (approximately 15% of the area). A Kenyan version of the Thai bucket costs Ksh 65 and also has an efficiency of about 30%. The jiko whose top

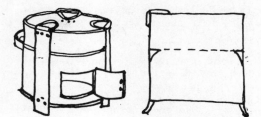

Traditional metal stove ("jiko").

Tradtional ceramic stove ("anglo")

Thai bucket stove.

Vermiculate liner stove.

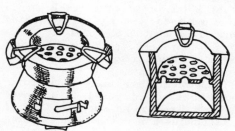

Ceramic liner stove.

Umeme stove.

is lined with vermiculate costs about Ksh 30 and has an efficiency of about 30-35%.

A more expensive jiko designed to have efficiencies in the region of 34-38% has also been developed by UNICEF in Kenya. It is called the Umeme stove and includes a pot that sits down into the stove. It contains a sloping inner combustion chamber made of metal that is insulated from an outer metal cladding by a layer of ash, sand or clay. It also has an air-intake slide. However, at Ksh 100 its cost may prevent its widespread use.

To reduce the price a simpler version the Haraka has been developed.

This stove involves a cone shaped grate, a double metal wall for insulation and a shielded cooking pot.

Work with the improvement of charcoal stoves is going on at several places, and it seems clear that efficiencies of 30-35% are possible to achieve, without making the stove too expensive. Cheap, improved Thai-bucket type stoves can reach efficiencies of and above 35%.

As mentioned above, there are many reasons why charcoal is usually preferred to wood in urban areas, e.g. transportability, convenience, cleanliness. However, even given the high efficiencies of charcoal stoves, the use of charcoal for cooking is always less energy efficient than cooking with wood. The reason is the large energy losses (some 75%) in charcoal making in the primitive earth kilns which completely dominate (see Chapter 6). Even compared to an open fire, a standard jiko uses about three times as much wood for the same end use. Seen from the wood-fuel point of view, a rural person moving into a city and switching over from wood to charcoal for cooking, therefore increases his wood requirements two to three times.

If charcoal were made in improved earth kilns and used in improved stoves, twice as much wood would still be used compared to an improved woodstove.

Cooking with a Umeme stove.

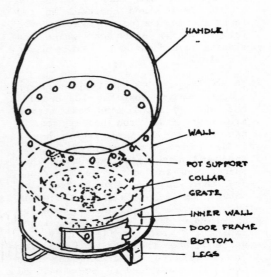
Diagram of Haraka stove.

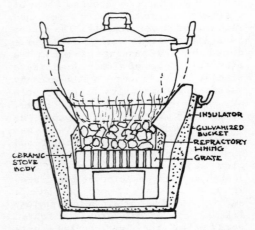
Improve Thai bucket for charcoal.

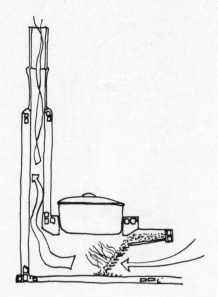

Rice husk stove.

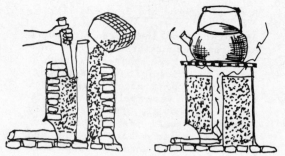

Packed stove for sawdust.

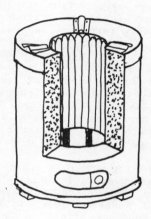

Chinese coal stove.

Coal briquette stove.

Nofflie "briquette" stove.

Other Stoves

Depending on the availability of special fuels (saw-dust, peat, coal, briquettes etc.) or the need to prepare special types of food etc. (baking, beer brewing), special stoves will have to be used.

The packed stove is designed to burn sawdust, rice hulls and similar fine grain fuels. They are usually made of ceramics or metal, and filled with fuel which is burned by regulating the air flow with small sticks making holes in e.g. the sawdust. Special stoves are also made for coal, similar to charcoal ones, but with thicker lining due to the higher temperatures. The problems with coal burning are that it is smoky and difficult to ignite. Peat burnt in stoves has similar characteristics.

The possibility of using briquettes made of wood, agricultural residues, peat etc. as a household fuel has aroused some interest recently (see Chapter 5). One advantage would be that briquettemaking would make available fuels that are otherwise normally wasted, such as rice hulls, straw, shells etc. Another advantage would be that briquettes are much more transportable than wood, due to their higher density. The "conversion" from biomass to briquette does not involve large energy losses, particularly as compared to charcoal. Good briquette stoves are now being developed, for example the "Noflie stove" in the Gambia. This stove works well with both briquettes made from peanut shells and with fuel-wood. Tests with this stove are now underway and results are largely positive, indicating relatively low smoke emissions, and a life lengths of about 2 years.

Cooking Methods and Hayboxes

Different cooking methods require different amounts of energy. In some cultures, where there has been a long tradition of fuel scarcity, cooking methods have developed so that they are very fuel-efficient. In China, for example, food is usually chopped in small pieces, beans are soaked in water before cooking and pots are often placed on top of each other so that several foods can be cooked at once, the ones that need most cooking at the bottom and the ones that need less cooking at the top. The foods are also often cooked or fried for a very short while. Another route to energy efficiency is in having a food industry distributing pre-cooked foods.

A different approach is to have better pots e.g. with tight fitting lids on aluminium pots that can be sunk into the flue from the fire. Other methods include pressure cookers, insulated pots or pots for special purposes like the modern samovar version that exists for boiling water in New Zealand, "Thermette". Water can be heated very efficiently in this with charcoal or wood but no other foods can be cooked in it.

An interesting method is the ancient "fireless cooker". It utilizes the principle of retained heat and may cut fuel needs at least in half. First, food is briefly cooked on a stove, much shorter than traditional cooking time. Then a thick insulation is put around the pot and the food spends time in this "fireless cooker" to complete the cooking process using the heat accumulated in the pot and the food. This time is longer than that required for traditional cooking time, but the pot does not need any fuel or supervision.

The simplest old-fashioned fireless cookers were made by filling a box with hay (this is why many people still refer to a fireless cooker as a haybox). A hole was left in the middle to make a "nest" for the pot and hay was piled on top of the pot in the box.

The stuffing material should be enclosed because this makes it easier to keep the cooker neat and clean. Any available material which is a poor heat conductor can be used to supply the necessary insulation around the pot: hay, straw, sawdust, dry feathers, cotton, wool, etc.

Stove Programmes in Developing Countries

In recent years there has been a considerable amount of acitivity in the area of stove design and im-

"Thermette" water heater.

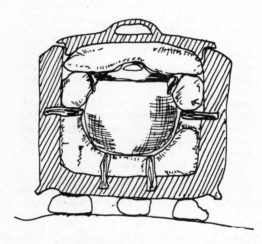

Haybox ("Fireless cooker").

proved stove promotion. In all it is estimated that about 100 stove-improvement and development programmes are currently underway largely in the following countries:

Guatemala: Following severe earthquakes in 1976 in the Guatemalan highlands, demand for wood as a building material caused severe shortages of fuelwood and a programme was started to improve stove efficiencies. The Lorena stove was developed by the Approvecho Institute of Oregon, U.S.A. and consisted of a solid block of sand and clay mixture from which was carved the internal fire chamber and three holes for pots. Added to the stove were a metal damper and a chimney. Since the design had a great deal of construction flexibility, and since it reduced smoke and raised the cooking level off the floor, it became popular and about 6000 such stoves were built up to 1983. Unfortunately the stove has a short life (about one year) after which time the construction crumbles and decays. It also takes a long time to heat up to an adequate cooking temperature and the chimney addition adds considerably to the stove´s cost.

Upper Volta: A great number of stove programmes have taken place in Upper Volta and the capital, Ouagadougou, now houses the headquarters of the Comité Inter-Etats de Lutte Contre la Secheresse au Sahel (CILSS) regional stove development coordination programme. In addition, the government agency, Foyers Ameliores, coordinates the great variety of stove programmes underway in the country. Specific stove programmes include a German design called the Nouna stove which consists of a long firebox with a chimney and two holes for cooking pots. The firebox is made from fire bricks and cement and the construction requires the services of a specially-trained craftsman. Although its life is approximately two years, it is a relatively expensive stove and is sold in the country at the heavily subsidised price of US$ 20, most buyers being from the country's more wealthy sectors. Over 1600 Nouna stoves have been built so far despite some problems with cracking. Another stove development in Upper Volta is the Kaya stove - a two-hole design made from concrete blocks - developed by a local private enterprise sponsored by the Dutch. Altogether about 5000 improved stoves have been installed in the country up to 1983.

Senegal: The Ban-ak-Suuf (clay and sand) stove programme was started in Senegal in 1980 and was based on the Lorena stove design. A smaller type called the Louga (women's) stove was also developed with the help of local Sengalese people. The design was largely the result of work undertaken by the Centre for Study and Research on Renewable Energy at the University of Dakar and took the form of a thick circular shield enclosing a single pot resting on three supports with a front opening in the shield to feed the fire. Of the 5,000 stoves built so far, about 30% have chimneys and a small number have two holes for pots. Once again, these stoves have short lives since they are particularly susceptible to heavy rain. In some cases, resin produced from boiling the leaves of the Baobab tree is added to the construction for rain protection.

Indonesia: Work on stove improvements was started in Indonesia in 1978 by the non-governmental organisation, Yayasan Dian Desa, and up to 1983 six generations of stoves had been developed. Taking the traditional stove as the base case design, the organisation then concentrated on a version of the Lorena stove, installing about 700 stoves throughout the country. Following this stage, work on a smaller, three-pot mud stove with a chimney called the Katasan stove was carried out and between 5000 such stoves were built. Unfortunately, poor construction led to

WOODSTOVES DISSEMINATION PROGRAMMES IN SELECTED DEVELOPING COUNTRIES

Year Initiate	Country	Materials/Types	Number Built	Outside Financial And Technical Support	Local Advisor Programme Execution
1977	Guatemala	mud massive	7000	ATI,IAF,FAO	ICADA,CEMAT, INTECAP,MCC,
		other types	100	USAID,CATIE	ICAITI,INTECAP, INAFOR,CII
1978	Indonesia	mud light	3400	FAO,VIA,NOVIB,OXFAM	UDD,BUTSI
		ceramic liner	6000	ITDG,MAF	YDD,PKK,BANGDES
	Sri Lanka	mud + ceramic	2140	VITA,HELVETAS,ARI	SSM
		ceramic liner	4000	ITDG,YDD	CISIR,CTR
	Upper Volta	brick + cement	4000	GFM/GTZ,PVC,EEC,AIDR	CILSS,FMK,IVE
	Nepal	mud light	1200	WB/UNDP/FAO,ADBN/UNICEF	CFDP,SFDP,RECAST
		ceramic liner	5300	USAID,STA/GTZ,ITDG	
1979	Niger	mud + banco	1000	CWS,VITA	AFN/CILSS
1980	Senegal	mud heavy small	5000	APROVECHI,USAID/PVD,FAO	CERER,CER
		mud massive	500	BOIS DE FEU,UNSO/DANIDA	ENDA & NGOs
	Ethlopia	pressed mud	2000	UNICEF,UNESCO	ENEC,BTC,REWA
1981	Kenya	mud ceramic	1000	APROVECHO,UNICEF,ITDG	MOE,KCWG
		metal + ceramic	5000	EDI,IDRC,VITA,CDP	KENGO,NGOs
	Mali	mud	1000	AETA	UNFM,IPR,CILSS

Reported number for 1983 = 50 000
Source: Caceres, A., 1985

the disintegration of most of the stoves within three months of construction. Following this, the Main Chong stove – a small two-pot Lorena type with a chimney – and the Tungku Lowon stove consisting of a simple mud construction without a chimney were developed with help from Britain's Intermediate Technology Development Group (ITDG). Most emphasis on stove development in Indonesia now centres on the sixth generation of stove, the Tungku Sae design incorporating a ceramic liner.

Most of the problems with stoves developed in Indonesia were the result of poor construction of the individual components which substantially reduced lifetimes of the stoves built. Part of the problem has been solved in the sixth generation stoves by prefabricating the liner and thereby reducing differences in liner quality.

IMPROVED WOODSTOVE DEVELOPMENT IN INDONESIA

Type of Stove Construction Materials	Research for Choice and Dissemination Strategies	Problems Encountered	Years Number Built
MASSIVE OR EGG LORENA	– Success in Guatemala, but volume was reduced – YDD members + villages	– Size too large and massive – cost too high – lighting was too slow – chimney was not necessary	1978–79 60
KETASAN sand/clay/ashes/cowdung	– smaller size was necessary and construction materials improved – YDD + cadre + autoconstruction	– third pothole was not used – good skills are needed for quality dissemination	1979–83 3,400
FAMILY/COOKER/MENCHONG sand/clay/ashes/cowdung	– need for room heat and improving flaking of – training of local cadres + simple information booklets	– accurate construction needed – highly skilled potters needed – chimney was expensive and unnecessary	1980–82 600
TUNGKU LOWON/LUNGKU SAE pottery liner	– massive dissemination easier high performance demonstrated and locally produced – production at small pottery factory and distributed through YDD network	– poor materials give short lifetime – transport of baked stoves could be difficult – standard size could be inappropriate	1981–83 6,000

Source: Caceres, A. 1985.

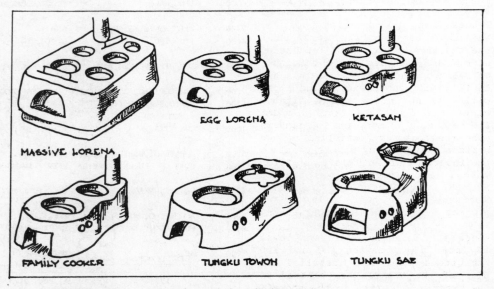

Improved Indonesian stoves for wood and charcoal.

Portable Magan Chula.

India: Stove development programmes have been continuing in India for many years and a wide variety of designs has evolved. For instance, the Magan-Chula stove is carved from a block of clay and incorporates a series of curved ducts, a metal grate and a ceramic chimney. This stove type was the direct precursor of the Lorena stove. A portable version of the Magan-Chula has also been developed which incorporates a number of ceramic parts which preceded development of pottery liners. Another Indian development is the HERL-Chula which is an L-shaped stove consisting of mud or brick in which there are three pot-holes and the system incorporates a chimney and a damper. The largest number of improved Indian stoves installed to date, however, use a design developed under the Safai Vidyala stove programme in Gujarat. The stove is made from burned brick with a precast concrete top and has two pot-holes and a chimney. There are an estimated 35 000 such stoves installed throughout India but, since construction of the stoves has generally been poor, about half the installed number have been abandoned and many of the remaining number in operation have been altered.

Recent Indian stove programmes of interest include the Indo-German stove project at Palambur in Himachal Pradesh and the development of the Nada Chula stove by Madhu Sarin under the auspices of the Ford Foundation near Chandigar. The stove developed at Palambur is known as a Dhauladhar Chula which consists of a mud stove with two pot-holes and a firebox under the first and baffles under the second, a chimney and two dampers (the front damper consisting of two horizontally-sliding panels to allow the insertion of long pieces of firewood). This programme was started in 1981 and since that time about 950 stoves have been built at the project site and 550 in other areas.

Since the Nada-Chula project began in late 1980, between 100 and 200 such stoves have been installed and part of the project has centred on the training of women in the stove's construction techniques.

Sri Lanka: The Sarvodaya Shramadana Movement in Sri Lanka has also developed six generations of improved stoves since 1979 with the aim of not only saving woodfuel but also reducing health hazards of traditional stoves and fires. Current emphasis is placed on the Tungku Sae stove with a ceramic liner. As in Indonesia, the liner is being mass-produced by potters following extensive training to maintain high liner quality. The liners are sold for US$ 1.5 each and are being distributed to households at the rate of 200 per month. Studies have found that the liner has had a marked effect on reducing cooking fuel and time.

Nepal: The Katmandu-based Research Centre for Applied Science and Technology (RECAST) began its stove development programme in 1981 under a contract between the government, FAO and UNDP and stove design and testing was carried out in conjunction with ITDG. Since work began, a number of different stove types have been developed including a double-walled, two-pot hole design with a chimney which is robust but difficult to build and expensive. In addition, pottery liners for mud stoves have also been developed consisting of four separate parts and a chimney. In this system, two pots, one of which acts as a fire chamber, are connected by a pottery pipe and the second is connected to the chimney. The complete unit is covered with clay. So far, 690 such stoves have been installed and have proved generally popular and efficient. It is intended that a further 15 000 stoves will be installed throughout Nepal in the 1983-85 period.

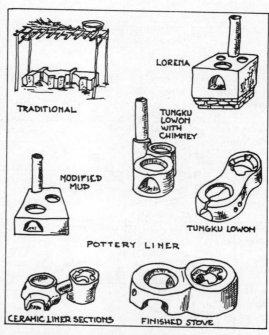

Six generations of improved woodstoves in Sri Lanka.

Dissemination

Stove programmes, to date have been slow in disseminating more efficient woodstoves. The Magan Chula has been making a slow but steady progress in Tamil Nadu for over 30 years. Safai Vidyalaya has had 35 000 stoves disseminated in Gujerat. Some 7000 Lorena stoves have been built in Guatemala, 5000 Luoga stoves in Senegal, 4000 stoves of several types in Upper Volta, about 6000 stoves in Indonesia, etc. A rough, optimistic, estimate would be that stove programmes to date, have contributed to the building of several hundred thousand new stoves. What is needed, however, are millions and not hundred thousands. One problem so far has been that the "stove problem" has been oversimplified into either an energy problem, a health problem, or a deforestation problem, or an economic cost versus benefit problem. There has also been relatively limited funding for past stove programmes, and the expectations for quick results have been much too large. The nature of stoves, the reasons for their acceptance and rejection are very complex. Successful programmes will necessarily take a long time and unrealistic expectations for quick solutions should be avoided.

Dissemination strategies must be based on the requirements and preferences of local stove users. Only recently has a realistic understanding begun to emerge about the importance of variety and alternative approaches in dissemination strategies. The options range from local artisans constructing in-situ custom-built stoves through an extension network, to prefabricated ceramic or metal stoves produced in small workshops and distributed through local markets. In the latter case dissemination must build on local initiative, entrepreneurs, and industrial considerations in the design, production, and marketing of cookstoves. In the earlier case dissemination must necessarily take place in close and active cooperation with the local women in developing, designing, building, and disseminating the stoves.

WOODSTOVES DISSEMINATION IN SELECTED DEVELOPING COUNTRIES

Material-Types Names	Countries/Number	Advantages	Disadvantages
Mud/sand heavy (Lorena, Karal, Coumbia Gueye or Ban ak suuf)	Guatemala (7000), Senegal (5500), Mesoamerica (300) Ecuador (200) tried in Indonesia, Senegal Nepal and others, Total = 13 000	– local materials – user-built – adapatable to conditions – keeps warm longer	– dissemination slow – follow-up necessary – lack of quality control – light-up slow
Mud/sand light (Ketasan, Karal Nepali Chulo)	Indonesia (2400), Sri Lanka (2200), Kenya (1000), Nepal (1200), Upper Volta (1600) Total = 8400	– smaller size – improved material – user-built	– third pothole not used – skilled dissemination
Mud/sand (Tunku lowon. Insert)	Indonesia (1000), Sri Lanka (825), Kenya (600) Total = 2425	– dissemination easier – improved materials – easy installation	– short lifetime – transportation difficult
Pottery mud/sand (Tungku sae Insert)	Indonesia (6000), Sri Lanka (4000), Nepal (5300), Kenya (1000) Total = 16 300	– dissemination easier – production line – easy installation	– short lifetime – transportation difficult
Pressed mud (Gulicha)	Ethiopia (2000) Total = 2000	– dissemination easier – several models to choose – use of a gypsum mould	– lack of quality control – chimney gets blocked – cracks in weak parts
Cement or banco (Kaya, Nouna, AIDR, Banfara)	Chad, Upper Volta (1000), Mali (1000), Niger (800), Gambia (500) Total = 3300	– longer lifetime – could have an oven – several models to choose	– dissemination slower – heavy
Cement + metal (Thai bucket)	Thailand (15 000), Kenya (500) Total = 15 500	– production line – fuel-efficient	– too heavy – parts break loose
Metal + ceramic (Umeme, Kimaki and other Jikos	Kenya (5000) Total = 5000	– dissemination faster – highly fuel-efficient – informal sector industry	– quality control in line

Source: Caceres, A., 1985.

Stove Economy

The economic attractiveness of improved stoves depends primarily on whether you live in a money economy where you have to buy your fuel or in a subsistence economy where you collect your fuel "free of charge". If fuel is free, the commercial incentive for consumers to invest in new energy-efficient stoves is virtually non-existent. And on top of that the small cash that may be available is handled by the men while the women are usually solely responsible for the fuel and cooking the food. In those cases, any improved stoves have to be more or less home-made, and easily built so that the time used building the stove is worthwile, because it gives you a cleaner and safer kitchen and uses less time collecting food and cooking.

The cost of building or buying a new stove varies widely, of course, from place to place, so the prices stated must be regarded with much scepticism. The examples provided below are only illustrative, but may provide interesting comparisons.

India: To build one stove of the Nada type takes some 2-6 hours. A woman typically charges some IRS for the construction work and some IRS 12 to provide the material as well. Dampers cost about IRS 10-18. The big cost is the chimney: it can be made entirely of mud, which is cheap but takes a lot of time. It can also be made of asbestos, cement, pottery or metal and costs between IRS 20-40. A complete stove thus costs between IRS 40-70.

If fuel is commercialised the situation becomes different. Here ready-made stoves can be considered. In India typical fuel costs per month can be IRS 80-100 which means that other stove types can also be economic provided they offer real savings in wood use.

Sri Lanka: Six generations of stoves have been tried here. The traditional stove takes 2 hours to build, costs very little and can be used for many years. A Lorena stove takes 20 hours to build, chimney and dampers cost 50 rupees, and life is about 2 years. A modified mud stove takes 12 hours to build, chimney and dampers cost 50 rupees, and life is one year. A Tungku Lowon with chimney takes 10 hours to build, chimney and dampers cost 50 rupees, and life is around 1 year. A chimneyless Tungku Lowon takes 8 hours to build, cost is negligible, and life 1 year. A pottery liner stove takes 4 hours to make, the liner costs 25 rupees, and life is about 1 year.

Kenya: A traditional charcoal-jiko costs Ksh 20. The UMEME-jiko costs Ksh 100, while the top-lined jikos cost between Ksh 30-35 but are more efficient.

Bourkina Faso: Cement stoves like the Nouna costs US$ 20 or more and a cast-iron top to be used on that stove costs US$ 20 by itself. This kind of stove is therefore restricted to wealthier people.

However, most poor people in subsistence economies cannot pay anything for a stove. There the home-made chimneyless mud stove is the only alternative to the open fire. Many poor people can afford a stove that costs around US$ 2-4, so that means that simple jikos and home-made chimney and damper stoves are a possible way to introduce efficient stoves if they are found attractive enough by the people supposed to use them. When fuel has to be bought, stoves that cost below US$ 10 may make sense even for poor people, but more expensive stoves can only be used by more affluent families.

The economic problem with efficient stoves is crucial; stoves have to be made that people can afford, that have a reasonable life, that are accepted and fuel-efficient. This is not an easy task.

In addition, in order to become accepted, it is obvious that stoves have to be made differently for different social categories. This shows once again that "the perfect stove" does not exist. One has to provide a stove that fits into the local economic situation, and this depends both on income and the prices of local materials and skills.

In this context it is of interest to observe how fuel use depends on how wealthy the people are. The poor use wood, especially in the rural areas, wood which is not bought. Charcoal is often used by the poor in urban areas, and mostly has to be bought. Kerosene is only used when you have extra money. To start with, it is used for shorter cooking tasks, like tea-making. As expected, gas and electricity are exclusively used by the rich.

Conclusions

The need for improved methods of burning fuelwood and charcoal is now more urgent than ever in the light of the diminishing areas of forest and fuelwood supply. However, following earlier disappointments with stove improvement programmes through the lack of understanding of the complete infrastructure surrounding the use and attributes of the traditional stoves and fires in developing world rural areas in particular, the drive to implement improved stove dissemination and use has made slow progress.

In order to make a noticeable impact on the energy problems, millions of efficient stoves are needed. Progress to date has been slow. The stove problem is more complex than expected and small amount of funding has been put into stoves up till now. Stoves are an important part of development and stove programmes should be encouraged, but the process is complex and will take a long time. This has to be understood so that unrealistic expectations are avoided. Fuel saving is only one of the possible benefits.

However, it appears that research organizations in this field are now far more aware of the problems and drawbacks than previously. Efforts are being made to produce improved stove designs far better matched to the requirements of the user than the earlier attempts. Development programmes are gradually producing stoves matched to local needs in terms of local production, effective operation, durability, reliability and cost.

Recent stove programmes have begun to understand these difficulties and work according to the complexity of the problem. Given adequate time, stove programmes can substantially alleviate the effect of fuel scarcity. In addition, they can improve domestic environment, encourage people's participation in development, reduce drudgery, give women time for other productive activities, and frequently promote their status in society.

After a decade of improved stove programmes, some of the conclusions one can make are:

Although domestic fuelwood consumption is contributing to some deforestation, the main forest depredators are extensive agro-business, expansion of the agricultural frontier, and abusive industrial forestry.

Within foreseeable time, even massive dissemination of woodstoves will only make a limited dent in the deforestation problem of the most affected areas, but it will certainly improve the household conditions, diminish the eye and respiratory diseases of women and children, save some of the time needed for wood collection or steady attention of open fire, and will make people participate in the solution of their own problems.

The promising second generation of improved stoves should be supported. It is essential in any national stove programme to develop a high-level of domestic competence in stove design, production, distribution, and evaluation. However, technical expertise is useful only when it has strong links of communication with local stove users and field personnel.

Stove designers need a strong background in combustion, heat transfer, selection of materials and other engineering aspects. Beyond this they should know the needs and priorities of the stove users, and understand the economic, social, and cultural context into which the stoves are introduced. It is also becoming clear that training of builders and users and follow-up are important parts in any stove programme.

For large-scale dissemination, a government role is often essential, whether by direct financing or by indirect programmes of public education and motivation. However, governmental organizations cannot effectively manage commercial activities, and should best leave this aspect of stove dissemination to the private sector. Non-governmental organizations play a very effective role in the introduction and initiation of stove projects and above all in training programmes. International agencies have a very significant role in providing financial support to such serious and locally based programmes that are now starting to emerge in several countries.

Bibliography

The Aprovecho Institute Newsletter
(1981-83) Cookstove News.
Aprovecho Institute, 442 Monroe,
Eugene, Oregon 97402, U.S.A.

Brunet, E. & Ashworth, J.H. (1984)
Technical Adaptation: BRET Wood-Con-
serving Metal Stove.
Associates in Rural Development,
Inc., 362 Main Street, Burlington,
VT 05401, U.S.A.

Caceres, A. (1985) Stoves and Kilns
- What has happened in the last de-
cade!
In, Bioenergy 84, Vol. 5. Bioenergy
in developing countries. Elsevier
Applied Science Publishers Ltd.,
Crown House, Linton Road, Barking,
Essex, IG11 8JU, U.K.

Clarke, R. (1985) Wood-stove Dissemination.
Proceedings of the conference held
at Wolfheze, The Netherlands. Intermediate Technology Publications,
9 King Street, London WC2E 8HW, U.K.

Foley, G. & Moss, P. & Timberlake,
L. (1984) Stoves and Trees.
Earthscan, 3 Endsleigh Street, London
WC14 ODD, U.K.

GATE Report S1/7 (1980) Helping People
in Poor Countries Develop Fuelsaving
Cookstoves!
German Appropriate Technolgy Exchange
(GATE), c/o GTZ, Postfach 5180, D-6236
Eschborn 1, West Germany.

ITDG Newsletter (1981-83) The Boiling
Point.
ITDG stoves project, Applied Research
Station, Shinfield Road, Reading
RG2 9BE.

Joseph, S. & Hassrick, P. (1984)
Burning Issues. Implementing Pilot
Stove Programmes. A Guide for Eastern
Africa.
Intermediate Technology Publications,
9 King Street, London WC2E 8HW, U.K.

Joseph, S. et al. ITDG Stoves Project
Technical Notes and Interim Technical
Reports.
ITDG, 9 King Street, London WC2E
8HN.

Krishna Prasad, K. (1983) Woodburning
Stoves: Their Technology, Economics
and Development.
ILO, Geneva, Switzerland.

Noflie II, A Portable Wood/Briquette
Stove (1984).
Project implemented by Department
of Community Development. Marina
Parade, Banjul, Gambia.
I.T.D.G. Ltd., 9 King Street, London
WC2E 8HW, U.K.

Sarin, M. (1984) Nada Chula - A Hand-
book.
Voluntary Health Association of India
(VHAI), C-14, Community Centre,
S.D.A., New Delhi-110016, India.

TATA Energy Research Institute, (1982)
Cookstove Handbook.
Bombay House, 24 Homi Mody Street,
Bombay-400 023, India.

VITA (1982) Testing the Efficiency
of Wood-Burning Cookstoves - Provisional International Standards.
VITA, 1815, North Lynn Street, Suite
200, Arlington, Virginia 2209-2079,
U.S.A.

8 BIOMASS COMBUSTION IN SMALL-SCALE INDUSTRY

Introduction

The combustion of wood and other biomass fuels has been used extensively in human culture for cooking, heating and other processes such as drying, crop preparation, forging etc.

Due to increased prices of petroleum products and the shortage or lack of purchasing power, biomass fuels will probably remain or become even more important for large groups of people in the near future. This applies not only to developing countries, but also to a large number of developed countries, where conditions for biomass fuels presently appear favourable.

This chapter will address the subject of the generation of process heat, mainly for small scale industrial or artisan use. Stoves for cooking are discussed in Chapter 7.

Much development of furnaces and other related equipment has been realized in developed countries. This technology, however, is often not suitable for developing countries because of the high levels of investment and mechanization required and problems relating to electricity supply. However, much of the knowledge gained during recent development should be applicable to conditions in developing countries. The main objectives of such a knowledge transfer would be: more efficient fuel usage; the possibility of using non-conventional biomass fuels, such as residues from agriculture; the possibility of choosing different fuels for the same furnaces or combinations of fuels for the same furnace, and finally, the reduction of environment problems associated with combustion.

A crucial factor when applying new technologies in developing countries, is the physical and economic availability of a reliable supply of electricity for the auxiliary equipment required for mechanical fuel handling, combustion air and flue gas fans, process controls and pumps, etc. Some furnace grates can be operated without the need for any electric auxiliaries but burners and fluidized beds will always require electric power for their operation. Most kinds of boilers require pumps and hence electricity for water circulation and steam boilers need extra capacity for maintaining high pressure.

Failure of the electric power supply can cause serious problems in large-sized boilers because of the heat stored in the fuel bed. If cooling of the furnace is not possible by air and water flow, overheating and damage may occur. Risk of explosion may also be induced because of unburned combustible gases produced from the hot fuel bed.

The Combustion Process

While the combustion of oil or gas is a homogenous process, the combustion of biomass fuels involves different processes. They are basically the same for all biomass fuels and combustion equipment, but the importance of each process will vary according to:

- moisture content
- heat value of dry substance
- content of volatile matter
- ash content
- ash melting point
- particle size and size variation.

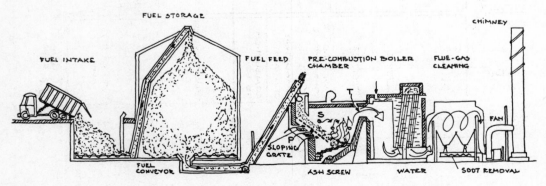

Principle of modern wood-fuelled industrial combustion system.

The combustion process for biomass fuels consists of three main stages (see Figure below).

1) drying, and 2) pyrolysis where heat is added to the fuel, and 3) oxidation, where heat is released.

1) Drying

The moisture content and particle size are important during the drying stage. High moisture content and large particle size require a long drying time due to the consequent lowering of the surface area to volume ratio of the fuel. The heat for drying is supplied by radiation from the flames and from refractory materials (eg. the furnace walls). It can also be supplied by preheated air of up to 450°C.

The need for refractory material and pre-heated air increases with the fuel moisture content. The maximum practical moisture content of wood fuels is about 65% of the total weight.

2) Pyrolysis

As the fuel becomes closer to bone dry its temperature is raised by the same heat transfer mechanisms as for drying. At about 200°C, the pyrolysis starts. The volatile matters consisting of combina-tions of H_2, C and O_2 then evaporate from the fuel. With further heating, the combustion of solid material, mainly coal to CO, starts. From now on the heat is supplied by the combustion itself. As combustion of the solid material will raise the temperature above the fuel bed, the volatile combustibles will also ignite. Biomass fuels have a high volatile content and it is important to mix these gases with air to ensure complete combustion.

3) Oxidation

The final combustion involves the smaller particles of charcoal which arise from the disintegration of the larger wood pieces. The surface area to volume ratio is high, causing raised combustion intensity and temperature. The temperature at which the ash gets sticky, and its melting point are important in the avoidance of slag formation and varies with fuel used as does the amount of ash produced. For some fuels and types of furnaces, ash will also be carried in the gas stream and if gas temperatures are high, the ashes can stick to the boiler tubes, causing reduced heat transfer.

Very dry fuels require cooling of the furnace so that the equipment is not damaged.

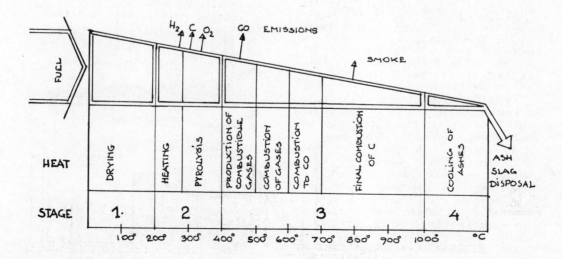

Outline diagram of the processes involved in biomass combustion.

Grates

The grate is a steel structure, on which the fuel rests and through which combustion air is supplied.

The simplest type is the flat grate. (See Figure).

To increase the movement of fuel and therefore increase combustion efficiency, the grate can be designed as a cone (see Figure). The declination of the grate spreads the fuel from the centre to the outer part of the grate, where only ash remains. Some ash will fall through the grate to a "box", from where it can be manually extracted.

The fixed sloping grate works by the force of gravity only. The angle of declination can vary along the grate to match the different properties of the fuel throughout the process of combustion, (see Figure). The fuels are fed either mechanically or from continuously filled chutes, according to the heat requirements of the boiler. The ashes are normally extracted at the bottom of the grate by means of a screw or scrapes into a water tank.

The moving sloping grate is a further development of the fixed sloping grate (see Figure). In this design the segments of the grate can move relative to each other and thereby feed the fuel down the grate. Normally, the grate is divided into sections that can be controlled separately according to the actual fuel being used. As the moving grate is more costly to operate and maintain, a compromise can sometimes be made by only making the lower part of the grate mobile.

The travelling grate is a widely applied design for solid fuel combustion (see Figure). A number of segments are connected to each other in the principle of a chain. Running over two axes, the upper part forms a horizontal grate. A bed of fuel is fed at a certain thickness on to the travelling grate at one end of the furnace. One possibility, tried successfully, is to feed coal and wood chips simultaneously in two layers, one over the other.

A briquette fuel source may have special properties that may require a moving grate. Due to the fact that a briquette is compressed during manufacture, the ash sometimes keeps the original shape of the briquette as it burns. This will cause some combustible material to be covered by ash and to not burn properly. The fuel bed should therefore be stirred to make the ash fall apart to allow complete combustion. It is also important to choose a grate that will

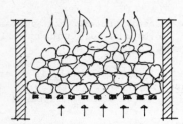

Horizontal grate, batch fed.

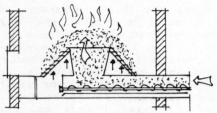

Cone grate, underfed.

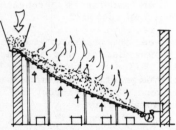

Sloping grate, overfed.

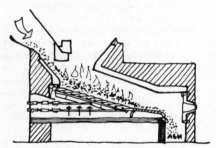

Moving sloping grate.

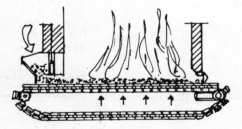

Travelling grate, sidefed.

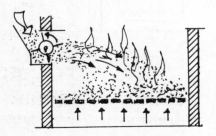

Spreader stoker.

not let the often large amount of
fines be lost by falling through
the grate.

Air Supply

Combustion devices with grates can
be supplied with air in a whole series
of different ways. The primary combustion air (P) is often complemented
by secondary (s) and even tertiary
(t) air flows which are usually introduced above the grate where most
combustion occurs in order to ensure
complete combustion.

In an open fire air is supplied from
the top, over fire (see Figure) and
the same thing happens in the simplest
furnaces. With a grate, air can be
supplied from below and pass through
the fuel under fire, (see Figure).
More efficient combustion can be
achieved if air passes through the
fuel from the top to the bottom reversed fire, (see Figure). This can
be achieved by putting the chimney
outlet under the grate.

With big logs the usual way is to
stack them in the furnace and let
the air pass through throughfiring
(see Figure). Secondary air is
supplied directly after the fire
to the flue gases. The primary air
can be supplied with a fan, turbofir-
ing, to increase efficiency.

If fired with logs or other large-
sized fuels, self induced draught
may be sufficient. As smaller sized
fuels create a bigger pressure drop
in the bed, they will require fans
for flue gas evacuation. Bigger sized
furnaces are mostly equipped with
combustion air fans.

In simple installations, where electricity is not available, fans can
be replaced by bellows or manually
operated fans.

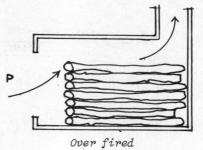

Over fired

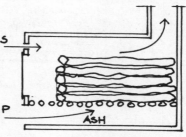

Under fired

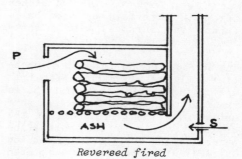

Reversed fired

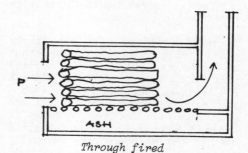

Through fired

Example of biomass fuelled underfed boiler system.

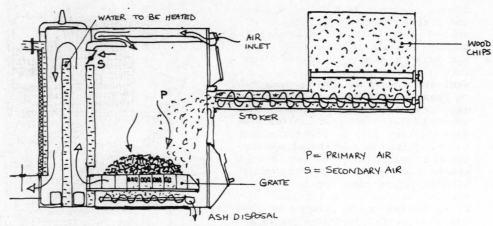

Example of biomass fuelled overfed boiler system.

Fuel Supply and Ash Disposal

Fuel may be introduced into a furnace from above the grate (over fed either manually or with some kind of stoker for bulk material) or from below (under fed furnace) with a screw feed and chipped wood fuels.

The spreader stoker (used in conjunction with the travelling grate) is a throwing machine that distributes the fuel through the furnace at some height over the grate. Some of the drying, pyrolysis and combustion occurs as the fuel is in suspension while final combustion is completed on the travelling grate. The spreader is a way of increasing the combustion intensity in the furnace.

A problem with most of the simple grate designs is that ash cannot be automatically extracted, meaning that firing must normally be reduced for about an hour every day to enable the manual extraction of ash and slag. The need for such a delay is avoided with more advanced grate designs which incorporate automatic ash extractors (eg. screw extractors).

In order to increase the flexibility of the biomass furnace for periods of insufficient biomass fuel supply, they can be fitted with a back-up oil-fired burner.

Heat utilization

There are several methods of utilizing the heat produced by a furnace. Flue gases can be used directly, (sometimes combined with air) or indirectly via a heat exchanger.

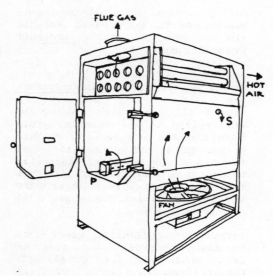

Hot air furnace, through fired with "Turbo" system.

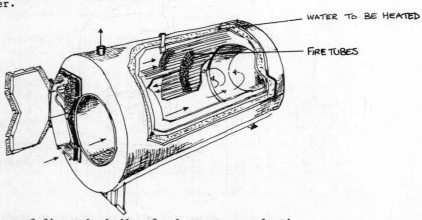

Diagram of fire tube boiler for hot water production.

Ovens for baking and other purposes can be heated directly by applying the fire or flue gases inside the oven. This heats the walls and enables baking using the retained heat. Ovens can also be heated indirectly by letting the flue gases pass in an enclosed space on the outside of the oven walls. These walls can be made from metal or brick; they get hot and heat the bread inside the oven. Drying can be done by mixing the flue gases with air and passing them through the matter to be dried. This approach should be avoided with foods and crops because of the risk of contaminating the food. Indirect drying can be done be letting the fuel gases heat the walls of a drying chamber.

There are many different designs of heat exchangers. These can be divided into two groups: fire-tube boilers (where the flue gases pass through a body of water or air in tubes), and water or air-tube boilers (where the flue gases pass outside the tubes that contain the water or air to be heated). A water-tube boiler can take the form of a combustion chamber whose walls comprise water tubes. Steam can be generated in similar ways.

More Sophisticated Combustion Systems

One of the most highly developed adaptations of the underfed grate furnace is the cyclone furnace. This is cylindrical in construction and features a high velocity, tangential secondary air flow over the fuel bed. This means that a major part of the combustion takes place in suspension.

Methods of combustion other than the grate are combustion in gas suspension and the fluidized bed. Biomass fuel can also be converted

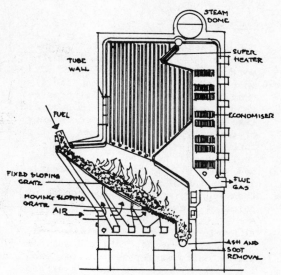

Biomass fuelled water tube boiler system.

to a combustible gas in gasifiers by thermal decomposition (pyrolysis) if heated in the absence or under reduced supplies of air.

Gas Suspension

Combustion of, for example, pulverized wood can be made in suspension in a gas stream. Small particle size enables the fuel to be more easily carried by a gas stream in a suspension. As particle size is small, the area to volume ratio is high. This means that the flame intensity will be high yet the flame length will be relatively limited. By adding swirl, a rotating air flow, the width of the flame can be increased and the flame length further decreased, thus allowing a more economically sized combustion chamber.

The advantage of pulverized solid fuels is that they can be burned in a boiler of the same design as for oil.

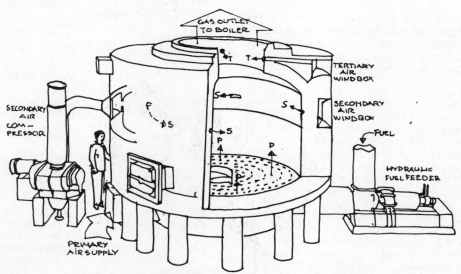

Diagram of biomass fuelled cyclone furnace.

Fluidized Beds

A fluidized bed is a vertical vessel with e.g. a layer of sand on the bottom which has a number of air inlets. When air is blown rapidly from below, the sand bed increases in volume and physically behaves like a fluid. Combustion occurs very efficiently in the sand bed as the turbulence and hence the mixing of fuel and air is good. The fluidized bed also allows for a high degree of cleaning of exhaust gases, making them attractive from an environmental point of view.

A development of the fluidized bed is the circulating fluidized bed. If air velocity is further raised, sand and fuel will follow the gas stream upwards and the mixture will more or less fill the volume of the combustion chamber. After the combustion chamber, a cyclone will separate sand and fuel from the gas stream to feed it back to the bottom of the chamber. Because of this loop the heat transfer in the bed and the possibilities of controlling the combustion will be increased. Also the circulating bed is less sensitive to variations in fuel qualities and different fuels such as oil, coal, gas, peat, wood and wastes can be used in the same furnace.

The fluidized bed systems are commercially available, but are still too complex to be recommended for widespread application in developing countries.

In more sophisticated systems the efficiency of combustion will generally reach 90% and often over 95% in modern equipment. In general terms an overall efficiency of 70-80% should be expected, due to flue gas and other losses.

Conversion of Oil-fired Boilers for Biomass Combustion

As far as combustion is concerned, most oil-fired boilers can usually be converted for a switch to solid fuels. Certain boiler types, such as pressurized boilers, are clearly unsuitable for such conversion. The most important constraint in this respect concerns the increased space requirements for manipulation, storage and combustion of the fuel. In addition, the boiler efficiency is reduced, since existing flue gas ducts and chimneys often have dimensions that are too small for firing with solid fuels. This means that the opportunities for conversion must be assessed from case to case, and no general rules can be laid down.

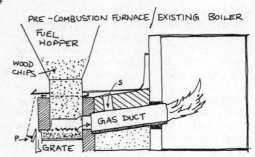

Example of biomass precombustion furnace.

The conversion can be carried out in various ways. The most common approach is to install a pre-combustion furnace (for moist fuel) in front of the boiler and connect it to the existing equipment with a gas duct. When dry fuels are to be used, the existing boiler is usually raised to make room for a grate and fuel feeding mechanism underneath.

The pre-combustion furnace consists of a sheet-iron jacket lined with insulating bricks and tamping clay. The grate may be horizontal or sloping. If the grate is horizontal, the fuel is often introduced by an underfeed screw stoker. With a sloping grate, an overfeed fuel conveyor may be used.

Flue-gas Cleaning

To achieve cleaning, cyclones (usually multiple cyclones) are usually used for dust separation. For large boilers, an electrofilter is also usually required. The separated dust should normally be collected in closed containers, and should not be recirculated through the furnace. The cleaned gases are evacuated through the chimney by means of a blower. These fumes are basic, so there is no acidic environmental impact, as is the case with oil firing.

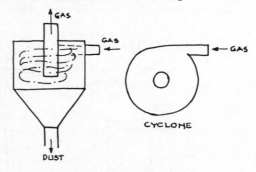

Cyclone smoke cleaner.

Ongoing Development

Intensive development of all the stages from preparation and conversion of fuels to combustion, is being carried out by official organizations and private companies in many countries.

The main motives for developing and finding equipment are:

-   reliable operation of fuel handling and combustion
-   low investment and operation costs
-   conversion of existing oil-fired units by refining the biomass fuel or mixing it with fuel oil
-   combustion in gas turbines, diesels, etc. for power generation
-   recapturing of heat by condensing moisture in the flue gases
-   combustion of wet fuels by special refractory kilns or super-charged combustion reactors
-   reduced emissions from combustion

It must be noted however that to date there have been no major attempts to adopt large- or medium-scale combustion equipment for usage under conditions prevalent in developing countries.

Local Adaptation

Several technical, economic and social factors may need special consideration when adapting combustion equipment for developing countries.

Technical factors may include:

-   availability and quality of fuels
-   availability of electricity
-   availability of infrastructure for maintenance, etc,

Economic factors involve:

-   availability of local capital or aid funds for financing
-   existence of monetary economy that can provide acceptable pay-back on investment
-   possibilities of local manufacturing
-   balance between manual and mechanized operation.

Social factors:

-   existing patterns of supplying fuel
-   willingness and motivation to adopt new methods for energy supply
-   availability of skilled labour

Bread Ovens

Bread which is an important constituent of the diets of most cultures is very difficult to prepare on a typical open stove. It must be baked using accumulated heat, usually stored in stones, bricks or clay, which must be evenly distributed over all sides of the dough.

In order to achieve fuel economies in the baking process, people have progressively abandoned small, individual stoves in favour of larger bread ovens. These may be shared by several families or run by village bakers working for a number of consumers.

The principal advantage of such ovens is that, once they reach the correct temperature, only small additions of fuel are necessary to maintain the desired heat level over long periods. To achieve optimum fuel economy, the oven should be well-insulated and opened as little as possible during use.

Bread ovens should be situated in bakeries which have facilities for storing flour and preparing dough. The oven or ovens heat the bakery and provide the steady temperature conditions necessary for the primary and secondary fermentation of the dough prior to baking. The building of the bread oven is only part of the task; what really matters is the production of good bread. It is, therefore, strongly recommended

Modern bakery oven of a type used in Botswana.

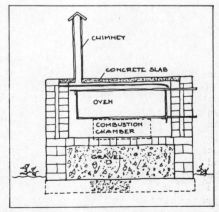

Principle of wood fuelled bakery oven.

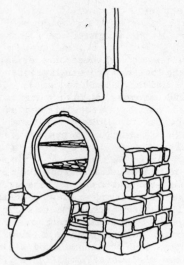

UNICEF oil drum bakery oven.

that matters are discussed with experienced local bakers before embark-
ing on construction. Bakers usually
recommend building two ovens in one
bakery; this facilitates work, increases the productivity of labour
and diminishes fuel consumption.
Experienced bakers know exactly how
to use ovens to their maximum
capacity. Apart from baking all sorts
of bread they roast meat and use
the accumulated heat to dry fruit
and vegetables. The remaining heat
serves to dry firewood, hardly any
is wasted!

Construction of the Bread Oven

The construction method described
below is based on the investigations
and long experience of Pierre Delacrétaz.

Mr Delacrétaz has been interested
in bread ovens and the baking of
bread for many years and is the
current Chairman of "L´Association
pour la maison du blé et du pain"
in Echallens near Lausanne. For a
better understanding of the construction and functioning of bread ovens,
readers are referred to Mr Delacré-
taz´s book, Les Vieux Fours à Pain.
Work on an oven starts with the levelling of the ground and the construction
of a plinth - the height of which
should be about 70 cm, which makes
it easy to maintain the fire, clean
the floor and insert the loaves. The
plinth consists of four walls constructed out of whatever material
is available, for example stones
or ordinary bricks. The space inside
the walls is filled with well-packed
earth, stones and sand. The surface
of the plinth should be larger than
the base of the oven and should be
covered with a layer of sand and
silt up to 20 cm thick which serves
as additional insulation. A final
layer of clay, which serves as the

floor of the oven, should be about
90 cm above ground level.

It is strongly recommended that there
should be incorporated into the clay
floor a flat stone or fire-resistant
brick at the entrance to the oven,
which is the area most exposed to
wear. As the floor is a vulnerable
part of the stove that must withstand
constant shocks due to the pushing
of fuel into different parts of the
oven, the raking out of embers and
the charging and removal of loaves. It
is advisable, whenever possible,
to build the floor with fire-resistant
bricks.

As soon as the floor is constructed,
the inner diameter of the oven must
be established. For family ovens
it may be fixed at 60-70 cm, while
for community use it may be increased
up to 140 cm. It should be noted
that an internal diameter of 140
cm gives a floor surface of 1.5 m^2
on which 40 loaves may be baked,
at any one time.

The inner proportions of the oven
are crucial for good combustion,
heat recovery and evacuation of
gases. If the vault is too high and
the opening too low, it will be
difficult to achieve good combustion. On the other hand, if the opening is too high and the vault too
low, heat recuperation will tend
to be poor. After carefully measuring
the proportions of a great many old
bread ovens, Mr Delacrétaz established
the optimal proportion of the inner
height of the vault to that of the
opening at 100 to 63.

Once the principal mould is ready,
the dome structure may be built around
it. The internal height of the vault
should be 10% less than the radius
of the inner circle. The resulting
structure thus resembles a slightly
flattened hemisphere.

Building the dome does not present
any particular difficulties. The

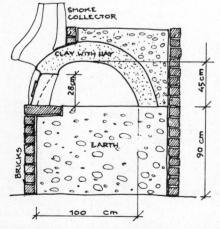

*Principles of traditional European
bakery ovens.*

lumps of clay used should be mixed with fairly long strands of hay and thoroughly kneaded to avoid air pockets. The final thickness of the walls should be about 20 cm. To strengthen the opening, bricks or stones may be positioned around the metal arch mould. A sheet metal door (2-3 mm thick) is used to close the opening. It is placed slightly obliquely and closes the opening by its own weight.

There is no chimney built into the oven. When the fire is lit, air enters through the lower half of the opening. Glowing embers heat the floor and the flames and hot gases strike the vault before being directed, by the curved surface, out of the oven through the upper part of the opening. As bakers say, the fire must "turn" in the oven.

If the oven is built in enclosed premises, it is important to form a smoke collector (built from clay or sheet metal) over the rim. The smoke collector will, of course, be linked to a suitable chimney. As soon as the structure is thoroughly dry (which may take up to two weeks), walls should be built around the plinth and the dome covered with a layer of sand and silt about 40 cm thick. This ensures that the oven is well insulated and keeps heat loss to a minimum.

The work of the oven builder thus comes to an end and that of the baker commences.

In places where fuelwood is scarce other forms of biomass can be used. A material which could be used to generate heat for baking is sawdust. In Sierra Leone work has been done on developing a baking oven fuelled with sawdust by using the "hole-through-sawdust" type burner to provide the heat required for baking.

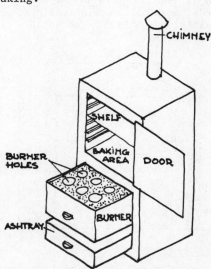

Sierra Leone saw dust bakery oven.

Production

Much of the equipment used for combustion and fuel handling is not complicated to produce once an appropriate design has been developed.

A large proportion of a furnace can be produced locally, while refractory materials for high temperature have to be imported. Grates and other metal parts such as furnace doors, etc. can probably be made locally from cast iron, but because the number of boilers and heat exchangers required in a given area will be limited, economic production will necessitate their importation, especially as much of the equipment can be obtained "off the shelf" from developed countries.

Auxiliary equipment is often "off the shelf material" that is cheaply produced in developed countries. However, they create a problem of spare parts when employed in developing countries. Consideration should therefore be given to the long term spare part availability problems that such equipment could create.

Environment

In many parts of the world, over-felling creates several problems of deforestation and erosion. A more efficient use of biomass fuels will decrease these problems through lower fuel consumption. Emissions from combustion will also decrease correspondingly. Efficient combustion of biomass in furnaces also implies that production of toxic matters such as unburned hydrocarbons will be further reduced. The main combustion product from biomass fuels is carbon dioxide. If re-growth occurs of the areas where biomass fuels are taken out, no additional carbon dioxide is added as with fossil fuels. The concentration of carbon dioxide in the atmosphere after combustion will thus be the same as if the biomass had been left for natural decomposition. Furthermore, in contrast to oil and coals, and excepting peat from some deposits, biomass fuels generally have a very low sulphur content. Nitrogen oxide emissions on the other hand have a tendency to be increased.

The environment close to a furnace can be protected against accidents and emissions by incorporating design features including cyclones and electrostatic precipitators for the cleaning of flue gas and improved methods of handling the fuel and resultant ash (which, when from biomass fuels, often makes good fertilizer).

Economy

The investment requirements and ex-pected rate of return of a biomass-fuelled furnace will vary substan-tially for different sites. Important cost factors are plant size, fuel quality, mechanization of fuel hand-ling and storage and plant loca-tion. Even in developed countries investment requirements will show surprisingly large variations due to different design philosophy of the suppliers and different needs related to the fuel deliveries and fuel quality.

The table shows an example from a developed country and has been chosen to illustrate the economy of a com-plete hot water boiler plant for chipped wood fuels in comparison with an oil installation. It should be noted that pay-back period of the additional investment of the wood-fired installation is 4 years. In-stallations in developing countries, can in most cases, be less complicated in design. In some cases, just the inclusion of a grate will enhance combustion efficiency thereby sav-ing money and/or valuable forest resources. In some cases, it is not the furnace, but the equipment using the energy produced that needs invest-ment in order to increase the overall efficiency. Other economic aspects, which are more important than invest-ment return, but which generally do not influence a private inves-tor are the saving of natural re-sources by the use of efficient equip-ment and the regional benefit of a better trade balance due to payment for fuel remaining within the area.

	Wood	Oil
Plant capacity	4 MW	4MW
number of fur-naces/boiler	2	2
Overall efficiency	75%	85%
Total investment	$800 000	$280 000
Number of full power hours h/yr	3 500	3 500
Fuel consumption tons/yr, m3/yr resp.	6 700	1 600
Fuel price	$40/yr	$260/m3
Fuel cost/yr	$270 000	$420 000
Labour cost/yr	$30 000	$8 000

Comparison of two 4 MW boiler systems.

Source: Jackie Bergman (1984), Background Report Commissioned by the Beijer In-stitute.

Dissemination

The introduction of new combustion technology to industrial users is probably less complicated than to households, provided there is a sub-stantial economic incentive. To change working conditions for employed people should not be too difficult, espec-ially if the working condition and the environment can be improved. Industrial buyers will also usually have more money available for invest-ment than private persons.

Information on new combustion tech-nology should first be developed practically for local conditions before it is widely disseminated. The training of local people can take from a few months to several years, depending on what kind of equipment is to be introduced and the avail-ability of suitable infrastructure. Government departments and branch organizations can be used as forum for information dissemination.

Conclusions

Equipment for the more efficient use of biomass is commercially avail-able, but adaptations of design to local conditions will be needed in order to establish technically, econom-ically and socially acceptable in-stallations.

Bibliography

Beagle, E.C. (1978) Rice-Husk Conversion to Energy.
FAO Agricultural Services Bulletin 31.
Food and Agricultural Organization (FAO), Via delle Terme di Caracalla, 00100 Rome, Italy.

Biofuel, an introduction (1983).
National Energy Administration, P.O. S-117 Stockholm.

Delacrétaz, P. Les Vieux Fours à Pain.
Edition de la Thièle, Yverdon, Switzerland.

Denton, F.H. (1983) Wood for Energy and Rural Development - The Philippine Experience.
Dendro Thermal Development Office. The Philippine National Electrification Administration (NEA) Manila, The Philippines.

Dickens, C.H. et al. (1983) Wood Fuel. Heat from the Forest.
The Swedish State Forest Enterprise (Domnänverket), 791 81 Falun, Sweden.

Egnéus, H. & Ellegard, A. (1984) Bioenergy 84.
Proceedings of the conference in Gothenburg, Sweden June 1984.
Elsevier Applied Science Publishers Ltd., Crown House, Linton Road, Barking, Essex 1G11 8JU, U.K.

Massaquoi, J. (1985) Development and Testing of Ovens Using Sawdust as Fuel.
Fourah Bay College, University of Sierra Leone, Freetown, Sierra Leone.

Micuta, W. (1985) Modern Stoves for All.
Revised Edition.
Intermediate Technology Publications, 9 King Street, London WC2E 8HW, U.K.

SIAR (1983) Dendro Thermal Project.
Wood-based energy production.
Scandinavian Institute for Administrative Research (SIAR), Stenkullavägen 43, S-112 65 Stockholm, Sweden.

Swedforest (1985) Seminar on Wood-Based Energy Production.
Swedforest Consulting AB, Box 154, 182 12 Danderyd, Sweden.

Tillman, D.A & Rossi, A.J. & Kitto W.D. (1981) Wood Combustion - Principles, Processes and Economics.
Academic Press, New York.

9 BIOGAS

Introduction

The technologies used for the conversion of organic materials to biogas have been in existence for many years in both the developed and developing countries, the gas being used either for direct combustion in cooking or lighting, or indirectly to fuel combustion engines delivering electrical or motive power.

The biogas production process involves the biological fermentation of organic materials such as agricultural wastes, manures and industrial effluents in an anaerobic (oxygendeficient) environment to produce methane (CH_4), carbon dioxide (CO_2) and traces of hydrogen sulphide (H_2S). In addition to the gases produced, the fermentation of these materials reduces them to a slurry containing high concentrations of nutrients making them especially effective and valuable as fertilizers. A further by-product of the process is its positive effect on public health. Bacteria harnessed in reducing the organic material to slurry and biogas kill pathogens (disease-producing bacteria) usually found at high concentrations in manures and which pose a severe threat to human well-being.

Because of the process's threefold advantages and the fact that it is particularly well-suited for use in agricultural areas where organic feedstock is available, biogas production would appear to offer great potential to energydeficient areas of the developing world. In most parts of the developing world there is a great biomass resource waiting to be tapped but there are also a number of problems associated with the introduction of biogas systems. The number of operating units may, however, differ substantially from the number of installed units, mainly due to maintenance problems.

Three countries have installed a large number of units. In numerical order these are: China (7-8 million digesters), India (100 000), and South Korea (29 000). The remaining countries, except Brazil (2900) have less than 1000 and usually the numbers are below 200.

Most of the developing countries utilize two basic designs: the fixed dome (Chinese) and the floating cover (Indian) concept. These designs have not been without problems, these being briefly that the Chinese design leaks and the Indian design is too expensive, but work is going to improve both of them.

In China and India the development and installation of the digesters has received substantial government support, taking some of the high cost out of systems now locally available. From the Chinese and Indian experience, it would appear that strong government support for biogas is a prime necessity for a successful biogas dissemination programme and one of the reasons why such systems are not widespread in other parts of the world is due to the low level of importance most governments attach to the potential of biogas.

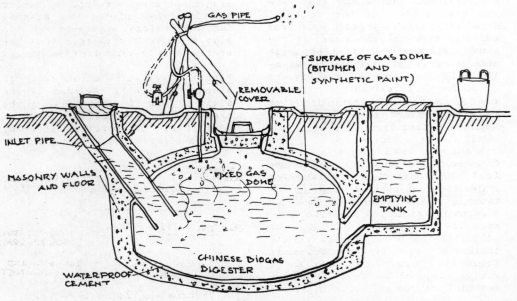

Chinese type digester with separate gas storage chamber.

The Biogas Process

Biogas is formed by bacteria during the anaerobic degradation of organic material. The degradation is a very complex process and requires certain environmental conditions as well as different bacterial populations. The degradation process also takes a long time. The more time that is allowed for the process to take place, the more complete the generation will be and the more biogas will be produced. The anaerobic process is also partly dependent on temperature. The higher the temperature, the faster is digestion and biogas production. Below 10°C biogas production occurs very slowly if at all, and above 65°C very few strains of bacteria can survive.

The size of the biogas digester is also critical. If it has too small a volume relating to the daily load there is a risk of washout of undigested material from the system. In such cases the growth of bacteria in the system is too slow to sustain balanced degradation. If the digester is too big it is a waste of money.

There are two optimum temperature ranges for methanogens (methane producing bacteria), one at 30-40°C and one at 50-60°C. These are known as the mesophilic and thermophilic ranges respectively. To operate at the higher temperature range a heating system must be employed. This makes the system more complex and expensive. The rate of methane generation approximately doubles for every 10°C rise. Temperature fluctuations should be minimized since they severely inhibit the methanogens. Toxic substances in feeds can severely inhibit the rate of digestion.

Different materials will give different gas yields. If more material is added, the biogas production from the digester will increase. At the same time, the gas yield per unit material will decrease as the retention time is shortened. A mixture of different materials will usually give a higher yield than the volume of gas produced from each one in separate digestion.

One commonly used parameter is the C/N (carbon/ nitrogen) ratio in the fuel stock materials. It has been found that the optimum ratio is between 20 and 30:1 and that too high a ratio restricts the microbial process due to a lack of nitrogen for cell formation while too low a ratio increases the shift towards ammonia toxicity. But there seems to be a wide range of suitable C/N ratios depending on the materials digested.

As temperature has a profound influence on the rate of gas production, insulation, composts around the digester and different heating systems (solar, biogas, or recycled effluent) have been used. Mixing systems can help production by maintaining uniformity, preventing scum formation and preventing solids deposition. Bio-degradibility can be increased by pre-treatment of the materials with chemicals, heat treatment or physical (e.g. shredding) means.

The addition of heating and heat retaining systems into the digester adds substantially to the cost of the digester thus, in most developing countries, digesters are unheated. With such systems it has been found that the digester needs to have a volume 40 to 60 times larger than the volume of material available daily as feedstock if sufficient biogas is to be produced to meet demand.

In developing countries the daily gas production volume per digester volume is usually 0.1 - 0.4 as compared to high rate systems employed in industrial countries where it may range from 0.7-3.

Another design parameter is the concentration of the feed, and a maximum total solids concentration of 9% is optimal in many cases. In general, any organic waste which is moderately bio-degradable can be digested.

Technical Description

In general, biogas digesters can be divided into two main types: continuous process type digesters in which material is added continuously and biogas production is uninterrupted, and batch type digesters in which material is loaded in one single operation and left to ferment until biogas production ceases. Continuous biogas supply is possible also with batch digesters if a gas storage tank is built into the system.

Batch fermentation of material to biogas is usually regarded as the simplest (and therefore usually the cheapest) biogas production system available. It involves charging an air-tight digester with a sub-

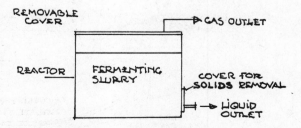

Principle of batch digester.

strate, a seed inoculum (a small amount of substrate containing bacteria from an earlier digesting process used to start a new digesting process) and, in some cases, a chemical catalyst to maintain a satisfactory pH value. Once loaded, the digester is sealed and allowed to ferment for 30–180 days depending on the material in the digester and the local heating conditions. Gas production builds up to a maximum during this period and then begins to decline as the material loses its ability to produce. Material in the digester can have a "normal" solids content of 6–10% or can be increased to 20% for "dry" fermentation. Although batch fermentation is simpler than fermentation with the continuous process type digesters, the process takes twice as long and is therefore better suited to fermentation of more fibrous materials.

Continuous fermentation. There are three main types of continuous process biogas digesters: the fixed dome (Chinese type), the floating cover (Indian type) and the plug flow or bag type.

The fixed dome consists of an airtight container constructed of brick, stone or concrete, the top and bottom being hemispherical and the walls straight. Sealing is achieved by building up several layers of mortar on the digester's inner surface. Gas leakage through the dome is often a major problem in this kind of design. The digester is fed on a semi-continuous basis (usually about once per day) and the gas produced rises to be stored under the upper dome. Gas pressures in the dome can reach 1–1.5 m of water pressure. Typical feedstock for these digesters is animal manure, "nightsoil" (human excreta) and agricultural waste. Gas production from the digester is commonly in the order of 0.1–0.2 volumes of gas to volume of feedstock per day, retention time in the digester being 60 days at 25°C.

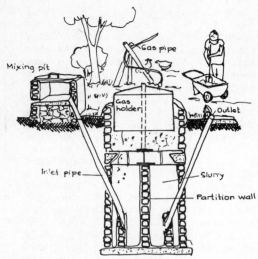

Indian-type floating cover biogas digester.

The floating cover digester design was developed by India's Khadi and Village Industries Commission (KVIC) and consists of a cylindrical container, the height to diameter ratio being in the order of 2.5–4.1:1, constructed of brick or concrete reinforced with chicken wire. Digester construction strength is not as critical as with the fixed dome type since the only pressure on the walls is the fermenting material, the gas collecting under the dome which floats upwards as the pressure increases. The cover is usually constructed of mild steel. Cost, corrosion and maintenance of the cover have been the main problems of this design. Typical feedstock is cattle manure which is fed semi-continuously into the digester, displacing an equal volume of slurry which is drawn out through an outlet pipe. Typical retention time for the feedstock is about 30 days in warm climates and up to 50 days in colder regions. Using cattle manure with 9% solids, gas yields of 0.2–0.3 volumes of gas per volume of feedstock per day have been regularly achieved. The pressure of the gas available depends on the weight of the gas holder per unit area, and usually varies between 4–8 cm of water pressure.

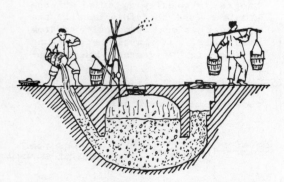

Feeding and emptying of a Chinese fixed-dome digester.

The bag digester type comprises of a long cylinder of either polyvinylchloride (PVC) or a material known as red mud plastic – developed in 1974 from the residues of bauxite smelted in aluminium production plants. Incorporated in the bag are inlet and outlet pipes for the feedstock and slurry and a gas outlet pipe. The feedstock inlet pipe is situated so that pressure in the bag is kept below 40 cm of water pressure. Gas produced is stored in the bag under a flexible membrane. A complete 50 m³ volume digester weighs just 270 kg and is easily installed in a shallow trench. Feed-

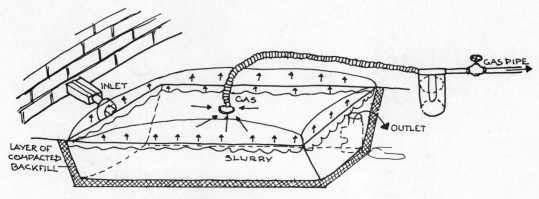

Bag digester developed in Taiwan.

stock is fed into the bag semi-continuously with the feed displacing
an equal amount of slurry removed
from the outlet. Typical retention
times vary for animal manure feedstock
from 60 days at 15-20°C to 20 days
at 30-35°C. Since the digester has
extremely thin walls, it can be solar-heated by direct exposure to
the sun. Gas production from the
bags has been found to vary from
0.23-0.61 volume of gas to volume
of feedstock per day depending on
local conditions and feedstock availability. If the red mud plastic
or PVC is unobtainable, the design
can also be constructed from concrete
with a flexible gas-collecting membrane situated at the top of the
container. Although the fabrication
materials are extremely strong, they
can still be damaged and are difficult
to repair, especially in remote rural
areas lacking access to repair facilities and materials.

Biogas stove.

Biogas Applications

Biogas produced from digesters can
be used for two basic purposes.
It can be burned directly for cooking
and lighting or indirectly in combustion engines to generate electricity or motive power. As a result,
it can be used to replace wood for
cooking and lighting and gasoline,
diesel or kerosene for lighting and
electricity generation. Between
1.3 m^3 and 1.9 m^3 of biogas (generally
containing 55% methane and 45% carbon
dioxide) is equivalent to one litre
of gasoline while 1.5-2.1 m^3 of biogas
is equivalent to one litre of diesel
fuel. Biogas will burn when mixed
with air with the biogas content
varying from 6% to 25%.

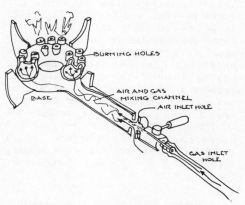

Principles of biogas stove.

area of burner parts to the area
of injector orifice should be between
225 and 300:1. Early stoves had
efficiencies in the 40% range but
newer designs, despite being more
expensive, have increased burning
efficiencies of about 60%.

Based on the fact that biogas has
different properties compared to
other commonly used gases, e.g. propane, butane, and is only available
at quite low pressures (4-8 cm of
water) stoves capable of burning
biogas efficiently are designed differently from propane gas stoves.
To ensure that the flame does not
"lift off", the ratio of the total

Gas pressures also affect the efficiency of stove burners, especially
with fixed dome types in which pressure can vary over time from just
a few cm water up to 150 cm of water.
To avoid the problems associated
with gas pressure differentials,
it is often advisable to incorporate
a simple constant pressure device
into the design.

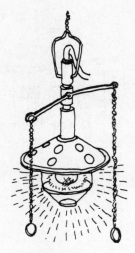

Biogas lamp.

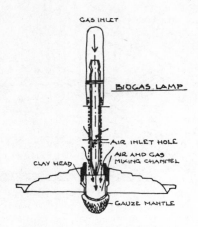

Principle of biogas lamp.

The same is true of lamps powered by biogas in that lamps should be of the mantle type if biogas is to be burned directly. Alternatively, the biogas can be used to power a generator to provide electricity for electric lighting and other electric appliances. In a gas lamp, it is important to maintain a gas pressure of 40 cm water and, since this is only possible with the fixed dome-type digester, gas lighting is often unsuitable for areas without this type of digester.

Besides maintaining constant gas pressures for cooking and lighting, it is crucial that sufficient organic waste and digester feedstock is available on a regular basis to produce sufficient gas for daily use. It has been found through trials that between 0.34 m^3 and 0.42 m^3 of gas are needed per person per day with about 0.15 m^3 of this needed per hour for lighting at a level of 100 candelas. For a six-person family requiring lighting for four hours, a total of 2.9 m^3 per day is required.

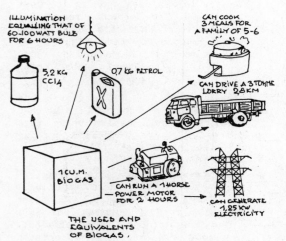

Energy equivalents of one cubic metre (m3) of biogas.

Biogas can also be used to fuel both spark and compression ignition engines. With spark ignition, an engine built to run on gasoline can have its carburettors modified to accept gas. Biogas can also be used in dual-fuel engines which are particularly useful in areas where the biogas supply is erratic. Such engines are best for static operation since fuel can be stored in quantity close by. Motive engines require transportable fuel which often rules out biogas since it is difficult and expensive to compress to an acceptable form. However, small farm tractors have been run for limited periods using gas stored in 1-2 m^3 capacity bags mounted on the roof of the tractor.

Size of Biogas Digesters

In terms of size, biogas digesters have been developed to cater for a wide range of biogas demand.

1. Household plants are small-scale plants of about 8-10 m^3 usually located in rural areas and depending on households for their inputs, control and utilization of outputs. This is the most common application of biogas technology and the one most referred to in this report.

2. Community-scale plants have capacities greater than 40 m^3, are generally found in rural areas and depend on the co-operation of a number of households for their inputs, control and output utilization.

3. Plants associated with intensive animal rearing (chickens, pigs or cattle), or institutional plants which have pre-existing management systems, e.g. prisons, hospitals, schools.

4. Plants associated with the disposal of industrial effluent (e.g. distillery wastes).

The prospects for plants located wholly in the monetized sector (types 3 and 4) seem on the whole to be favourable, whereas plants located wholly or partly in the subsistence sector (type 1) sector appear to face particular problems.

The benefits of community-sized plants include: lower capital cost per m^3 installed due to economies of scale, better operation and maintenance since it is possible to employ a full-time operator, ability to use organic wastes other than manure due to the larger size, generation of sufficiently large volumes of gas so that engines can be installed, and the possibility of achieving a more equitable distribution of benefits to the poor in the village. Furthermore, it is likely that higher gas yields could be expected due to addition of other organic wastes, better mixing and heating and closer operational control. At the moment there are only a handful of community-sized digesters in developing countries.

In general, it can be concluded that the community plants are usually more efficient than the household-scale plants due to their economies of scale and better associated operation and maintenance. However, such plants can be negatively affected by social problems relating to: rights to the biogas in relation to the organic material input from individuals in the community; rights to the slurry; the reduction in the amount of freely-available organic materials to the poorer sections of the community who depend on such materials for fuel and may have less access to the biogas than the richer sections; and difficulties in rapidly altering the lifestyles of a large group of people who will almost overnight have to change from using traditional fuels to a new and relatively unknown fuel. Such a plant could also bring monetarism to areas which were previously subsistence-oriented and immune from monetised life-styles – a factor that could ultimately exclude the community's poor from the benefits of gas supply. However,

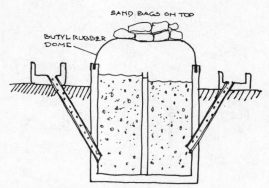

Biogas digester with a rubber dome.

evidence from some of the few community plants now operational, especially from a case study in Nepal, suggests that these problems can be successfully overcome.

Future Developments

As previously outlined, there are still a number of problems to be overcome with biogas digesters before it can be concluded that the technology is proven, feasible and appropriate to the developing world. Some of the problems, such as gas leakage and corrosion of the floating cover, are of a technical nature whilst others, such as the settlement of rights over the ownership of the gas and slurry from the digester, are social.

In both cases, a relatively high level of research and development work is proceeding and certain solutions have been put forward. The Chinese, for example, are now working on digesters combining both the Chinese and Indian systems by adding a separate gas container in conjunction with fixed dome digesters (see Figure). China has also started work on developing a cheap floating cover for the Indian design using low-cost concrete. The Indians, meanwhile, are working on covers constructed of ferro-concrete (see Figure) and fibre-glass and the Swedes have experimented with butyl rubber domes in Zambia and Tanzania.

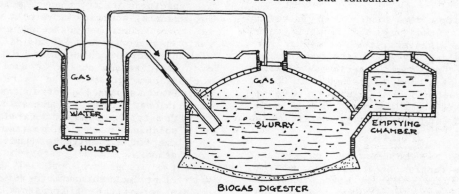

Principle of Chinese fixed-dome biogas digester.

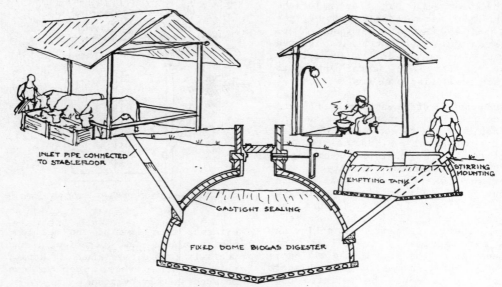

Principle of improved fixed-dome digester used in Tanzania.

The Centre for Agricultural Mecanization and Rural Technology in Tanzania (CAMARTEC) has worked together with German Appropriate Technology Exchange (GATE) on an improved fixed dome biogas plant in Arusha.

The big advantage of the fixed dome Chinese biogas plant is that almost no steelwork is involved, this does not only reduce the cost for material but also makes the setting up and the supervision easier. The modifications in the Tanzanian design involve a hemispheric shape and a narrow manhole which makes it easier to get the biogas plant gastight. Fixed underground piping and an extended neck reduces the risk of blocking the gaspipe. The plant does not have a mixing chamber, instead the inlet pipe is connected directly to the concrete stable-floor. This makes feeding easier and all the urine and dung from the animals can be utilized. The plant operates with a long retention time. Finally, the biogas-plants of this design are standardized to 3 fixed sizes, 8, 12 and 16 m^3. The results are largely successful and as a result, many similar biogas plants are now being built around Arusha.

Besides the interest in improving the fixed dome and floating cover type digesters, the Chinese in particular, have shown great enthusiasm for the red mud plastic bag digester and such bags, made cheaply in China, are now becoming widely accepted due to their low cost, ease of handling and durability (estimated at up to 20 years).

Work has concentrated on digester construction materials and, so far, relatively little effort has been made to rationalize the designs, reduce their costs and to increase gas yields. This is due primarily to the lack of sound technical data and the comparatively weak research and development capabilities of the countries concerned in these design developments.

An improved data base would enable existing and promising designs to be assessed more accurately and would also identify with greater precision the niches where biogas plants could become both socially and economically acceptable. Such information would also throw new light on methods to improve digester performance by incorporating simple heating, material pre-treatment and mixing techniques.

A further deficiency in existing biogas digester developments is in the shortage of effort concentrating on "unpackaging" recent advances made in developed countries (such as plug flow digesters, anaerobic filters and anaerobic baffler reactor designs etc) for use in the developing world. All these designs show considerable promise for digesting wide varieties of feedstock to produce higher gas yields at lower capital cost.

The anaerobic filter (see Figure), for example, consists of a tall digester container (height to diameter ratio of approximately 8-10:1) filled with a material such as river pebbles or plastic refuse acting as nucleii for organism growth. The waste to be treated is passed upwards through this filter with the gas exiting through a syphon. Due to its design, only soluble wastes can be treated in this form of digester but the loading rate can be increased and retention times are just 1-2 days with gas production reaching 4 volumes of gas per volume of feedstock per day.

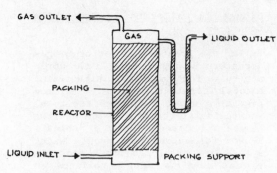

Principle of anaerobic filter.

The process is still at the pilot development stage, requiring further relatively high technological input, and has yet to be fully tested in the developed world before it can be used in developing countries. Anaerobic filters in particular, offer some potential for treating sewage and industrial effluents.

Production of Biogas Digesters

By and large, the fabrication of biogas digester components and the construction of simple devices is possible locally, although it has proved important in China and India in particular to have high- level government interest in system development and dissemination before wide-scale biogas digester construction and use comes into effect. With systems like the Chinese fixed dome digester, most of the materials needed for construction – bricks, rock and mortar – are available locally as is sufficient labour. This approach gives low costs, but sometimes gives problems concerning quality control. With the Indian floating cover design, although local workshops are generally capable of fabricating the cover, such fabrication is generally carried out at the more centralized engineering plants and not at the actual location of use, so transport costs need to be added to the relatively expensive initial fabrication costs.

More advanced biogas digesters, such as the plug flow and bag designs, use materials not always readily available in the developing world, so local production of these systems is only possible in a few countries. China has started fabrication of the bag digesters as has Taiwan.

Biogas-burning stoves and other appliances have been in use in a number of developing countries for many years, although only more recently have designs reached efficiencies of 60% or more. Many of the earlier models had efficiencies generally lower than 40% but such appliances could be made in local workshops. Parts needed to convert gasoline engines to biogas are not generally readily available or easy to fabricate in the developing countries and hence often carry heavy import duties.

Economics

Despite the ability to construct biogas plants locally, such construction can still be relatively expensive and the initial cost will often put biogas-producing systems out of reach of the poor farmer.

Having said that biogas digesters are expensive, it must also be noted that the actual economics of such systems are difficult to define since a problem arises when deciding against which of several different parameters the value of such digesters should be measured.

If it is required to make a financial analysis, market prices must be examined. However, inputs and outputs are often produced and consumed at home rather than purchased or sold in the market and also a biogas plant provides both fuel and fertilizer. There is still no consensus on methods of analysis on the economy of biogas. The problems encountered are:

1   What values should be placed on the gas and the fertilizer? This problem is compounded by the fact that there are differences in output (nitrogen content of the composted slurry and methane content of the biogas) associated with variables such as quality of feedstock and temperature.

2   What is the value of the key inputs? e.g. cost of dung production cost of its purchase or collection, or its opportunity cost (its value in the best alternative use). Opportunity costs must also be assessed with the inclusion of factors such as labour (especially family labour), land, water, and local raw materials.

3   Biogas is generally not traded and therefore has no market price. Hence it has to be valued according to equivalents that are traded. What equivalent should be chosen: dung, firewood, charcoal, other gases, electricity or kerosene?

4   The value of biogas also changes with different end-uses and biogas usually has a mix of end-uses like cooking, lighting or substituting for diesel in irrigation pumps.

5   Another problem of economic evaluation is how should the marginal utility output be valued. To assure the farmer, as is often done, that he will be able to use addi-

tional energy has to be justified in terms of need, and any additional costs have to be accounted for.

6 Finally, there are two points relating to social analysis that have created problems in economic evaluation: how to evaluate forest degradation and soil condition, and how to evaluate the health benefits from biogas because of the elimination of pathogens in human waste?

However, the social worth of biogas programmes is closely linked to the weight attached to these secondary benefits. In China improved sanitation has been a principal objective, in promoting biogas. Likewise, a major drive behind renewable energy research has been the effects of deforestation and the search for firewood substitutes. Even with these secondary benefits taken into account, in the case of India, biogas digesters still do not represent on attractive economic option for household use, whereas in China the costings are far more attractive. A family-sized biogas plant in India costs in the region of US$ 300 whereas this is reduced to about US$ 100 in China.

At the other end of the scale, large biogas production systems appear more cost-effective through their economies of scale. However, until large numbers have been installed, there remains much uncertainty surrounding the economics of such systems, although with such systems the social and organizational problems appear to outweigh those of finance.

It is hoped that future developments, once in large-scale production, might significantly lower the cost of digesters and increase biogas production efficiencies. It is claimed by the Chinese, for example, that the digester bag made of red mud plastic can be produced in the country for as little as US$ 8 per m^3 (compared with the cost of the Indian floating cover design at about US$ 40 per m^3 and about US$ 12 per m^3 for the Chinese fixed dome) where as (in Taiwan) the cost is reported to be around US$ 30 per m^3. In addition, (in India) the cost of stoves with efficiencies greater than 60% has declined recently to US$ 12-15 which should make the cost of using biogas marginally more attractive.

The financial viability of biogas plants depends very strongly on whether their outputs in the form of gas and slurry can substitute for fuels, fertilizers or feeds which were previously purchased with money. If this is so, then the cash earnings can be used to repay the capital and to cover maintenance costs.

Biogas Installation in Developing Countries

By far the greatest number of biogas digesters in any country in the world are to be found in China which claims a total installation of seven million digesters up to 1982 with the total increasing at the rate of one million a year. Although part of China's reason for encouraging such a widespread use of biogas stems from the urgent need to reduce reliance on commercial fuels and wood (to reduce the pressure on China's financial resources and on its threatened forests) an equally significant part of the reasoning was provoked by a desire to improve sanitation and health in rural areas. Although it is difficult to incorporate the health improvement factor in an economic analysis of the cost-benefits of biogas production systems, this reasoning can often be of prime importance in convincing individuals and communities to invest in such a plant.

China has by far the most sophisticated biogas development programme in the world with substantial support from the state for research and development, system implementation, installation, financing and publicity of available systems and their benefits. This is due to a number of factors, among them China's historical involvement with the technology which first started in the 1920s. In 1958, Chairman Mao generated a resurgence of interest in biogas following a decline in use but, even so, up to 1972, the country had just 1,300 digesters installed and many of these were inoperational due to neglect.

The real revival of interest started in Sichuan province in 1970 when a group of farmers attracted state support for their work and, by 1982, with seven million digesters installed the greatest proportion are still to be found in this province.

Without state support, it is doubtful whether biogas in China would have developed so fast and the involvement of a great number of research and development organisations has been particularly instrumental in bringing proposal designs into full-scale use. China's primary research organizations involved in this field include the Chengdu Institute of Biology, the Chengdu Biogas Research Institute, the Beijing Institute of Solar Energy, the Agricultural University of Shejiang Province, the Shanghai Institute of Industrial Microbiology, the Sichuan Biogas Extension Office and the Guangzhou Institute of Energy Conversion.

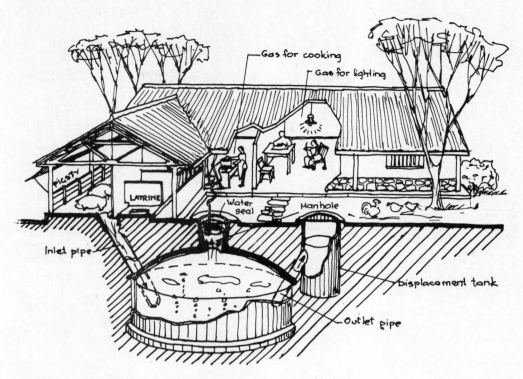

Typical set-up for use of biogas in China.

The results of work at such institutes are widely disseminated throughout China and implementation is encouraged by a network of biogas extension offices which hold meetings to explain the use of biogas systems, initiate training courses and publish biogas information in the press and advertize its use through radio and television. Such activities are coordinated through the Beijingbased National Office for Biogas Development and Extension and by the State Office of Biogas Utilization and Expansion in Sichuan Province.

The country's main policy has been to concentrate biogas utilization on areas with wood shortages, severe deforestation rates and a high incidence of the disease schistosomiasis. The main emphasis has been on household-sized digester installation and subsidies for digester installation are made available to potential users.

With such great numbers of digesters in operation, to keep them functional (and in order to avoid the damaging publicity of having many faulty or inoperational systems installed) the state has expended a great deal of effort on training technicians and maintenance personnel. Without such maintenance teams, China's biogas production programme would probably fall into sad decline as was experienced in its previous incarnation soon after the 1970 revival when too-rapid an expansion policy and uncontrolled implementation resulted in several million leaking sys-

tems. Even now, several reports indicate that a sizeable number of existing digesters are out of work due to technical problems caused by faulty designs. Unfortunately, reliable figures on the life-length of Chinese digesters are not yet available. According to some figures, only about 3 million digesters are now in operation, many of them re-designed to avoid leaking. This is mainly because they were so crudely built that most of the methane gas that they generated escaped. But the government plans to make them popular on farms again.

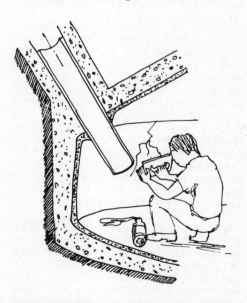

Repairing cracks in gas-tight dome.

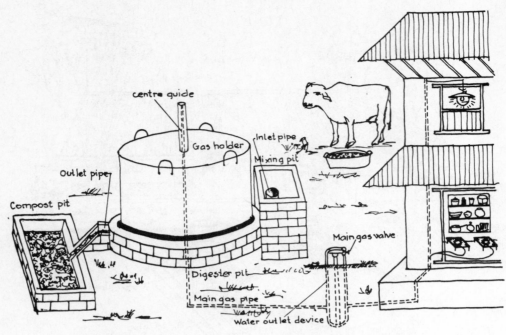

Typical set-up for use of biogas in India.

In many respects, the same situation has occurred in <u>India</u> where a rapid biogas digester implementation policy exceeded the capabilities of India's research and development organizations to produce reliable designs and to optimize digester efficiencies. As a result, early digesters in the country were expensive and inefficient. This situation has been remedied somewhat and India now has about 80 000 biogas digesters installed. However, as the link between research and development and practical implementation has been at best tenuous, new developments and designs are not being incorporated as rapidly as they might, and co-ordination and feed back require much improvement if development is to be achieved.

Since India's first biogas digester was installed at the Mantunga Homeless Lepers Asylum near Bombay in 1897, the country's interest in biogas systems has gradually increased. In 1960, the Gobar Gas Research Station was set up to investigate biogas production and, in 1962, the KVIC became involved in the field. This involvement led to the installation of some 6000 biogas plants in India by 1974 and, a year later, the government's Department of Science and Technology initiated the All-India Coordinate Biogas Programme involving KVIC, the Indian Agricultural Research Institute, the Planning Research and Action Division of the government of Uttar Pradesh, the Structural Engineering Research Centre, the Indian Institute of Management, the National Environmental Engineering Research Institute and the Department of Physics at Lucknow University.

This joint research programme produced the floating cover design digester which is the most common system in use in India today – most having a capacity of 8 to 10 m^3 for household usage. However, to be successful, constant information feedback on system performance was required and, despite the efforts of the central biogas system information dissemination office – the All-India Committee on Biogas in New Delhi – a great deal of vital information was not fed back into the system mainly because R & D and implementation work was generally run on separate lines.

The poor performance of earlier biogas digesters can also be attributed to poor back-up services – a situation which is still largely prevalent – which led to a relatively high breakdown rate. This was and is compounded by poorly coordinated

Maintenance of the floating dome gasholder.

biogas dissemination programmes resulting in a considerable waste of both financial and material resources. These deficiencies have been recognized by India's biogas controlling organizations and steps are being taken to remedy the situation.

Despite having the world's second largest number of installed biogas digesters, India's biogas programme has mainly concentrated on fairly expensive systems capable of being installed only by the wealthier inhabitants of rural areas. Three or four cattle are required to produce the necessary substrate and as a result, only about 10% of the country's farmers have the resources to be able to invest in the available systems. Work is progressing, however, to encourage the use of community-sized digesters but, so far, such systems have not proved altogether successful.

Nowhere throughout the rest of Asia and indeed the rest of the world has any country embarked on such large-scale biogas implementation programmes as China and India - although a number of countries have shown considerable interest in the technology, and implementation programmes are underway. In South Korea the government's office of rural development originally started large-scale implementation in 1969 and, by 1975 29 000 household-sized units had been installed. However, it was found that performance was severely restricted by the country's cold climate and work was largely discontinued except for the development of village-scale units.

In the Philippines, Maya Farms pioneered the implementation of batch digesters in 1972 and, in 1976, the government established a non-conventional energy programme with heavy emphasis on biogas system development. However, by 1979, only 340 digesters had been built and only about 70% of them were operating. Current research work centres on the activities of the National Institute of Science and Technology, the University of the Philippines, the Central Luzon State University and Maya farms.

Thailand had about 300 digesters installed up to the end of 1981 but about 60% of these have ceased to function effectively. Work continues in this field at the Asian Institute of Technology in conjunction with the Applied Scientific Research Corporation. The situation is slightly better in Taiwan where an estimate of the number of digesters installed is about 1000. Most work has concentrated on the development of the red mud plastic bag digester carried out by Union Industrial Research

Laboratories and the Industrial Technology Research Institute at Ksinchu.

Nepal is reported to have about 400 digesters installed and to have a relatively efficient R & D and dissemination capability. This encourages the continual re-evaluation of designs and dissemination mechanisms and has resulted in the constant evolution of biogas technologies. Most biogas system development is carried out by the development and consulting services of the Butwal Technical Institute, the Energy Research and Development Group of Tribhuvan University and the Gobar Gas Company, a private concern set up to commercialize biogas.

In Sri Lanka, about 150 digesters have been installed by the end of 1981, about 45% of which were of the Indian floating cover design, the remainder being fixed dome types. In Sri Lanka, work relating to this field is carried out by the Ceylon Electricity Board and the Industrial Development Board.

By 1980 Pakistan had installed about 60 digesters and 50 more were under construction. The main responsibility for biogas development lies with Pakistan's Appropriate Technology Development Organization in Islamabad and with the Energy Resources section of the Ministry of Petroleum and Natural Resources. Bangladesh is thought to have about 10 digesters installed.

By far the greatest implementation of biogas systems in Latin America has taken place in Brazil which now has 2 300 systems installed and a well-organized and effective national biogas co-ordination programme in effect. Although the country only managed to construct 1,600 of the 6 000 digesters originally intended for installation, due to financial constraints, the Brazilian biogas programme looks like becoming successful. If biogas does take off in the country on a wide scale, it will be largely through the concerted efforts of the Brazilian Enterprise for Technical Assistance and Rural Extension (EMBRAPA). EMBRAPA works through 25 state biogas managers, each of which utilizes the capabilities of local agricultural extension workers (of which there are 30 000 in the country).

In addition to the EMBRAPA work, the Brazilian Enterprise for Agricultural Research (EMBRATER) has carried out important R & D, concentrating on systems suited for use on small and medium-scale farms. The programme recommends biogas use for electricity generation through combustion engine-powered generators in order that electric power should be brought to rural areas. During 1980-81, the national biogas programme distributed 500 000 information folders and participated in 700 radio and 65 television programmes and EMBRATER held 128 courses to train 2000 biogas system technicians. With such a widespread dissemination of information, systems and servicing skills, it can be expected that Brazil's programme will avoid many of the pitfalls experienced by China and India in their early biogas dissemination years.

Elsewhere in Latin and Central America, about 150 digesters have been installed throughout Mexico and about 110 digesters are thought to be operating in Guatemala and Costa Rica.

In Africa, biogas digesters have yet to be used in any quantity although Kenya is reported to have 150-200 such units, many of which may be inoperational due to neglect and lack of service back-up facilities. In Tanzania, about 120 units had been installed up to 1979, 40% of which had ceased to function by 1980. Ethiopia is said to have about 100 digesters installed and Egypt has 20. However, with three national research organizations currently involved in biogas system research and implementation, it is thought that Egypt's plan to install 1000 digesters over three years could well spur a larger use of biogas systems in the country.

In Sudan, a project to utilize water hyacinth failed recently due to lack of technical support whilst, in Upper Volta, work continues in developing a dry batch digester to process agricultural wastes and cattle manure to produce fertilizer. Elsewhere in Africa, very little biogas production system development is progressing and there is an urgent need to develop indigenous technical expertize and strong national biogas programe coordination before any form of widespread dissemination of systems can be attempted.

Such strong national coordination appears to be a necessary prerequisite for the successful and large-scale dissemination of biogas digesters anywhere in the world if the experiences of China and India are anything to go by. Without the government support provided in these two countries, it is unlikely that such large-scale use would have occurred despite the advantages of biogas production in energy supply, nutrient recycling, waste treatment and improved health.

Dissemination

Considering the numerous advantages of biogas and the fact that in only two developing countries (China and India) has it achieved even a moderate degree of penetration, raises the question as to which factors are important in its dissemination. Besides the technical considerations, there are a number of other social, institutional, financial and political factors - all of which are strongly interrelated. An important factor inhibiting technology diffusion in the rural Third World are factors associated with the causes and consequences of poverty. Many of the needs of rural people are often poorly translated into "effective demand" (whereby people have cash to convert their needs into purchases). Just as production of more food does not necessarily lead to increased food consumption, so the availability of energy production technologies does not necessarily lead to their use. Where cash is scarce and non-cash fuels such as wood, crop residues and dung remain, a technical change will require the use of government funds and, almost certainly, a delivery and maintenance system that does not depend on profit margins supplied by customers.

Another factor is the extent to which the new technology meets the intended user's needs. Rural societies are increasingly recognized as being very complex systems and the failure of much technical intervention appears to arise from misunderstanding by outsiders of what rural people see as their problems. An estimation of the balance of benefits and burdens using biogas is particularly difficult in the subsistence sector or in circumstances where market prices are considered unreliable indicators of value.

On a more specific level, there are a number of social factors that influence the diffusion of biogas. Firstly, in some societies there exist social taboos about the handling of excreta, especially human wastes (in Muslim societies, the handling of swine wastes). Secondly, the division of tasks: women collect fuelwood and cook while men in most developing countries make the main decisions. The men may not be prepared to invest the capital since it does not directly impinge upon their area of responsibility. Thirdly, there seems to be a strong correlation between the educational level of the plant owner and acceptance of biogas technology.

Ideally there should be a strong national interest in the promotion of biogas, as in India and China, and a national biogas coordination committee should be created which is charged with coordinating all associated biogas activities within the country, e.g. research and development, implementation and financing. Without such a committee a great deal of duplication and waste of resources results. The committee should ensure that related functions proceed hand in hand so that no bottlenecks occur in diffusion. Unfortunately, in many countries, there has been a strong pressure to "be seen doing something". Hence, in many of these cases implementation proceeds faster than the capacity of R & D institutes to provide technical advice and for banks to finance such plants. The net result is to build plants which fail due to technical, social and financial difficulties.

Besides technical and economic considerations, there are a number of other factors which influence the dissemination of biogas. There has to be sufficient intensity of land usage and animal husbandry, so that raw material (manure, night soil and agricultural residues) are available. There has to be a need for both energy, fertilizer and improved public health, and people have to be acquainted with the practice of composting. There has to be a sufficient degree of monetization so that there is an economic incentive for digester construction, and there has to be a sufficiently high technical capability to build and operate a digester. Where these factors exist, biogas seems to be a promising technology.

Bibliography

Borda Biogas Team, Biogas Plants Building Instructions.
German Agency for Technical Cooperation (GATE), P.O. Box 5180, 6236 Eschborn 1, Federal Republic of Germany.

Buren, van A. (1980) Transferring Biogas Technology to the Rural Areas of the Third World.
International Institute for Environment and Development (IIED), 3 Endsleigh Street, London, WC1H ODD, U.K.

Buren, van A. (1979) The Chinese Development of Biogas and its Applicability to East Africa.
International Institute for Environment and Development (IIED), 3 Endsleigh Street, London WC1H ODD, U.K.

Buren, van A. (1979) A Chinese Biogas Manual.
Intermediate Technology Publications Ltd., 9 King Street, London WC2E 8HN, U.K.

Ellegard, A. & Jonsson, A. & Zetterqvist A. (1983) Biogas - not just Technology!
Swedish International Development Authority (SIDA), 105 25 Stockholm, Sweden.

Ellegard, A. & Egneus, H (1984) State of the Art on Biogas Technology Transfer and Diffusion.
Swedish International Development Authority (SIDA), 105 25 Stockholm, Sweden.

Kellner, C. (1985) The Diffusion of Family-size Unit Biogas Plants in suitable Areas in Tanzania - The Development of an Extension Strategy.
Paper presented to International Conference on Research & Development of Renewable Energy Technologies in Africa, Mauritius March-April 1985.
Centre for Agricultural Mechanization and Rural Technology (CAMARTEC), P.O. Box 764, Arusha, Tanzania.

Stuckey, D.C. (1983) Technology Assessment Study of Biogas in Developing Countries.
International Reference Centre for Waste Disposal (IRCWD), Ueberlandsstrasse 133, CH-8600 Duebendorf, Switzerland.

TATA Energy Research Institute (1982) Biogas Handbook.
Documentation Centre, Bombay House, 24 Homi Mody Street, Bombay 400 023, India.

United Nations (1980) Guidebook on Biogas Development.
Energy Resources Development Series No. 21.

10   DRAUGHT ANIMAL POWER

Introduction

Animals have been used for centuries throughout both the developed and developing worlds to either replace or supplement the need for human energy expended in transportation and the production of food. However, despite the vast amount of accumulated experience in this field, this area of renewable energy technology has been seriously neglected in recent years.

Draught animal power is an outstanding example of a mass level application of appropriate technology to millions of small farmers in the Third World. For both agriculture and transport, the renewable draught animal – integrated with the popular milk and meat systems based on its own species – has no equal.

There are approximately 400 million draught animals in the developing world. They are valued at US$ 100 billion, would cost US$ 250 billion to replace by petroleum based fuels and provide the people of developing countries with over 150 million horse power. During their working lives such animals in addition to providing this level of motive power also contribute fuel for rural communities, and fertilizer for the soil. When their useful working lives end, animals are still useful in that they provide food and a number of other valuable products such as skin and bone.

As they are the main production means of the small farmer, live stock are more evenly distributed than any other "energy source". 85-90% of Asia and Africa depend on manual and draught power whilst the benefits of mechanization serve barely 10%.

Such contributions are, and will continue to be for the foreseeable future, of major importance to the millions of people in the developing world.

Draught animal power can only fulfill its true potential if conscious attempts are made to upgrade it. Upgrading draught animal power systems does not necessarily mean that more animals have to be brought into use, although this is required in a number of countries, particularly in Africa. Only a well-equipped infrastructure can upgrade draught animal power and emphasis should be placed on raising output from the same system by:

a   increasing the amount of time for which draught animals are utilized

b   better feed and health care

c   better implements and carts

d   better recovery and use of by-products during the working life and after the death of the draught animal.

e   genetic improvement and breeding programmes

Draught animal power should not be seen as a backward technology in conflict with mechanization and modernization. On the contrary, considering that it will be the only feasible and appropriate form of energy for large groups of people in many countries for years to come, it should be given high priority in the country´s policies on agricultural production, rural development, employment, small transportation, etc.

Chinese horse cart

Since the bulk of animal work is carried out in rural areas, most technological improvements will directly affect the efficiency of agricultural practices. As was recently pointed out by the FAO, "It is being increasingly recognized that the core of the energy strategy for increasing farm productivity in the developing countries depends mainly on the appropriate combination of human, animal and mechanical power for specific situations within the country including the technical suitability and economic and social objectives".

Draught Animal Power Resource

Evidence showing the development of the use of draught animal power in improving work efficiency is found throughout history. Since animals were first domesticated and used for work in around 8000 BC, the number of animals and the variety of species have increased substantially throughout the world. For example, the earliest reindeer sledges, were probably used in Northern Europe around 5000 BC, while evidence of ox-drawn sledges has been found in Mesopotamia dating back to 3500 BC. The first known wheel, dated around 3000 BC, was discovered in Southern Russia.

The increased use of animals in providing motive power has culminated in a total current draught animal population of about 400 million which includes horses, mules, donkeys, bullocks, buffalo, yaks, camels, llamas, elephants, dogs, reindeer, sheep and goats. About 80% of these draught animals are oxen and other bovine species with most of the remaining 20% being of equine (horse) species. World populations of all draught animal types are given in the Table below.

India has the largest draught animal population of all the developing countries with 70 million bullocks, 8 million buffalo, a million horses and a million camels being used to provide motive power. Almost 60% of the energy used in the agricultural sector of the country is provided by draught animal power.

India has fully recognized the importance of animals in its development programme and has set up several breeding stations to produce ideal draught animal breeds, as well as a number of centres which are working on the development and production of improved implements suited to animal power. Nevertheless, the majority of farmers still rely on traditional agricultural tools and, overall, the quality of livestock in India is deteriorating. In an effort to stem this decline in standards and to improve equipment designs, the Government plans to introduce an Animal Energy Board to oversee the development of equipment and to generally improve animal power efficiencies.

In China the policy is to place equal emphasis on animal and mechanical power in order to improve the efficiencies of both sectors in parallel. China has about 50 million draught animals and half of the country's 100 million ha of agricultural land are cultivated using draught animal power. Mechanical power in the form of tractors is used to cultivate the remainder.

No other country uses its animal population more effectively than China which uses each animal on average for about 300 days/year for ploughing and transport. In comparison few other countries manage to use each animal for more than 150 days per year for these purposes.

(All figures are given in millions)

	Milk + Total world	Meat + Developed (all)	Draught a/ Developing (all)	Possible draught animal power in developing countries	
Cattle and yaks b/	1212.0	425.0	787.0	246	61%
Buffaloes	130.6	0.9	129.7	60	15%
Horses	61.8	22.4	39.4	27	7%
Mules	11.6	0.7	10.9	10	3%
Donkeys	42.8	2.0	40.8	40	10%
Camels	16.8	0.2	16.6	16	4%
Llamas	1.4	-	1.4	1	-
Elephants c/	-	-	-	-	-
Total	1477.0	451.2	1025.8	400	100

a/ Packing and logging included
b/ Yaks of China have been included under cattle
c/ There may be about 20 000 elephants engaged in logging work.

Source: U.N. (1981).

In order to achieve this work rate, China has made major progress in sustaining the health of its livestock, and has placed great emphasis on developing efficient implements matched to the animals' working capabilities. The country is currently raising 11 million scientifically-bred donkeys for use in the transport sector.

In other Asian countries, the situation is mixed. In Bangladesh and Pakistan, a great deal of use is made of draught animal power but in Indonesia, the Philippines and Thailand their use is declining. Elephants have proved especially useful for many years in the logging industries of Burma, India, Sri Lanka and Thailand.

In Africa, the use of animals varies from country to country with Botswana, Egypt, Ethiopia, Madagascar and Tanzania, all having long histories of animal power use, whereas Ghana, Kenya, Nigeria and Zambia use animals relatively little. In the Francophone (French speaking) African countries including Gabon, Mali and Bourkina Fasso and also in Benin, Guinea, the Ivory Coast, Liberia and Sierra Leone the use of animal power has only been introduced relatively recently with some success. However, of all developing countries, the African countries require the most support from external sources in increasing their use of animal power.

In Latin America, the horse, mule, donkey and llama have been in use for centuries. In recent years their use has declined with the increasing application of mechanized transport and machinery, but rising fuel prices are forcing many Latin American countries, including Brazil, Colombia and Mexico, to reassess the potential of draught animals. Peru has always depended on llamas for high altitude work.

In terms of work efficiency, the majority of draught animals produce 0.4-0.8 horse power on a sustained basis and, if one assumes a mean of 0.5 horse power from approximately 300 million draught animals, it can be seen that the developing world derives in the region of 150 million horse power from this source. There appears to be a "natural average speed" of movement for each breed of animal, which they will tend to adopt wherever possible. The average working speed of such animals is in the order of 2 to 5 km/h.

Draught animals do have the important characteristic that they can generate substantially higher draught forces and power outputs over a short period of time. This allows draught animals, for example, to pull a plough through a difficult piece of ground, to accelerate a loaded cart from rest and haul it uphill, or to uproot a tree. However, using draught animals in this way, with intensive short bursts of effort, does reduce their overall working efficiency over a normal working day.

The working performance of a draught animal is a function primarily of its weight, provided that this consists of muscle rather than fat:

Oxen in good condition can reasonably be expected to exert a draught force of about 10% of their body weight.

Horses have a better output: weight ratio than oxen and can reasonably be expected to exert a draught force of about 15% of their body weight. Using two or more animals harnessed together results in a relative efficiency loss of about 7.5% for two, and 15% for three. However, it is often necessary to combine the pulling power of two or more animals in order to carry out particularly heavy tasks. Draught animals are capable of working efficiently for between three to six hours a day.

WORKING PERFORMANCE OF DRAUGHT ANIMALS

Animal	Average weight kg	Approximate draught kgf	Average speed of work m/s	Power developed kW
Bullock	500-900	60-80	0.6 - 0.85	0.56
Cow	400-600	50-60	0.7	0.35
Buffalo	400-900	50-80	0.8 - 0.9	0.55
Horse	400-700	60-80	1.0	0.75
Donkey	200-300	30-40	0.7	0.25
Mule	350-500	50-60	0.9 - 1.0	0.52

Source: H.J. Hopfen. Farm Implements for Arid and Tropical Regions FAO 1969.

Traditional western harnessing devices.

Harnessing Devices

The efficiency of draught animal power application is dependent on the harnessing device, which is the link between the animal and the implements or cart. The object of the harnessing device is to ensure that the draught power is utilized to the maximum extent. In addition it should be comfortable and not injure the animal, be easy to put on and take off, be easy to repair, and possible to manufacture with local materials. It should also be cheap enough for purchase by farmers and carters.

Harnesses for horses and mules have been developed to a high standard in the developed countries and many designs for maximizing the efficiency of converting animal power to useful work have been developed and are suitable for developing country usage. These developments include breastband harnesses for light work and full collar harnesses for heavier tasks. When two or more horses are needed for a certain task, it has been found that independent hitching of the animals to the farm implement is best for maximum power conversion

Traditional plowing with oxen.

efficiency. It is important to disseminate this knowledge throughout
developing countries and to modify
the harness designs so that local
materials can be used. For bovines
the traditional harnesses used in
developing countries are the neck
yoke and the head yoke. These harnesses have a number of disadvantages. The neck yoke (see Figure)
does not fully utilize all the
strength of the animal's legs, shoulders and back. It can cause injury
and discomfort because the load is
applied over a small area, when the
throat strap is pulled against the
animal's windpipe. Finally, the
attachment point of the implement
on the animal is high, which is inefficient since only a proportion of
the animal's input is converted into
useful draught force. The head yoke
also has a number of substantial
disadvantages: it does not utilize
the full power from legs, shoulders
and back, it can cause injury, the
animal cannot swing its head, the
attachment point is high, it causes
discomfort and can only fit a certain
size of animal.

Improving these harnessing techniques
in order to convert more of the animal
power to useful energy not only increases the efficiency of the animal
and implement but also improves the
animal's working capacities since
a more comfortable harness often
extends an animal's working day and
indeed its working life.

For single ox-drawn implements, four
improved harnesses have been developed. The Chinese V-yoke is a simple
neck yoke which offers improved comfort and better use of the animal's
strength; the flexible harness has
the same effect but does so without
the use of rigid parts (this is not
suitable for pulling wheeled vehicles); the collar harness offers
the greatest increase in efficiency
but is relatively complex and expensive, and the Japanese back harness
is the best suited for load-carrying
animals.

Double ox harnesses can be improved
by carving or padding the double
neck yoke to make it fit the animals
better, or by using rigidly linked
collar harnesses. This latter arrangement appears to offer the greatest
efficiency increase. It has also
been found that poor harnessing sometimes results in the need to use
two animals for a particularly heavy
task, whereas improved harnessing
can improve the animal's pulling
performance by up to 60%, thus reducing the need to use a second animal
in certain circumstances.

Neck yoke

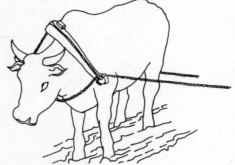

Chinese V-yoke

Flexible harness

Collar harness

Japanese back harness

Farming and Farming Tools

At present there is a shortage of power supply for the existing level of world farming requirements let alone for increasing cultivated area and yield. In Africa and the Far East, mechanized power is expected to contribute only 8-9% of the total power input by year 2000. While the overall share of mechanical power could rise 8% in 1975 to 19% in 2000, the bulk of power needs would still have to come from humans and animals. In Africa human energy would still comprise as much as 82% of the input in most areas and in the Far East. This proportion is likely to be around 66%. To the largest possible extent this labour should be transferred to draught animal power. There is no conflict or competition between mechanized and draught power - each source can complement and supplement the other. China uses both machines and draught animal power for cultivation, but for other operations it largely uses the abundant supply of manual labour.

It is important that the tools are suited to the soil, the draught animal, to the users themselves, and that they are light, simple and robust. There are several major factors which should be borne in mind whilst attempting to improve tools and implements for use with draught animals. They must be:

1   adapted to allow efficient and speedy work with the minimum of fatigue

2   not injurious to man or animal

3   of simple design, in order to encourage local manufacture

4   made of easily available materials

5   light-weight for easy transportation

6   ready for immediate use without loss of time through preparatory adjustments

In rain-fed cultivation, the main applications of tools are soil preparation (using ploughs and harrows), sowing (especially where the agricultural season is short), weeding (this is undoubtedly the operation which gains most from mechanization, since it involves a veritable race against the weeds), harvesting and transport (which accounts for a high proportion of farming activities).

In irrigated farming, especially with rice, implements are needed for the levelling and building of permanent bounds (levellers, ridgers and diggers), for ploughing and puddling.

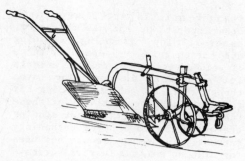

Typical West African plow

Typical West African harrow.

There are also multi-purpose tools (toolbars) that can be used for a limited variety of purposes. These typically consist of a frame with a wheel and handles onto which can be fitted a plough, cultivator, hoe, lifter, harrow or seeder. Experience in some African countries (e.g. Senegal and Gambia) has led to a recommendation for the standardization of equipment in order to facilitate production, maintenance and distribution of spare parts.

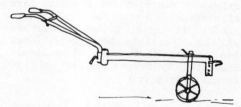

Multipurpose toolbar

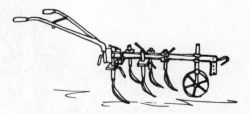

Toolbar used for harrowing

Toolbar used for plowing

Draught animals are also used in other operations such as threshing, water-lifting, grinding, crushing and cotton-grinding, etc. In terms of volume, the use of draught animals for such work is declining, even though there is certainly scope for increasing the productivity of these tasks. It is essential that studies are conducted to improve technology currently in use and locate new uses for draught animals in rural areas.

If they are to be appropriate for developing world use, the essence of these improved technologies must lie in their simplicity. Simple design is an essential pre- requisite for local manufacture which is, it- self, essential in reducing system costs and making repair and servicing possible.

Transportation

For relatively short distances, small loadings of up to 2 tons and in opera- tions involving long loading times, animal-drawn vehicles provide an attractive alternative to motorized transport. To be economic, motor vehicles often need to be used for 250-300 days/year, whilst animal-drawn vehicles can repay their low capital costs if only used for 50 days/year. In addition, when little agricultural activity is possible, the farmer can earn extra income from the hire of the vehicle for local transporta- tion of goods and people. In many areas of the developing world, ani- mal-drawn vehicles are the only form of transport suited to local condi- tions. In India, for example, half of the country´s 500 000 villages are situated in remote locations connected by tracks suited only to animal transport.

Grinding with draught animal power.

There is tremendous scope for in- creasing the productivity of the animal-drawn vehicle system in many developing countries by:

1   improving the design of existing animal-drawn vehicles

2   increasing the utilization of these vehicles

3   increasing their numbers and varie- ties, and diversifying their uses

4   introducing suitable animal-drawn vehicles where appropriate

5   improving breeds and numbers of draught animals suitable for pull- ing vehicles

6   creating and improving existing infrastructure for the development of animal-drawn vehicles.

In traditional existing designs of vehicle, much of the draught effort is wasted unnecessarily in overcoming the friction of the rough and crude bearings and in hauling very heavy platforms. Some improvement can

Traditional ox-cart in India.

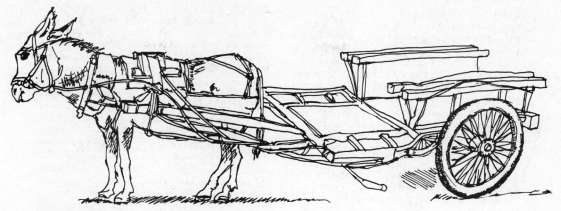

Modern donkey-cart in China.

be made by using worn out truck wheels
and tyres, but these are also very
heavy. The basic design for an improved cart should include pneumatic
tyres which should be of a special
design so as to cope with slow movement and rugged terrain. They should
not be easily punctured and they
should be light, cheap and easy to
maintain. Special tyres for use
in conjunction with animals are made
in India with smooth bearings (these
can be ball or roller bearings).
Animal-drawn vehicles should be fitted
with slow speed, cheap bearings (not
costly truck bearings meant for heavy
loads and high speeds), lighter platforms (the introduction of steel
sections has made the carts lighter),
shock absorbers, a brake (the provision of a breaking device is essential; a simple wooden pole, acting
against the rim will often suffice)
and improved harnesses. Through
these improvements a cart's capacity
can be increased from one to three
tons on level roads (if two animals
are used). Whenever the higher capacity is not required, a single animal
can pull one or even 1.5 tons easily.
With an improved vehicle, the carters
are able to transport in a single-ani-
mal cart as much as they did previously in a double-animal cart thereby reducing costs.

These improvements also cause less
damage to road surfaces and thus
extend the period of the year during
which tracks can be negotiated, especially in areas with high seasonal
rainfall whose rough roads deteriorate
quickly during heavy rains if damaged
by heavy carts with iron-rimmed wheels
and no suspension.

Bullock-drawn carts are slow and
reduce traffic capacity. Horses
and mules are better draught animals;
they can haul heavier goods at a
faster pace. It is proposed, therefore, that horse and mule-drawn carts
be encouraged for urban transport,
most of which is by professional
carters, but that rural transport
should be integrated with the draught
animals used in agriculture.

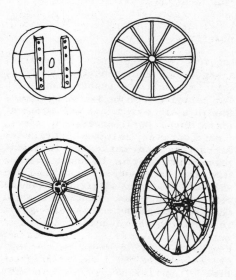

The development of wheels.

Draught Animals in Forestry

In some industrialized countries,
well bred work horses are still used
in the process of logging. The technology, equipment and harnessing
devices, etc, are being documented
and much of this information might
be of use in the developing countries
where the practices are still inefficient. For instance, throughout
Latin America logging oxen are head
harnessed. This method presents
many inconveniences and should be
revised. In Burma, India, Sri Lanka
and Thailand, most logging is carried
out using elephants and such practices
are now being tried in the forests
of Zaire.

The FAO examined the use of elephants
for logging and suggested a number
of steps to improve the system.
Random observations indicate that
several types of equipment, used
in the advanced countries, will be
of great help in improving the efficiency and reducing the cost of
elephant logging operations.

Winter forest logging with horse in Sweden.

Logging by horses in Sweden employs a variety of equipment for handling, skidding, rolling, pushing, dragging and carrying the logs. Sweden holds annual shows to demonstrate the best varieties of work horses, harnesses, equipment, etc. Pedigree certificates for horses, as well as medals and prizes, are awarded in order to provide the necessary incentive. Activities of this nature to upgrade the technology and techniques could also be applied in the case of logging by cattle and elephants.

In forestry operations, besides improving harnessing techniques, a number of log transport system improvements are possible. Sledges, two-wheeled carriages, four-wheeled carriages and six-wheeled crawler vehicles (for heavy loads and difficult or soft terrain) have all been developed to a high standard in the developed world, and log-handling equipment including winches, loading cranes and loading lifts are all relatively simple tools which could easily be applied in developing countries.

Although it is important to ensure that effective tools are utilized,

Simple tools for handling heavy logs.

it is the whole process of work, i.e. the combination of man, animal, tools, vehicles and mechanization that is critical. Different working processes ar suited to different terrain and wood types. In Scandinavian countries the renewed interest in horses for logging has been brought about because it was shown that they were more economical than tractors in certain circumstances — when the area to be cut is small, when the terrain is difficult or if the soil is fragile. One of the main advantages is that less damage is caused to soil, roots, and trees than by heavy machinery. There are about 30 000 horses used as draught animals in Sweden and in 1972-73 they logged about 2 million m^3. Draught animals have an important role in logging in developing countries. Training is needed in horse-driving and loading and tools; vehicles and work processes have to be developed for the local conditions.

Animal-powered Pump Technologies

Water-raising systems using animals to provide the motive power have been in existence for centuries but only recently has it been recognized that animal-powered water pump efficiencies can be increased with the use of improved system components and designs.

1   The sack and rope water-lifting system is commonly used in the Sahara region and utilizes the motive power of a camel to lift individual sacks of water from a well. About 2000 litres of water/h can be lifted from wells 6 m deep, and often this method is used to lift water from far greater depths.

2   The Sakia is often used in Egypt to collect water from wells up to 1.5 m deep and consists of a vertical wooden wheel with clay pots mounted on the wheel's periphery. The wheel is driven by a single animal or pairs of animals walking in a circle around the well. Power is transferred to the pump by means of a wooden right angle gear drive. The system has recently been modernized with the inclusion of galvanized steel sheet to replace the wooden wheel, and cast iron replacing the wooden gear drive. In this updated system, as the steel sheet wheel rotates, water collected at the periphery gravitates to the centre of the wheel where it is discharged at ground level. With an efficiency of about 60%, the Sakia system can supply about 30 litres of water per second from wells up to 1.5 m deep with a single ox providing the motive power. Typi-

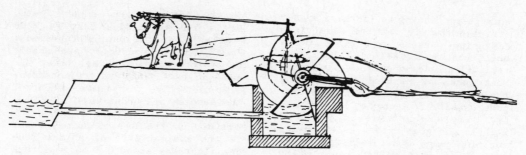

The Egyptian Sakia pump.

cal operation involves the use of two oxen, each working a total of five hours to lift water for 10 hours/day. The system is estimated to cost in the region of US$ 34/ha of land requiring irrigation.

3   The chain and washer pump, or Paster Noster pump, is found mainly in China and operates on the principle of a continuous looped chain and washer being pulled through a closely fitting pipe, over a geared wheel and back to the foot of the pipe to pass once more up the pipe with its next load of water. With the inclusion of a cast iron right angle drive, the system can be adapted for animal power. The maximum practical lift is 15-20 m and when driven by a pair of bullocks, about 14 litres of water per second can be supplied from a 4 m well, dropping to 9 litres per second from a 6 m well. The system is estimated to cost in the region of US$ 44/ha of land requiring irrigation.

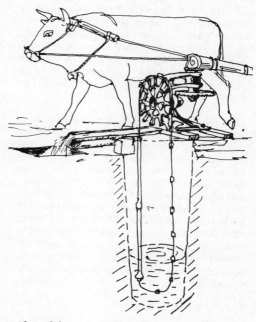

The Chinese chain and washer pump.

4   The Mono Pump, unlike the traditional animal-powered pumps and water lifting systems, incorporates a gearbox and a rotating pump. The Mono Pump was developed in the UK and uses a positive displacement, helical rotor/stator pump and a gearbox. With this system, the turning speed is directly proportional to the water flow, but the output pressure of the raised water remains virtually independent of that speed. The system can be used in wells of up to 110 m deep and deliver 1200-4600 litres of water/h depending on the availability of animal power. Another system made by the Bunger company in Denmark, is claimed to have a production capacity of 8000-10 000 litres of water/h from wells 7-8 m deep. The pump costs approximately US$ 2400.

Animal-powered pumps may have a great potential in some areas, but they are not always the most economic solution. Economic analysis has shown that, for irrigation, animal-powered pumps could sometimes be the best option for small farms of 1 to 5 hectares.

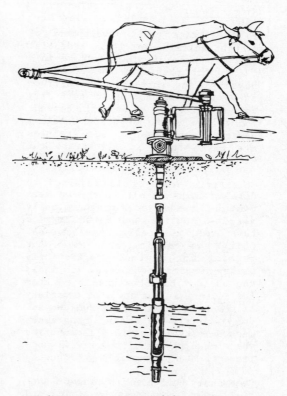

Modern rotary pump with gearbox.

Work is continuing to increase system efficiency while, at the same time, designing systems capable of being built and installed using local materials and components at low cost. Most work has centred on improving the efficiency of converting motive power to kinetic energy by incorporating simple gearboxes in place of right angle gear drives and by using rotating pumps as water-lifting mechanisms. Wood is often replaced by metal in more modern designs.

So far, dissemination of the pumping systems has been slow despite the cost advantages for small farms. Once again, dissemination has been hampered by ignorance of the merits of the proposed systems and by a shortage of funds to pay for installation.

If such systems are to become widely used, like the traditional pumps, they must be simple, durable, reliable, cheap, easy to maintain and should contain components made in local workshops using locally-available materials. Imported materials and components add significantly to the system costs. If these criteria can be met, it is considered likely that improved versions of animal-powered water pumps could prove competitive and become more widely used in areas where the supply of other sources of energy for pumping is costly or difficult.

The Infrastructure needed for Draught Animal Power

To facilitate rapid development of draught animal power it is essential that a number of programmes be initiated for providing the necessary physical and soft-ware infrastructure. Requirements would naturally vary from country to country, depending on the state of draught animal power and environments. Some examples are given below to illustrate the kinds of measures required.

Scientific breeding is required to develop farm animals which can meet draught needs yet be economically viable. Europe and North America have produced excellent work horses. China has ongoing programmes for raising work animals particularly donkeys, mules and horses. In India top quality bullocks have been bred for centuries, but now, for a variety of reasons, enthusiasm has waned. Comprehensive breeding programmes must be established in all countries where draught animal power will continue to be of economic significance.

Veterinary health services are essential. Many countries in Africa and Asia still have the problem of typical livestock ailments such as foot and mouth disease, rinder pest, etc. Funds and facilities must be made available for eradication programmes. Some African countries are unable to promote draught animal power because of tsetse fly and trypanosomiasis, but active programmes to control the disease are under way.

A limit on the use of animal power is the availability of animal fodder. It takes about 0.5 to 1 hectare to produce enough fodder for 1 animal, but in many places grazing on non-agricultural land and using waste products makes the support of a sizeable number of animals virtually cost free.

Inadequate feeding practices for draught animals are extremely frequent and these lower the levels of animal health and thus their efficiency. There are various native plants in many countries which could be grown and used for animal feeding. Pasture improvement on marginal lands still offers good prospects. If special fodders cannot be grown, straw can be supplemented with molasses and urea, which add to the nutrient requirements. The use of agricultural wastes and industrial by-products should be promoted.

In some countries extension of credit and the formation of co-operatives have been part of the measures designed to bring institutional credit and organized modernization to rural areas. A larger flow of credit can be achieved by combining the efforts of rural banks, co-operative credit societies, rural branches of commercial banks, etc. However, much greater efforts are still required to meet the needs of the farmer. Insurance facilities should be extended to all sectors of draught animal power - the death of an animal may completely disrupt a small farming operation. Studies in various parts of the world have shown how pro-

Animal keeping.

ductivity can be increased with better equipment. Improved implements and harnessing devices must be developed, produced and serviced in the countries which use draught animals. The rural artisan has an important role to play in the making, provision and maintenance of equipment.

The slaughter system, starting from the collection of animals, transportation to slaughter, stunning methods, slaughter itself, recovery of by-products and their processing, etc. must all be modernized. Fattening the animals for a period of three to six months before slaughter would considerably enhance the quality and quantity of meat.

Information must be distributed on all these aspects. Extension workers should be trained in all aspects of draught animal power and supplied with literature and other materials through which they can create awareness amongst farmers and carters about the benefits of modernizing draught animal power. Extension must be at farm level. On the regional level, centres for draught animal power could conduct socio-economic studies, research and development work, as well as providing education, training, and extension.

Only an adequate infrastructure can upgrade draught animal power. Such an infrastructure is lacking or missing in most developing countries, but it could be implemented and strengthened rapidly with relatively little input.

Economic Aspects

A small subsistence farmer is only likely to invest in new techniques if he considers that they will be profitable. In view of the limited financial resources of the farmer, the purchase of equipment means that most have to rely on a national credit agency for financial support. In order to encourage investment, such an agency must not charge a high rate of interest and must only demand modest guarantees from its borrowers.

The benefits of using animal rather than human labour were particularly well illustrated in recent field trials in Sierra Leone. It was found that maize cultivation by hand, including hoeing, seeding and weeding took 160 man days/ha/year whereas this was reduced to only 50 man days with the use of an ox. Ground nut cultivation which took an average of 182 man days/ha/year and by hand was reduced to 30 days when using an ox, hand-cultivation of cow peas taking 240 man days/ha/year was reduced to 112 days, and swamp rice cultivation by hand took 197 man

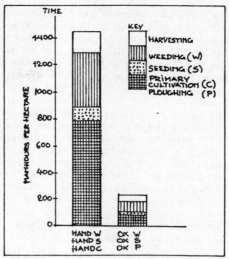

Diagram showing time budgets for groundnut cultivation, using manual labour and draught animals. (Source: Starkey, 1981).

days/ha/year as opposed to 82 man days with the use of an ox. In addition it was shown that with the use of animal power a farmer could plan to plant and harvest more accurately and could carry out these activities at the most appropriate times thus producing up to 20% increases in yield and better product quality, adding significantly to his crop value. The study also revealed that the farmer now had the ability to intensify cultivation, use multi-crops, grow inter-season crops and cultivate green fertilizer.

Naturally the value of this extra produce and its greater quality is offset against the cost of the ox and the associated equipment.

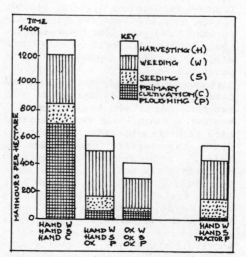

Diagram showing time budgets for maize cultivation using manual labour, draught animals or tractors. (Source: Starkey, 1981).

The cost of oxen varies with season, weight, ability to bargain, location and country. However, in most cases, it appears possible to buy an ox and resell it for a profit even after some years use as a draught animal.

Village labour rates, which generally include food as well as cash, also vary greatly and specific contract work is more common than a daily wage. A farmer who uses family labour for tending grazing cattle or for crop cultivation may not consider the labour as a real cost at all.

Profit can be achieved by increasing productivity through

1   reduction in work hours needed per hectare

2   increase in unit yields

3   intensification of cultivation by having enough time for additional cultivation activities

4   improvement in the quality of the produce

The mechanization of agriculture must be directed at increasing the cultivated area and the substitution of manual power by animal drawn implements.

It has been found in Sierra Leone that the capital investment could be repaid quickly through an increased and higher quality of crop production. This was calculated on the assumption that the animals were used for a minimum of 28 days/year, that three ha of land were ploughed, and 4 ha of land were seeded. The ox-cart needed to be used at least 50 times per year to become an economically viable additional expense. If these figures are a true reflection of the economic benefits of substituting animal power for human labour, then it would appear that such a capital outlay in the developing countries is not only a very attractive proposition for the individual farmer, but also to the developing countries as a whole as they will benefit substantially from increased agricultural production if improved farming practices are encouraged.

Conclusions

It has been found from a number of studies and field trials that the use of draught animal power can and does have a significant impact on the increased production and improved quality of agricultural products, and that animal power can greatly benefit the transportation sector of most developing countries, especially in rural areas. In addition, it has been shown that animal power can be used at acceptable cost and that it is often far more economic than the use of mechanized agricultural equipment and motorized transport.

Since few developing countries are likely to experience substantial economic growth in the short term, it is thought that for the foreseeable future, the use of animal power represents the most economic and appropriate form of technology improvement. By transforming agricultural practices from human labour- based operations to those based on the use of draught animals, improved crop production and quality will aid economic growth without which developing countries cannot develop.

Unfortunately the infrastructures necessary to make these proposed improvements a reality are often missing. Thus, before a country can consider trying to educate farmers to make greater use of animal power, it must first install the dissemination components necessary in order to make such a programme a success. However, there remains a severe shortage of initial investment capital in the developing world, and countries in such a position can do little to alleviate a situation in which the potential for self-improvement exists but where improvement programme activation is impossible through the lack of funds. It is felt that this is an especially important area to which the aid agencies should pay increasing attention in terms of providing sufficient capital for system technology improvements and enabling animal breeding and health facilities to be established. Aid agency capital is also important for providing farmers with loans for farming improvement measures.

Bibliography

Barwell, I. & Ayre, M. (1982) The Harnessing of Draught Animals.
Intermediate Technology Publications Ltd., 9 King Street, London WC2E 8HW, U.K.

ESCAP (1983) Draught Animal Power.
Renewable Sources of Energy, Volume V. ST/ESCAP/270.
Economic and Social Commission for Asia and the Pacific (ESCAP).
The United Nations Building, Rajadamnern Avenue, Bangkok 102 00, Thailand.

FAO (1982) Animal Energy in Agriculture in Africa and Asia.
Technical papers presented at the FAO Expert Consultation held in Rome, 15-19 November 1982.
Food and Agricultural Organisation (FAO) of the U.N., Via delle Terme die Caracalla, 00100 Rome, Italy.

FAO - C.E.E.MAT (1972) The Employment of Draught Animals in Agriculture.
By Arrangement with Experimental Centre for Tropical Agriculture Equipment (E.E.E.M.A.T.) France.
Food and Agricultural Organization (FAO). Via delle Terme di Caracalla, 00100 Rome, Italy.

Hamer, M. (1982) Animals Still Full of Pull.
Article in New Scientist, No. 18, November 1982.

Hopfen, H.J. (1969) Farm Implements for Arid and Tropic Areas.
Food and Agricultural Organisation (FAO) of the U.N. Via delle Terme die Caracalla, 00100, Rome, Italy.

Lundell, S. et al. (1983) Swedish Forestry Techniques with possible Applications in the Third World.
The Forest Operations Institute, Box 1184, S-116 13 Spaanga, Sweden.

National Research Council (1981) The Water Buffalo: New Prospects for an Underutilized Animal.
National Academy Press.
Commission on International Relations (JH-217), National Research Council, 2101 Constitution Avenue, Washington, D.C. 20418, U.S.A.

Ramaswamy (1980) Draught Animal Power.
Paper No. 8. Panel of High-Level Experts.
World Bank - UNDP joint study. Financial Requirements of Supporting Actions and Pre-investment Activities in Developing Countries.

SAREC (1980) Workshop on Water Buffalo Research in Sri Lanka.
R:1-3 1982.
Swedish Agency for Research Cooperation with Developing Countries (SAREC), S-105 25 Stockholm, Sweden.

Starkey, P.H. (1981) Farming with Work Oxen in Sierra Leone.
Sierra Leone Work Oxen Project, Njala Unversity Collage, Private Mail Bag, Freetown, Sierra Leone.

U.N. (1981) Report of the Ad Hoc Working Group on Draught Animal Power.
U.N. Conference on New and Renewable Sources of Energy Nairobi 1981 A/CON.100/PC/39.

Section II

BIOMASS ENGINES AND BIOMASS ENGINE FUELS

11 PRODUCTION OF BIOMASS ENGINE FUELS

Introduction

Mechanical energy in the form of shaft power is urgently needed for the development of industry, agriculture and transportation systems in developing countries. In fact, improvement of the standard of living in many developing countries is only possible through an increased productivity, requiring increased industrialization and modernization of the agriculture and the transportation sectors. This will inevitably lead to an increased need for shaft power.

This power can be generated in many ways. In both industrialized countries and developing countries, internal combustion engines using petroleum fuels are commonly used for vehicles and agricultural machinery. For stationary applications, however, the different infrastructures of industrialized and developing countries leads to different solutions for the supply of shaft power. In the industrialized countries with well developed electric grid systems supplied by large hydro-electric, coal-fired or nuclear power plants, electric motors are often used to provide shaft power. In the developing countries, where electric energy is often not available where power is needed, internal combustion engines using a petroleum fuel are frequently used in stationary applications.

The increasing prices of petroleum fuels and the growing awareness in non-oil producing countries of the potential risks associated with a very large dependence on imported petroleum fuels have led to a rapidly increasing interest in alternatives to the petroleum fuelled internal combustion engine for shaft power production.

The Figure below gives an overview of these alternatives. The possible or feasible power range for each energy type is also shown. As indicated in Table 1, some of the alternatives are restricted to stationary applications, others require special climatic or geographic conditions for siting.

The most versatile shaft power generators are those which use chemical energy in a fuel for conversion to mechanical energy (i.e. shaft power). There are three demonstrated principles for this:

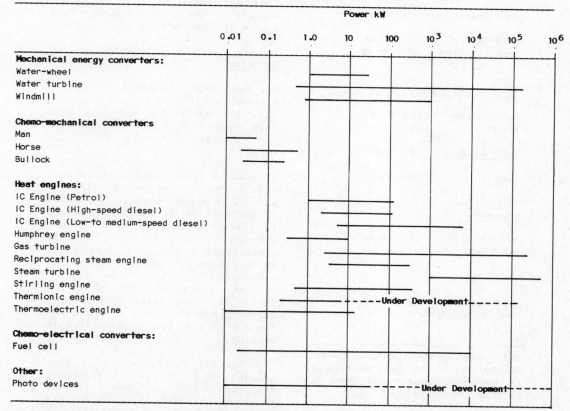

Energy converters and their power range

131

1 Heat engines where the fuel is combusted to generate heat and the heat is converted to mechanical energy. Diesel engines and steam engines are examples of heat engines.

2 Chemo-electrical converters where the chemical energy is converted to electric energy which can then be used to generate shaft power by an electric motor. The fuel cell is an example of a chemo-electrical converter.

3 Chemo-mechanical converters where the chemical energy in the fuel is converted directly to mechanical energy. Muscles are examples of chemo-mechanical converters.

Direct conversion of chemical energy in biomass to mechanical power by use of animals is discussed in Chapter 10. Engines based on this principle have been demonstrated, but only for very special chemical fuels. The use of biomass as fuel in such engines does not appear to be a realistic possibility for the next

Table 1.* Power requirements for shaft power applications in developing countries

Application	Typical power requirement
Stationary[1]:	
Water pumping	
Small farm	2 – 10 kW
Large plantation	30 – 250 kW
Grinding, milling	
Small village	5 – 10 kW
Region	50 – 200 kW
Small workshop	2 – 30 kW
Sawmill	30 – 200 kW
Electric Power generation	
Village level	10 – 100 kW
Towns and regions	500 – 10 000 kW
Mobile:	
Road vehicles	20 – 150 kW
Small fishing boats	5 – 50 kW

[1] It is claimed by Beale (1981), see Chapter 13, that 95% of the shaft power needs in developing countries is found in the power range below 10 kW.

* For reasons of clarity, Tables in Chapters 11-14 are numbered.

10 or 20 years. In any case it will not be discussed further here.

Chemo-electrical conversion in fuel cells is much further developed and is applied to specialized uses such as power supply in space craft. Demonstration plants for utility, domestic and industrial electricity production are in operation using clean fuels e.g. natural gas. Biomass can be used to produce suitable fuels for fuel cells. The fuel cell technology is fairly expensive though, and even if research and development bring the costs down, it cannot be expected to be competitive during the next decade and thus will not be discussed further here.

It must therefore be concluded that the alternative to petroleum fuelled internal combustion engines must be looked for amongst the other types of heat engines. Such alternatives are further reduced by the limitations in power range of some of these heat engines and by the limited possibilities of operating them with other fuels than gasoline, kerosene or gasoil, e.g. biomass fuels.

Most of the applications for shaft power in developing countries can be found in the power range below a few hundred kW, see Table 1. The following discussion will therefore concentrate on this power range. The discussion will be limited to engines operating on biomass or a fuel derived from biomass, since biomass is the main indigenous fuel source in most developing countries.

Overview

For each type of engine system the state of development of the technology will be described. Also the main data needed for an assessment of the economic feasibility will be given.

Biomass is seldom available in such a form that it can be used directly as fuel in a heat engine. Some kind of preparation of the biomass fuel is therefore often required. Such preparation may include: physical processing e.g. drying, size reduction or densification; thermal processing e.g. carbonization, or chemical processing e.g. the production of alcohol fuel. Different forms of fuel preparation and the costs involved for these will also be discussed in this section.

The Section concludes with a discussion of factors which influence the choice between the different types of available biomass engines with some examples of three selected applications.

Properties of Biomass Fuels

As schematically illustrated in the Figure on the top of next page the main constituents of a biomass fuel are:

- water

- minerals (i.e. ashes)

- hydrogen

- carbon

- oxygen

If all water and minerals are removed, the basic composition of most types of biomass does not differ greatly. When biomass is burned, the hydrogen, together with oxygen in the air, forms water vapour whilst the carbon combines with atmospheric oxygen to give carbon dioxide.

The maximum amount of heat which can be obtained by combustion of water and ash-free biomass (the so-called higher heating value) is also not much different for different types of biomass, see Table 2. This does not imply, however, that all

types of biomass are equally suitable as an energy source, as in practice the biomass is neither water nor ashfree. The importance of the water and ash content depends on the type of technology used for energy conversion. This will be discussed in a later section of this chapter. One important consequence of the water and ash content is, however, that the amount of heat which can be extracted from each kilogramme of biomass by combustion is reduced. Often the latent heat in the water vapour generated by the combustion cannot be utilized. The amount of heat which can be obtained by combustion under such circumstances from a biomass which contains water is called the effective heating value. The bottom figure on next page shows how the effective heating value varies with the moisture content for wood, which can be considered as a typical biomass fuel.

It should be observed that the water content can be affected by drying the biomass. This is discussed briefly in the section on fuel preparation. It should also be observed that the ashes partly consist of minerals included in the structure

Table 2. Composition and properties of some types of biomass

Biomass	Proximate analysis (dry basis)			Ultimate analysis (dry, ash-free basis)			Higher heating value (dry, ash-free basis) MJ/kg
	Volatile matter %	Fixed carbon %	Ash[2] %	Carbon C %	Hydrogen H %	Oxygen O %	
Wood							
Birch			0.6	49.9	6.1	44.0	20.4
Coconut palm	79.7	19.3	1.0	49.1	6.2	44.7	19.2
Douglas fir			0.8	52.7	6.4	40.9	21.2
Ipil Ipil (Leucaena)	82.6	16.6	0.9	49.0	6.1	44.8	19.1
Agricultural residue							
Coconut husk			6.0	46.8	6.9	46.3	19.3
shells	78.9	20.3	0.8	50.2	6.4	43.4	20.3
Coffee hulls			1.8				18.6
Corn cobs	80.3	18.1	1.6	48.4	6.1	44.9	18.9
Cotton gin trash stalk	63.6	18.8	17.6				19.9
Peach pits	73.9	25.2	0.9				23.2
Rice hulls	68.4	24.9	1.2				19.9
Sugar cane bagasse			15-20	45.6	6.7	47.7	18.9
Wheat straw			1.9	46.9	6.7	46.4	20.5
Other							
Bamboo			4.5				20.1

NOTES:

[1] Variations between different species exist. The age of the plant and the conditions under which it has been grown may also have some influence.

[2] The soil conditions and the age of the plant have some effect on the ash content. Handling after harvest may have a great effect.

Compiled from various sources.

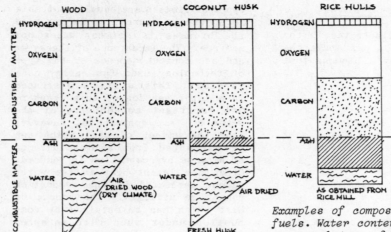

Examples of compositions of biomass fuels. Water contents are given for a range between fresh and air dried material.

of the biomass, and partly of soil which has been picked up during handling and transportation after the harvesting. Biomass to be utilized with a conversion technology which does not tolerate a high ash content should therefore be handled with some care after harvesting.

A property of the importance for transportation and handling of biomass fuels is its bulk density. This and the physical size of the pieces of biomass are also of great importance for the design of combustion chambers, combustion equipment and gasifiers. The bulk density, as well as the size distribution, can be adapted to suit the conversion technology chosen, as is briefly discussed in the section on fuel treatment. Table 3 gives some data for bulk densities for different types of biomass.

For the design of gasification and combustion equipment the amount of volatile matter (i.e. matter leaving the biomass as vapour when the biomass is heated) is of importance. For most types of biomass 70-80% will

be volatilized when the biomass is heated to 950°C.

For conversion to an alcohol fuel, the contents of simple sugars, starches, cellulose, semi-cellulose and lignin are of importance. This is discussed in the section on fuel preparation.

When the use of biomass as an energy source is considered it is important to observe that different types of biomass may require different types of energy conversion technologies. Modifications and adaptations of the biomass as well as the conversion technologies are possible to some extent, to make fuel properties and technology requirements match. There are, however, often economic constraints on this and the properties of the available types of biomass are therefore of great importance in governing the ways in which biomass can be used to replace petroleum fuels.

Table 3. Bulk density for different types of biomass fuels

	Moisture content of the sample (kg water per kg wet fuel)	Bulk density kg/m³
Wood		
Sawdust	about 10%	180
Wood chips	about 10%	170
wood blocks	about 20%	260
Agricultural residue		
Corn cobs	about 10%	300
Peach pits	about 10%	470
Rice hulls, loose	about 10%	about 100
Wheat straw, loose	about 10%	about 80
bales	about 10%	about 320
briquettes	about 10%	about 700

Compiled from various sources.

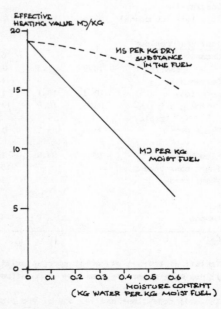

Effective heating value of the wood as a function of the moisture content.

Preparation of Solid Biomass Fuels

The need for fuel preparation

Heat engines work on the principle of combusting fuel to generate heat which is then converted into mechanical energy.

Depending on the design of the heat engine system there will be different demands on the fuel so that it can be handled in the system and properly combusted. Some designs will only operate on a liquid fuel and for those it will be necessary to convert the biomass into a liquid with the correct properties. Other systems can handle solid fuels but require a certain maximum or minimum size of the fuel particles or a certain maximum moisture content.

Biomass is seldom available in the form required for utilization in the heat engine system and some kind of preparation of the fuel is therefore often necessary.

For most types of heat engines there is a trade-off between costs for the engine system and costs for fuel preparation so that designs which require less (and cheaper) fuel preparation cost more, whereas cheap heat engine systems require more expensive fuel preparation. The application and load situation determine which option is the most favourable from an economic point of view, as will be shown later in this chapter.

Physical treatment of solid biomass fuel

Size reduction   Some types of equipment can only handle solid fuels if the pieces of fuel are within a certain size range. This applies to gasifiers in general, many types of combustion equipment and most devices for the automatic handling and transportation of the fuel.

Suitable technology for size reduction and the costs involved, depends on the original form of the biomass and the requirements imposed by the equipment for the handling or combustion of the biomass fuel.

Equipment suitable for chipping wood is commercially available in a large range of capacities, starting from about 15 cubic metres per hour. The cost for such a chipper would be about US$ 2500. It can be operated by one man. The resulting cost for the prepared fuel depends obviously on the utilization of the equipment. Such wood chippers are extensively used for instance in Sweden and Finland by farmers using wood chips for heating their homes.

Densification   Gasifiers and some types of combustion equipment will not work properly with biomass of a low density or in the form of small particles. Such biomass can be converted to an acceptable form by briquetting or pelleting. The technology for this is commercially available. The

Wood-chips

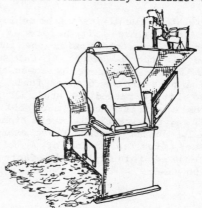

Chipping machine

Tractor mounted chipper

cost for briquetting or pelleting can be estimated to be about US$ 1 per gigajoule of prepared fuel (see Chapter 5).

Drying   Some types of gasifiers and combustion equipment will not operate properly on raw biomass fuel and require a fairly dry fuel, typically with a moisture content of less than about 20%. Drying of biomass fuels also leads to an increase of the available energy, and to less rapid decomposition, i.e. reduced losses during storage.

Raw biomass fuels will typically have a moisture content of 40-50%. Biomass will dry naturally to a moisture content in equilibrium with that of the air if stored under cover with adequate air circulation. The Figure to the right shows the equilibrium moisture content of wood as a function of the relative humidity in the air and the air temperature. Similar curves are valid for other types of biomass fuel. Natural drying may require a long time (several months) and other methods of drying are often used.

Cold air drying, where the ambient air is blown past the fuel by means of fans, is fairly effective when the relative humidity of the air is low – the drying time can range from a few days to a week. Electric energy requirements of the fans can be 150 to 200 kWh/kg of water removed from the fuel.

More effective drying can be achieved by heating the air before it is brought in contact with the fuel. The air can be heated by the sun, see Figure below, or by combustion of a part of the biomass fuel. If the latter method is used, drying from a moisture content of 50% to 20% may require 15-20% of the biomass fuel. Drying systems where exhaust gases from an engine or a boiler are used for drying the fuel are also used.

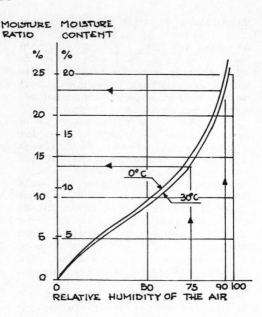

Equilibrium moisture content of wood at ambient temperatures 0°C and 30°C.

Solar drying has been estimated to lead to an added fuel cost of about US$ 0.3/GJ. The cost for drying using the biomass fuel as a heat source will cost about the same. Drying using exhaust gases from the plant where the fuel is burned may cost less than 30% of this, i.e. less than US$ 0.1/GJ.

Carbonization   of the biomass, in order to eliminate volatile matter, is necessary for use in some types of gasifiers which convert biomass to a combustible gas which can be used as fuel in an internal combustion engine. This will be explained in the section on producer gas. Technologies for carbonization, i.e. the manufacture of charcoal, are described in Chapter 6, although it should be mentioned here that different uses of the charcoal imply different quality requirements. Charcoal suitable for cooking may not necessarily meet the requirements of producer gas generators.

Liquid Fuels from Biomass

There are several possibilities for the production of liquid petroleum substitute fuel from biomass.

The easiest route to a liquid fuel from biomass is to use a sugar-bearing feedstock (like sugar cane) for fermentation to ethanol. Other types of biomass can also be used for production of alcohols (i.e. ethanol or methanol), but the technology is more complicated. Some seeds and nuts (e.g. sunflower and coconuts) can be used for production of vegetable oils which may be used as a substitute for gasoil.

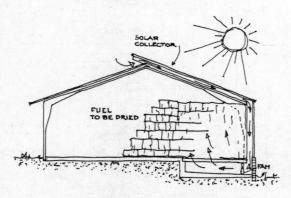

Diagram of solar drying system.

Liquid fuels can also be produced from biomass by treatment at high temperature. Such pyrolytic treatment will give a liquid similar to crude oil, which may be further up-graded to engine fuels. This possibility is being studied in Sweden, Finland, Canada and the USA. It is not believed that this technology can be used for commercial production within the next ten years, therefore it will not be discussed further.

The other options will be discussed below with emphasis on short term applications in developing countries, feedstock limitations and fuel properties.

Ethanol Production Processes

The principal production scheme for biomass ethanol involves pre- treatment of the feedstock, fermentation and ethanol distillation.

In the case of sugar cane the cane is first washed, crushed and filtered to separate the bagasse. The sugar juice is further concentrated and sterilized before fermentation.

For starch-containing raw materials the pre-treatment also involves hydrolysis of the starch molecules by enzymes into fermentable sugar. Once the fermentable sugar is formed, processing is identical for sugar and starch materials. Conventional ethanol technology uses batch fermentation with common strains of yeast to produce an 8 to 10% alcohol solution after 24 to 72 hours of fermentation. The ethanol solution is then distilled in a multistage distillation column to a concentration of about 95%. If anhydrous alcohol is the desired product, which is the case when ethanol is used for blending with gasoline, benzene is added to a third distillation column to split the azeotrop ethanol and water forms at a 95% concentration of ethanol. The anhydrous ethanol contains 99.8% ethanol. A process flow diagram for a cassava-based plant is shown below.

Feedstocks for Ethanol Production

Ethanol can be produced from three main types of biomass feedstock:

- sugars (sugar cane, molasses)

- starches (cassava, corn)

- celluloses (wood, agricultural residues)

Sugar-bearing feedstocks are attractive for ethanol production, since they already contain the simpler sugar forms such as glucose or fructose, which can be fermented to ethanol directly. Starches contain carbohydrates of greater molecular complexity and therefore have to be broken down to simpler sugars by a saccharification process. This means that another process step has to be added, which increases capital and operating costs. Carbohydrates in the cellulosic materials have an even greater molecular complexity

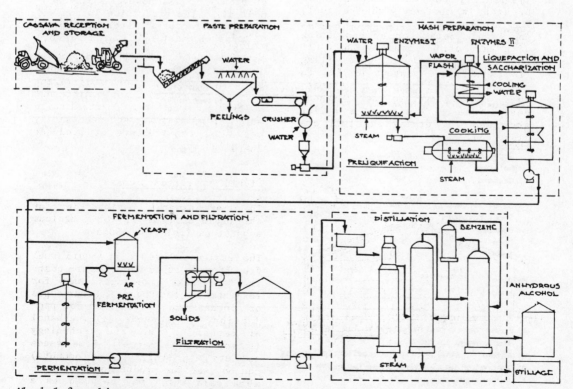

Alcohol from biomass: process flow diagram for cassava based plant.

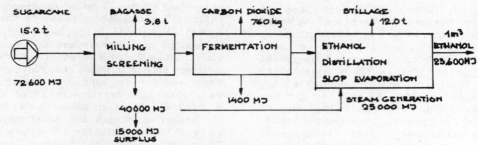

Flow diagram for sugarcane ethanol plant.

and have to be converted to fermentable sugars by acid or ensymatic hydrolysis.

Most sugar-bearing crops will generate residues which can be utilized as fuel for supplying energy to the process. Sugar cane for instance, generates bagasse, that can be used to provide energy for generating the steam and power needed for crushing, fermentation and the distillation process.

Other interesting feedstocks are sweet sorghum, cassava and corn. (See Table 4).

Sweet sorghum has a short growing season of 3 to 4 months and can be grown on sugar cane land when the land remains fallow between the last harvest and the next planting. The availability of another sugar material during the period when cane is not available would allow an extension of the production season and reduce the capital charges per unit of alcohol produced. Sugar cane distilleries normally operate for 160 to 180 days per year.

Cassava is a root crop grown extensively as a subsistence crop in a large number of developing countries. It is suitable for production only in the tropical regions. Cassava can be grown on marginal land and offers the potential of yielding a high volume of ethanol/ha. Efforts are under way to develop higher yield and more disease-resistant varieties of cassava but before it can be considered a major source of energy, these efforts have to be further intensified.

Corn, which can also be used for ethanol production, is an important food in many developing countries and large-scale utilization of corn for ethanol production would directly reduce food supplies. From a social viewpoint, but also based on technical and economic aspects, alcohol production from corn is not likely to be very attractive for developing countries unless the production is based on corn which is free from alternative uses.

Molasses, which is a by-product of cane sugar production, is also a potential feedstock for ethanol production. It is, however, only economical if the production is annexed to a sugar factory. Alternative uses of molasses exist and it may therefore

Table 4. Yields of alcohol fuels

Raw material	Ethanol yield liters/ton	Raw material yield[1] ton/ha	Alcohol yield liters/ha/yr	Energy yield GJ/ha/yr
Ethanol				
Sugar cane	70	50.0	3 500	1 350
Molasses	280	NA	NA	–
Cassava	180	12.0	2 160	324
		(20.0)[2]	(3 600)[2]	(540)
Sweet sorghum	86	35.0[3]	3 010[3]	945
Sweet potatoes	125	15.0	1 875	405
Babassu	80	2.5	200	67.5
Corn	370	6.0	2 220	162
Wood	160	20.0[4]	3 200	540
Methanol				
Wood		20[4]		394

NA = Not Applicable

1) Based on current average yields in Brazil, except for corn, which is based on the average in the U.S.

2) Potential with improved production technology.

3) Tons of stalks/hectare/crop. Two crops per year may be possible in some locations.

4) Current average yield in Brazil.

Source: Proceedings of the V International Symposium on Alcohol Fuels (1982).

be economically more attractive to sell the molasses on the world market than use it for ethanol production.

Development of Ethanol Production Processes

Competition between land use for food and energy production would be a less difficult problem if economic methods for producing ethanol from cellulosic material such as wood could be developed.

Wood consists of cellulose, hemicellulose and lignin. The latter cannot be broken down to sugar by hydrolysis. Cellulose and hemicellulose decompose by hydrolysis mainly to hexoses and pentoses. This can be done chemically by acid or biologically with enzymes. The hexoses can be fermented by yeast, while the pentoses cannot.

The yield of ethanol from wood is highly dependent on the conversion rate both in the hydrolysis and the fermentation step. Two major problems have to be solved before cellulose-based ethanol will be economically attractive. First the sugar yields have to be improved. Enzymatic hydrolysis seems to be a promising way of doing this but so far it has not been proven on a larger scale. Acid hydrolysis is a proven process and a number of plants are operating in the Soviet Union. Its economics are, however, poor and yields have to be improved. The other problem concerns pentose fermentation. Large efforts have been made during the last few years to solve this problem.

Today's level of development for hydrolysis processes and pentose fermentation implicate that the technical and economic risks of building commercial-scale plants will be significant. A commercial-scale ethanol production in the developing countries is therefore likely to be based on starch and sugar crops as feedstocks for at least the next 10 years.

The technology improvements for the commercially available ethanol processes aim at continuous fermentation processes and higher energy efficiency. Such development involves microbiological research work to yield liquids with a higher alcohol concentration, and more efficient distillation and heat recovery using engineering concepts commercially proven in other chemical engineering industries. Furthermore, ethanol concentration can be increased through absorption, vapour recompression, and/or multiple effect evaporators, but these techniques would require considerably more effort in research and development to develop into commercial practice. Other methods of alcohol separation being investigated include crystallization, use of molecular sieves and reverse osmosis, all of which will have the advantage of reduced energy requirements. These improvements would generally be at the expense of added capital cost.

The feedstock cost represents a large portion of total revenue requirements for ethanol production from both starch and sugar crops. Technology development effort is also needed in the agricultural area to improve yields of both food and energy crops, develop optimum crop rotation patterns and convert some existing subsistence crops (e.g. cassava) into commerical energy crops.

Methanol Production Processes

For production of methanol, high cellulose content materials such as wood and agricultural residues are suitable. This implies less competition between use of agricultural products for food and energy purposes.

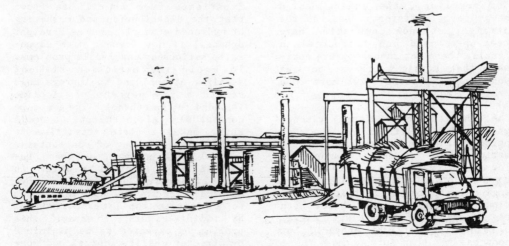

Alcohol distillery.

The production technology for methanol is totally different from the biological processes for ethanol production. In principle the process is conducted in two steps. First, the raw material is converted into a gaseous intermediate from which methanol can be synthesized. The methanol synthesis step is well-known and commercially available, while the gasification step is still under development. Such research is carried out in several countries with large wood resources e.g. Sweden and Brazil. Demonstration and commercial plants are expected to be built in the late 1980s or more likely during the 1990s. This means that a transfer of the technology to developing countries will probably not take place before the year 2000.

For economic reasons methanol will probably be produced in large plants. The minimum economic capacity for a methanol production plant using biomass as feedstock is seldom considered to be lower than 300 000 tonnes of methanol per year. This is very different from ethanol production where the "economy of scale" is at all not so pronounced.

Properties of Alcohol Fuels

Some properties of alcohol fuels are given in Table 5. In most respects ethanol and methanol are fairly similar to gasoline.

Gasoline or diesel fuel substitution in internal combustion engines can be accomplished by blending or by the straight use of alcohol as fuel.

No adjustments to the engines are required for up to a 20% ethanol blend in gasoline. Straight ethanol on the other hand has significantly different combustion properties to gasoline. To take advantage of the specific properties of ethanol (e.g. a higher research octane number than for gasoline), the engine design needs to be different. Engines for straight ethanol combustion have been developed and introduced in Brazil, where car manufacturers estimate that straight ethanol burning vehicles cost about 5-10% more than ordinary models.

The substitution of diesel presents greater difficulties since the properties of ethanol like high vaporization heat, low cetane number and low viscosity make the fuel unsuitable to use directly in existing diesel engines. Different technical solutions have been presented for both the blending of ethanol with diesel oil and for straight ethanol, but none has yet been introduced.

Methanol like ethanol, can either

Table 5.   Properties of fuel alcohols

Property	Fuel alcohols		Gasoline (for comparison)
	Ethanol	Methanol	
Density kg/m^3	789	793	720-750
Heating values:			
higher MJ/kg	29.7	22.3	46-47
lower -"-	27.0	19.7	43.6-43.9
lower MJ/m^3	21.3	15.6	about 32
Stockiometric air/fuel ratio kg/kg	9.0	6.5	14.6
Heating value for stockiometric air/fuel mixture		2.68	2.71
Boiling temperature at 1 bar °C	78.5	65	30-225
Heat of vaporization kJ/kg		1110	400
Vapour pressure at 38°C kPa		32	62-90
Viscosity c stoke		0.58	0.6
Octane number: research	106	112	91-100
Motor	89	91	82-92
Cetane number	15	10	

Compiled from various sources.

be used straight or as a blending agent in gasoline. Difficulties exist, however, in blending more than a few percent of methanol with the gasoline without having to add higher molecular weight alcohols. This is due to molecular weight phase separation effects that otherwise appear. Ethanol-blended gasoline does not require such an addition.

Experience from Brazil has shown that the distribution and marketing of hydrated ethanol (used as straight ethanol) is not fraught with especially serious or unsolvable problems. To avoid the explosion risk of straight ethanol and to improve cold starts a few percent of gasoline is added to the ethanol. The gasoline content also gives the fuel a smell which makes it less attractive to drink. The drinking of concentrated ethanol, has however, not been regarded as a serious problem in Brazil. Service station operators may be faced with the temptation to adulterate the hydrated ethanol by adding water. In Brazil this problem is expected to be minimized by stricter quality control and more frequent unannounced inspection. Other problems that have arisen e.g. corrosion, appear receptive to relatively easy solutions.

The Energy Balance

The cost and amount of energy inputs to produce the alcohol is of huge importance to its economic viability as a fuel.

Controversy has persisted regarding the energy balance in the production of ethanol from agricultural crops. Much of the controversy has revolved around the question of whether ethanol production from biomass in fact generates a surplus of commercial energy carriers.

The energy balance for commercial energy is usually expressed as a net energy consumption ratio (NER) which is defined in the following way:

$$NER = \frac{\text{total commercial energy consumed} - \text{byproduct energy}}{\text{energy in ethanol}}$$

The total commercial energy consumed includes energy contents of fuel and chemicals (farm inputs) consumed in growing and harvesting the crop as well as commercial energy inputs in the alcohol production process. Solar energy stored in the crop is not included.

The energy in ethanol is equivalent to its heating value. The evaluation of the by-product energy depends on the use of the by-product. If the by-product is used directly as a fuel, the energy is taken as the heating value of that fuel. If the by-product is used for some other purposes, like cattle feed, the energy is taken as the amount of energy required to grow or otherwise produce an equivalent material.

The energy balance for commercial energy is positive if NER is less than 1 and negative if NER is above 1.

Table 6 shows the estimated net energy ratio for different feedstock under Brazilian and U.S. conditions.

Sugar cane shows a positive net energy balance for commercial energy, derived wholly from the availability of bagasse. As long as a surplus of bagasse is produced there is less incentive to improve the energy balance for sugar cane based ethanol until an alternative economic use is found (such as external power generation and/or multiple feedstocks in the alcohol plant).

Cassava has a modest energy balance when wood fuel for process heat is assumed to be purchased. If a wood plantation is included in the agricultural system the net energy ratio improves substantially.

Under US conditions, see the production of ethanol from corn is a net energy consumer because the processing of the corn is relatively energy-intensive and the corn stover (stubble) is not commonly available at the plant to be used as fuel. Even if an "energy conservation" design is employed it is still a net consumer of commercial energy. The production of ethanol from corn could be made a net energy producer if nearly all the stover could be made available at the alcohol plant. However, it may not for different reasons be feasible to collect, transport and store that much of the stover for use at the plant.

The main energy losses in methanol production, based on naturally grown wood or agricultural residues as feedstock, are associated with the generation of synthesis gas and the actual methanol synthesis. The estimated efficiency for methanol production is in the range of 50-55%.

Table 6.

Brazil		United States	
	NER (kcal/kcal in ethanol)		NER (kcal/kcal in ethanol
Sugar cane	0.12	Sugar cane	0.33
Cassava without Tree farm	0.93	Corn-traditional	2.21
Cassava with Tree farm	0.13	Corn-energy conservation	1.20
(The stillage is used as fertilizer and for irrigation purposes.)		(The stillage is used as animal food after drying.)	

Source: World Bank, 1980.

The Economics of Alcohol Production

Cost estimates for ethanol production vary considerably. This is true in particular for cost estimates prepared outside Brazil. Cost estimates for ethanol production in developing countries presented in the literature during 1982, range between US$ 0.17 and US$ 0.55/1, i.e. US$ 8 to 26 per GJ. The lower figures are usually calculated for Brazilian conditions and the higher for other countries. US$ 0.55/1 (US$ 26/GJ) has been estimated for Thailand, which could be compared with the ex-refinery gasoline cost of US$ 0.25 to 0.30/1, i.e. US$ 7.7 to 9.2/GJ.

The reason for the differences in cost estimates is partly the lack of actual experience in most countries of constructing and operating this type of plant, partly because of different assumptions about the feedstock cost which have a large impact on the overall economy, and partly due to different estimates for the capital investment required for the process equipment. Brazil has its own manufacturing industry for ethanol plant equipment which gives them advantages regarding capital costs in comparison with other countries. Other countries are to a greater extent dependent on engineering companies from developed nations where the cost levels are high. Up to 70% of the plant cost may have to be spent outside a country on equipment, goods and services.

Capital costs for a 120 000 1/day distillery based on sugar cane have, in Brazil, been estimated at about US$ 9 million. Based on an analysis made by the World Bank the sugar cane cost must not exceed US$ 14/ton at the factory gate if the ex-refinery gasoline price is US¢ 27/1 (US$ 31 per barrel of crude oil) and the desired economic rate of return is 10%. For other developing countries the cost of a distillery of corresponding size has been estimated at US$ 30 million, which demands an even lower feedstock price in order to be viable.

Calculations made for South East Asian conditions, show already that the feedstock prices for ethanol production are about equal to ex-refinery cost of petroleum fuel. The cost of ethanol production is, as mentioned above for Thailand, more than double the ex-refinery gasoline cost. The import savings of gasoline that can be made by converting agricultural products to ethanol indicates that there is no economic advantage in reducing exports by diverting sugar, molasses or cassava to ethanol production, especially if a large portion of the capital cost is spent outside the country and chemicals have to be bought from overseas.

Without government subsidies ethanol fuel cannot presently compete with gasoline. Both in Brazil and the Philippines the ethanol price is linked to the price of gasoline. Thailand is a net exporter of carbohydrates and could therefore become a potential user of ethanol fuel. The government, however, does not seem to have any plans to subsidy the ethanol price, but prefers to leave it to market forces and the private sector for implementation.

The advantage of the economy of scale for ethanol distilleries diminishes rapidly above the capacity of 300 000 1/day. However, taking into account other factors that have larger influence on the production cost such as feedstock cost, operating days per year, etc., the optimum size for most developing countries is likely to be between 120 000 and 240 000 1/day for sugar-cane based distilleries. In some cases such as with an isolated location and expensive petroleum products, even smaller plants could be viable.

Mini distillery.

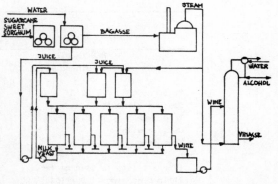

Flow diagram of small scale distillery, using surplus bagasse as fuel.

The existing technology for ethanol production was developed for applications where the cost of production and energy consumption were less important than today. A Brazilian study, presented by the Instituto Mana de Technologia, has compared the cost for ethanol when produced by the current available technology, with future process improvements and a switch to new raw materials. The following alternatives were analyzed:

1   Current technology

2 a Use of sugar cane having higher fermentable sugar content

    b Use of improved extraction processes

    c Inactive distillery period minimized by use of concentrated sugar cane juice

    d Use of an optimized fermentation process

3 a A pre-treatment of the cassava to obtain a starch suspension free from fibrous material

    b Saccharification and fermentation steps to give higher productivities for cassava than those usually mentioned in the Brazilian literature

4 Both sugar cane and cassava are used as raw materials. The process has the same characteristics as 2 and 3

5 A wood process using diluted sulphuric acid for cellulose hydrolysis

The results are presented in Table 7.

Improved sugar cane technology (alternative 2) as well as increased operating time by using cassava during the sugar-cane off season (alternative 4) show the possibility of a significant increase in the volume

Table 7. Cost estimates for ethanol production in Brazil

ALTERNATIVE NO	1 Cane Current	2 Cane Improved	3 Cassava	4 Cane+Cassava		5 Wood
Agricultural Productivity (ton/ha/year)	60	80	17.5	80	17.5	15
Total Sugar Content (kg/ton)	150	170	330	170	330	330
Industrial Yields						
Preparation (%)	91	97	98	97	98	72
Fermentation (%)	83	90	90	90	90	90
Destilation (%)	94	98	98	98	98	98
Losses Factor	0.93	0.97	0.96	0.97	0.96	0.99
Overall Industrial Yield (%)	66	83	83	83		63
Ethanol volume per ton of Biomass (liters/ton)	64	92	177	92	177	134
Days of Operation	200	300	300	200	100	134
Nominal Capacity (liters/day)	120 000	180 000	120 000	180 000		120 000
Annual Production (m³/year)	24 000	54 000	36 000	54 000		36 000
Needed Harvested Area (ha/year)	6.250	7.337	11.622	10.702		17.910
Overall Productivity (liters/ha/year)	3.840	7.360	3.098	5.046		2.010
Specific Investment (US$/m³/year)	353	218	245	220		669
Cost (US$/m³)	169	120	184	130		258

Source: Proceedings of the V International Symposium on Alcohol Fuels (1982).

RET-F

of ethanol produced with a decreased production cost. The disadvantage with cassava (alternative 3) is the demand for a larger area for agricultural production. In Brazil, however, this is a less important matter compared with other developing countries since it has quite a large area of land suitable for cultivation. A potential increase in the cassava productivity far beyond the value presented seems also to be possible. In the economic calculations for a wood alternative (5) it must be noted that by-products were not credited.

The economics of methanol production in developing countries are difficult to estimate. Capital costs are likely to be equal for both developed and developing countries, while the feedstock cost will probably be different. Investment costs for a 300 000 ton/year methanol plant based on High-Temperature- Winkler gasification technology have been estimated to be about 230 million for developed countries. The total production cost depends, however, on feedstock cost, salaries and other conditions that are country-specific.

Past and Present Use of Alcohol Fuels

Alcohols have been used as engine fuel since the beginning of this century. The interest in alcohol fuels has fluctuated with the prices of petroleum fuels and has peaked in times of shortage of such fuels. The rapid increase in petroleum prices during the last decade has resulted in renewed interest. Brazil has been producing and using ethanol as a blending agent in gasoline since the 1920s and is the leading country in the introduction of alcohol fuels. Some 500 000 cars are operating totally on ethanol. All other cars use a blended fuel consisting of 20% ethanol in gasoline. The original reason for Brazil´s ethanol production was the low world sugar prices which forced the sugar producers to promote the use of alcohol as a motor fuel. By 1975 the world sugar price was beginning to decline again and the oil prices had already been escalating

Table 8

Market Penetration Barriers to Ethanol Fuels In Selected
Developing Countries

	ARGEN-TINA	BOLI-VIA	BRA-SIL	COSTA RICA	PARA-GUAY	REP. DOMINI-CANA	KENYA	MALAWI	ZIM-BABWE	INDO-NESIA	PHILI-PPINES	THAI-LAND
Institutional Constraints												
· national program	–		+	–	±		–		o	o	±	–
· consumer attitude	o	o	±	–	±	o	o	o	o	o	o	o
· automaker, motor trade and allied Industries	±	o	+	–	+	o	±	o	o	+	o	o
· oil marketers/retailers	+	±	±	+	+	o	–	+	+	o	o	o
· refineries	±	±	–	+	±	+	±	o	o	o	o	o
· ethanol producers	+	o	+	–	+	+	+	+	+	o	+	o
· farmers	±	±	+	–	+	±	+	+	o	+	+	±
· sci/tech community	±	–	±	–	–		±	o	o	±	±	±
Economic/Financial Constraints	±	±	±		–	±	–		±	±		±
Pricing Constraints	+	–	±	±	±	o	±	+	o	–		±
Tech. Transfer Constraints	±	–	±	±	–	–	–	±	±	±	±	±

Legend: o unknown; + affirmative; - negative; ± uncertain

Source: Trindade, Implementation Issues of Alcohol Fuels: an International Perspective. In Proceedings VI International Symposium on Alcohol Fuels Technology, 1984

for a couple of years. The National Alcohol Programme (Proalcohol) was then established with the main objectives to:

- reduce balance of payment deficits by decreasing the demand for imported petroleum fuels

- increase the security of Brazil's energy supplies

- utilize excess capacity for sugar cane production

- increase income and create jobs in rural and underdeveloped areas.

The original goal was to increase production of ethanol to replace 75% of the total gasoline consumption by 1985. The production goal of 10.7 billion litres per year has now been extended from 1985 to 1987. For this production about 3 million ha of land will be required to grow the necessary sugar cane. In 1981 about 1 million ha were used for ethanol production and 1.7 million for sugar production.

In the USA, a blended fuel consisting of 10% ethanol in unleaded gasoline, "gasohol", is marketed. This fuel can be used in most spark ignited engines designed for unleaded gasoline. Test fleets of cars using straight alcohol fuels or blended fuels are operated in many other countries.

Other countries, already utilizing or setting up alcohol-programmes and building ethanol plants for replacing gasoline in the transportation sector are the Philippines, Thailand and some African countries e.g. Kenya and Zimbabwe. With varying economic success the basic technology for the production of ethanol is well known and can easily be transferred to most developing countries. See Table 8.

A National Fuel Alcohol Programme was launched by the government of the Philippines in 1980 with the aim of blending 15% ethanol in all gasoline by 1985 and 20% by 1990. In Thailand a Power Alcohol Committee was set up in 1979. The private sector has applied to set up fuel alcohol distilleries from a variety of agricultural raw materials. In Zimbabwe and Kenya alcohol plants are already under construction or completed. One must however be aware that ethanol is no general solution to the energy problem of the developing countries. The main concern over expanded use of biomass-based fuels is the potential conflict between land use for energy and food production. The lack of fertile land will set a limit to any large substitu-

tion of petroleum world-wide. If the entire world production of sugar cane, molasses, corn and sweet sorghum for which commercially-proven technology is available, were converted today, the total energy production would substitute for only about 20% of the total gasoline consumption. The opportunity to cultivate more land does not always exist. It is therefore of great interest to compare the yields of alcohol fuels produced from different feedstocks. Such a comparison is made in Table 4. It appears that from a land use point of view, ethanol production from wood is almost as effective as the most efficient food crops and more effective than ethanol production from, for instance, corn, sweet potatoes and cassava.

The implication of this is that since wood can be grown on land where food production is not suitable there is not necessarily a need for land use competition between food crops and feedstock for alcohol production. There are, however, other important uses for wood for instance as fuelwood. This must be taken into consideration when the resource base for alcohol production is assessed.

Environmental Impacts of Alcohol

Production

During the fermentation and distillation of ethanol the following by-products are produced:

- carbon dioxide, which is usually discharged into the atmosphere

- fusel oils (mainly amyl-, isoamyl-alcohols and glycerol) which can be collected for sale or blended into ethanol as a fuel denaturant

- stillage

Stillage is the liquid effluent from the distillation system and its disposal can be a major problem. It is produced in large quantities, about 10-13 times the volume of the alcohol produced. The two main potential uses for stillage are animal feeds and fertilizer. In Brazil the stillage is applied directly on to the soil by trucks or it is pumped to the top of neighbouring hills and gravity fed to irrigation systems for surrounding fields. Cane yields appear to be increased substantially on such land, due to both the fertilizer and irrigation effects.

Production of Vegetable Oil Fuels

Feedstocks for the production of vegetable oils

There are a number of crops which can be used for the production of vegetable oils. Table 9 shows the world production of the most important of these. Other important oil producing crops are sesame, sunflower and linseed.

Most oil crops give 300-800 kg of oil/ha. Higher yields are possible, for instance with the African oil palm which might give up to 3,000 kilograms of oil per hectare. The yields vary with the species, but also with climatic, soil and irrigation conditions. Average values are not very useful when estimating the production potential for a specific location.

There are no differences between the growing and harvesting of seeds for edible oil production and that for the production of vegetable oil fuels. The available experiences from production of vegetable oil for food can therefore be applied directly to fuel production.

Oil Extraction Technology

There are two alternative extraction technologies in use, namely mechanical pressing and solvent extraction. Mechanical pressing is the simpler way to extract the oil. Screw-presses are most common in modern practice. In such units the feedstock is normally pre-heated and then fed into the press which consists of a worm shaft rotating within a barrel. As the material

Screw press for vegetable oil production.

passes through the press, the pressure increases and oil is extracted continuously. The meal will leave the press with up to 20% of the original oil still in it. A second pressing will increase the extraction efficiency, leaving about 5% of the oil in the cake. A typical throughput for an oil production plant with mechanical pressing is between 60 to 100 tons/day. There are also small presses marketed with a throughput of about 40 kg/h. The oil residue in the cake is often higher for such presses, typically about 14%.

The energy requirement for the pressing varies according to the feedstock, the amount of preheating, the size of the press etc. Figures between 0.1 and 1 kWh/l of oil have been reported. A reasonable assumption is that about 5% of the energy value of the oil output has to be consumed as mechanical energy for the press. If this mechanical energy is generated in an internal combustion engine using the oil as fuel, about 20% of the oil produced would be needed for the operation of the press.

Table 9.   World production and production cost of main vegetable oils.

Oil crop	World production 10^6 ton 1978/79	Higer heating value MJ/kg	Bulk price[1] USD/ton 1981	Energy price[1] USD/GJ
Coconut	2.6	38.8	575	14.8
Cotton seed	2.9	39.3	530	13.5
Ground nut	3.3	39.7	546	13.8
Palm	3.8		585	
Rapeseed	3.7	39.7	478	12.0
Soy bean	13.2	39.4	539	13.7
Sunflower	4.6	39.6	526	13.3

[1]   Compare with Rotterdam price for gasoil as of March 1985: 235 USD/ton, i.e. 5.5 UDS/GJ.

Compiled from various sources.

The most effective method for vegetable oil production is solvent extraction in which a chemical solvent such as hexane is used to extract the oil. This will leave only about 1% of the oil in the residue.

Solvent extraction requires relatively large units and sophisticated technology and thus is not too well suited for local application in developing countries.

Past and Present Use

The production of vegetable oils for edible purposes or as fuel for lamps dates far back in history. World production in 1978/79 was more than 36 million tons of which about 87% was used for consumption purposes and the balance as an industrial raw material or for lighting.

Experiments using vegetable oils as engine fuel were carried out as early as the 1930s and have aroused new interest recently. There appears to be no regular use of vegetable oils as fuel at present apart from tests conducted with single vehicles or tractors.

Energy Balance

The energy balance for vegetable oil production is without doubt positive. The main energy needs are fuel for agricultural machines, estimated at 6 percent of the oil production, and energy at about pressing, estimated to about 20 percent if the oil is used as fuel for an internal combustion engine operating the press.

Properties of vegetable fuel oils

Table 10 shows detailed comparison between the properties of sunflower oil and diesel fuel. Other vegetable oils have properties which are similar to those of sunflower oil. The heating value of vegetable oils is generally somewhat lower than for ordinary diesel fuel e.g. gasoil, but this is of little significance. The most important difference is the much higher viscosity which leads to significantly different spray patterns from regular injector nozzles. This causes problems as is discussed further in the section of engine performance.

Table 10. Fuel properties of sunflower oil and diesel fuel measured in accordance with ASTM standard testing procedures.

Properties	Diesel fuel[*]	Sunflower oil: crude degummed
Gross heat value, MJ/kg	45.93	39.38
Specific gravity	0.84	0.93
Viscosity at 37.8°C, mm^2/s	3.90	34.70
Cetane number (typ.)	47–48	37
Flash point, °C	55–77	321.00
Cloud point, °C	-0.60	-6.60
Carbon residue, %	0.15	0.42
Ash weight, %	0.01	0.04
Distillation 90% °C	335.00	355.00
Sulphur %	0.25–0.29	0.12
Copper strip corrosion	No. 1	No. 1B

(Source: Quick, 1980).

[*] Australian automotive distillate.

Modification of Vegetable Oils for Fuel Purposes

It is possible to bring the properties (e.g. distillation curve) of vegetable oils closer to those of gasoil and at the same time bring about a significant reduction of the viscosity to approximately the same order as for gasoil by changing the oil into an ethyl or methyl ester.

Blending gasoil with vegetable oil is another method of making vegetable oil fuel more similar to gasoil. A significant reduction of the viscosity can be obtained by mixing vegetable oil with 50% gasoil. Kinematic viscosity at $37.8^{\circ}C$ of this mixture is about 10.9 mm^2/s compared to 39.4 for the sunflower oil and 3.9 for gasoil. This will reduce or perhaps eliminate operational engine problems but leads, of course, to a smaller saving on gasoil.

The Economics of Vegetable Oil Production

Taking the bulk prices into account, it seems unlikely that vegetable oil fuels will have a near-term impact on commercially well developed fuel markets, either in industrialized or in developing countries. As shown in Table 9, the bulk prices per energy unit in 1985 were more than twice the Rotterdam price for gasoil. The use of these fuels will more likely start as local initiatives on farms or in regions where there is an abundance of vegetable oil and gasoil is expensive. A typical example would be a tropical island which was a long way from the nearest petroleum depot and which produced copra and palm oil. Local extraction presses would have to be available and traditional trading patterns for the oil-seed might have to be changed. It would be necessary to have a local consumption of the meal-cake, which might not always be easy, due to the quality and oil content of the locally produced cake, availability of cattle and other domestic animals, etc.

An economic analysis should be made before taking any decisions to introduce vegetable oil fuels. Three factors will determine the outcome of this analysis: the local price for gasoil, the local selling price for the oil-seed, and the cost of local extraction of oil. The value of the meal-cake as a cattle feed, may also influence the economy. A general economic assessment of vegetable oil fuels will be of very limited value for decisions regarding investments in individual cases. Specific studies taking the local conditions into account are required for such decisions.

Bibliography of Biomass Fuels

Celis, R.V. et al. (1982) The Foreign Trade Deficit and the Food Crises, Anticipated Results of an Aggressive Program of Alcohol Fuel Production in Costa Rica.
"unpublished material" Discussion Paper D-73H.
Resources for the Future, 1755 Massachusetts Avenue, N.W., Washington, D.C. 20036.

FAO Agricultural Services Bulletin 46 (1981) Energy Cropping versus Food Production.
Food and Agriculture Organization (FAO), Via delle Terme di Caracalla, 00100 Rome, Italy.

Hall, D.O. (1982) Food versus Fuel, A World Problem.
King's College London, 68 Half Moon Lane, London SE24 9JF, U.K.

Kovarik, B (1981) Fuel Alcohol, Energy and Environment in a Hungry World.
Earthscan, 3 Endsleigh Street, WC1H ODD, London, U.K.

(1980) Proceedings of the IV International Symposium on Alcohol Fuels.
Sao Paulo, Brazil, October, 1980.
Instituto de Pesquisas Technolgicas, University of Sao Paulo, Brazil.

(1982) Proceedings of the V International Symposium on Alcohol Fuels.
New Zealand.
Ministry of Energy, P. Bag, Wellington, New Zealand.

(1984) Proceedings of the VI International Symposium on Alcohol Fuels.
Ottawa, Canada, May 1984.
Transportation Energy Division, Ministry of Energy, Mines and Resources, 580 Boath Street, Ottawa, Ontario K1A OE4, Canada.

Quick, G. R. (1980) Farm Fuel Alternatives.
Article in Power Farming Magasine No. 2. 1980.

Shell Briefing Service No. 5 (1982) Alternative Road Transport Fuels.
Shell International Petroleum Company Limited, London, PA 012, U.K.

van Swaaij, W.P.M. & Baenackers, A.A.C.M. (1982) The Methanol from Biomass Pilot Plant Program of the Commission of the European Community.
Twente University of Technology, P.O. Box 217, 7500 AE Enschede, The Netherlands.

World Bank (1982) Emerging Energy and Chemical Applications of Methanol: Opportunities for Developing Countries.
World Bank, 1818 H Street, N.W., Washington, D.C. 20433, U.S.A.

Bibliography of Ethanols

Carioca J.O.B. et al. (1981) Cassava-based minidistilleries in Brazil.
Biomass Vol. 1 No. 2.

de Groot, P. (1985) Breeding Sugar Cane for Energy.
New Scientist 18 July 1985.

Mathewson S.W. (1980) The Manual for the Home and Farm Production of Alcohol Fuel.
Ten Speed Press, Calif. U.S.A.

Ministry of Industry and Commerce (1981) Assessment of Brazil's National Alcohol Programme.
Secretariat of Industrial Technology, Ministry of Industry and Commerce, Brasilia, Brazil.

Monier-Williams G.W. (1922) Power Alcohol.
Oxford Technical Publications, London.

O'Keefe, P. & Shakow, D. (1981) Kenya's Ethanol Program: Is it Viable?
Ambio Vol. X No. 5 1981, Royal Swedish Academy of Sciences, Box 50005, 104 05 Stockholm, Sweden.

Pereira, A. (1983) Employment Implications of Ethanol Production in Brazil.
International Labour Review, Vol. 122, No. 1, Jan.-Febr. 1983.

Stetson, F. (1980) Making Your Own Motor Fuel With Home and Farm Alcohol Stills.
Gardon Way Publishing, Charlott, Vermont, U.S.A. 1980.

Stuckey, D. & Juma, C. (1984) Power Alcohol in Kenya and Zimbabwe: A Case Study in the Transfer of a Renewable Energy Technology.
Report for the U.N. Conference on Trade and Development, Geneva.
Department of Chemical Engineering and Chemical Technology, Imperial College, Prince Consort Road, London, SW7 2BY, U.K.

World Bank (1980) Alcohol Production from Biomass in the Developing Countries.
World Bank, 1818 H Street, N.W., Washington, D.C. 20433, U.S.A.

Bibliographies of Vegetable Oils and Producer Gas are given in Chapter 12.

12 BIOMASS FUELS IN INTERNAL COMBUSTION ENGINES

The Possibilities of Using Biomass Fuels in Internal Combustion Engines

Internal combustion piston engines are commonly used in developing countries for water pumps, mills and grinders, threshers, small electric power plants and, of course, for cars, buses, trucks etc.

Spark ignition engines operating on gasoline and kerosene are often preferred for applications where low weight is an advantage and where the shorter lifetime (typically less than 5000 hours) and lower efficiency (typically about 25% at full power) can be accepted from an economic point of view. This situation often applies when the annual number of operating hours is low, say less than 500.

Compression ignition engines operating on diesel fuel, are heavier, are generally designed for lifetimes of more than 20 000 hours, and show a higher efficiency (typically about 35% at full power). They are preferred for applications where weight is less important and where the gains on lifetime and fuel economy compensate for the higher price of the engine.

Both types of engine can be used with biomass derived fuels, although the technologies are somewhat different, as are the changes in engine performance in comparison with operation on petroleum fuels.

There are, in principle, three quite different ways in which biomass can be utilized as engine fuel. Most simple for the user is to substitute the gasoline or diesel fuel by a liquid fuel derived from biomass.

Spark ignition engines can be used with ethanol or methanol. These fuels can also be used in compression ignition engines but this requires more complicated measures such as the addition of ignition improvers to the fuel or dual fuel operation. For such engines, vegetable oils appear a more suitable liquid substitute fuel.

Another possibility is the conversion of the biomass to a combustible gas, "producer gas", which can be used as fuel in the engine. Spark ignition engines can use producer gas without major modifications. Compression ignition engines will not operate on this gas only, but can be run on a mixture of gas and suitable liquid fuel (for instance a vegetable oil). Finally, it appears possible to use biomass in the form of a fine powder directly as fuel for compression ignition engines. This technology cannot yet be considered sufficiently developed for practical operation and will therefore not be discussed further below.

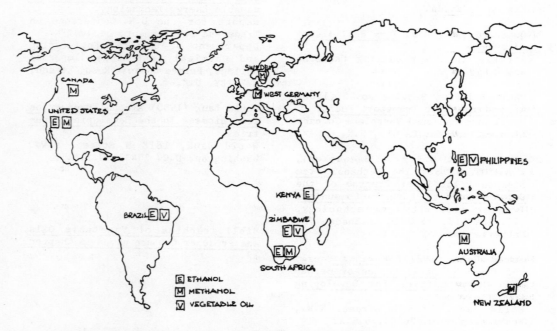

Alternative biomass engine fuel programmes in the World.

150

Alcohol Fuels

Production and Properties of Alcohol Fuels

The production technology, choice of feedstoves, production costs and properties of alcohol fuels have been treated earlier in Chapter 11.

Engine Performance with Alcohol Fuels

In most respects ethanol and methanol are fairly similar to gasoline and therefore they are more suitable for engines working with spark ignition than with compression ignition. The necessary modifications to a spark ignition engine depend on whether the engine will operate on straight alcohol, or on a blend of alcohol and gasoline. For blended fuels with less than 5% methanol or 10% ethanol, no modifications to an engine designed for gasoline are normally required and the power and the efficiency of the engine will not be affected. With 15% methanol or 25% ethanol, minor parts in the fuel system may have to be changed (or other additives added) because alcohols are more aggressive to some polymers and metals, which may have to be changed. There will be no power loss and often a slight improvement in the efficiency of the engine.

For operation on straight alcohol fuel the minimum modification necessary to operate the engine is to adjust or replace the carburettor nozzle in order to allow double the fuel flow. To simplify cold starts, it is also advisable to maintain the possibility for operation on gasoline. This means in practice that two separate fuel systems will be used, one for alcohol and one for gasoline, including either two carburettors or a special carburettor with an additional nozzle for starting fuel.

With these changes the advantages of alcohol fuels are not fully utilized. With the optimization of the engine for alcohols, increased power output and a higher efficiency can be achieved. The changes necessary to reach this include increasing of the compression ratio and replacing the inlet manifold. The effect of this may be a power increase of some 5% and an efficiency improvement of 15 to 20%.

The use of small amounts of ethanol (ie. 5 to 10%) in diesel oil is possible in compression ignition engines without any modifications to the engine. To use only alcohol fuels in compression ignition engines is not possible without addition of "ignition improvers" to the alcohol.

This is because the alcohol/air mixture will not ignite spontaneously when it is compressed in the engine. The "ignition improver" may be a special substance, for instance hexyl nitrate, but it is also possible to operate on an emulsion of alcohol and gasoil.

Dual fuel operation is also possible either by admission of the alcohol air mixture to the compression ignition engine through the air inlet and injection of a small quantity of gasoil to bring about ignition, or by the use of double injection systems, one for the alcohol and one for the gasoil. Practical tests with buses in Sweden using this latter system for methanol, shows a saving on gasoil of 80%. The efficiency of the engine was not changed.

It is also possible to introduce to some compression engines a spark ignition system, or a glow plug ignition system. The MAN factories in Germany have developed a system for spark assisted ignition in direct injection engines. Glow plug ignition systems are being studied in Brazil. These systems give higher efficiencies than for compression ignition operation on gasoil. An improvement of up to 20% has been reported.

Economic Considerations

There is no reason why engines designed for alcohol fuels should be significantly more expensive than petroluem engines operating with the same compression ratios. The cost data collected in Table 11 which is valid for gasoline and diesel engines, should therefore also roughly apply to alcohol engines.

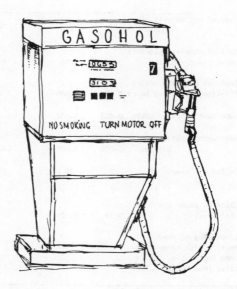

Alcohol-gasoline mixtures are sold as "Gasohol" in the United States.

The cost for the conversion of a spark ignition engine to alcohol fuel will depend on whether minimum modifications are made or whether the engine is optimized for the new fuel. Minimum modifications can be made for less than US$ 100.

Maintenance costs and engine lifetime can be assumed to be the same as for petroleum fuels.

Fuel costs depend, of course, on the price paid for the alcohol fuel,(see separate section on the production of such fuels), and the efficiency of the engine. As mentioned above, the efficiency of engines converted on a minimum modification basis can be assumed to be the same as for petroleum fuels, whereas a 10 to 20% improvement of the efficiency can be expected for optimized, specially designed alcohol engines.

Environmental Impacts

The emissions of carbon monoxide and hydro-carbons will be less for properly adjusted alcohol engines than for engines operating on petroleum fuels. However, the emissions of aldehydes are higher and catalysts must be used to bring these down to the same level as for operation on gasoline.

Development Needs

It appears that alcohol fuels can be used in spark ignition engines with fairly simple modifications to the engine. A large number of such alcohol powered engines are in use in Brazil. Research and development is still warranted for spark ignition engines in order to improve the basis for optimizing the engines to alcohols.

With respect to compression ignition engines, it is still uncertain as to which of several possible adaptations of the fuel or of the engine is the best for a given purpose. Continued testing will be necessary to answer this question.

Dissemination

Gasoline substitution on a local level using ethanol and spark ignition engines with modified fuel systems should not present any particular introduction problems. In most countries the same is probably true for the introduction of fuel blends with a low alcohol content (less than 5% methanol or 10% ethanol).

Introduction on a national basis of a pure alcohol fuel or a fuel blend with a high alcohol content will require replacement of some corrosion resistant coatings and

Table 11. Collection of data for economic assessment of internal combustion engine systems using liquid fuel.

Power output kW	5		10		50		200	
Type of engine	Spark Ignition	Compression Ignition	Spark Ignition	Compression Ignition	Spark Ignition	Compression Ignition	Spark Ignition	Compression Ignition
Fuel	Gasoline or Alcohol[2]	Gasoil or Vegetable[3] oil	Gasoline or Alcohol[2]	Gasoil or Vegetable[3] oil	Gasoline or Alcohol[2]	Gasoil or Vegetable[3] oil	Gasoline or Alcohol[2]	Gasoil or Vegetable[3] oil
Cost of engine only[1] UDS/kW	110	250	90	200	50	100	–	50
Cost of complete electric[1] power plant – USD/kW	610	750	450	560	180	230	–	160
Expected economic engine lifetime, operating hours	5 000	30 000	5 000	30 000	5 000	30 000		30 000
Lubricant costs, USD/kWh	0.002	0.002	0.002	0.002	0.002	0.002		0.002
Maintenance costs, UDS/kWh	0.008	0.008	0.008	0.007	0.008	0.005		0.004
Engine efficiency %	23	28	25	30	27	33		34
Electric power plant efficiency % (depending on load factor, high value Load factor 1.0, low value load factor 0.25)	9–19	12–24	11–22	13–26	12–24	15–30		15–31

[1] Cost level Jan. 1984
[2] Engine costs may be slightly higher than for gasoline, difference probably less than 5%. Efficiency may be 1-3 percentage units higher than for gasoline.
[3] Engine cost uncertain, may be 5-10% higher than indicated.

some rubber and polymer components in the fuel storage and handling systems. Also, to make full use of the potential advantages, specially designed and manufactured engines must be used. This implies that such a programme will be easier to carry out in a country like Brazil, where the manufacture of cars and engines is within the country. Many car manufacturers have undertaken research and development for special alcohol engines and it is not unlikely that such engines will be commercially available in a fairly short time if the requests from the users are strong enough.

Vegetable Oils

Production and Properties of Vegetable Oil Fuels

The production technology, choice of feedstucks, production costs and properties of vegetable oil fuels have been treated earlier in Chapter 11.

Past and Present Use

Experiments using vegetable oils as fuel for compression ignition engines are reported from the 1930's. There has been a renewed interest during recent years, but there appears to be no regular use of vegetable oils as engine fuel except in a few test vehicles.

Engine Performance with Vegetable Oils as Fuel

As the properties of most vegetable oils closely resemble diesel fuel they appear to be potentially suitable only for use in compression ignition engines.

As shown in Table 10, the viscosity of vegetable oils is, however, much higher than for diesel fuel. This means that the spray pattern will be different when standard injection nozzles are used, giving considerably larger droplets (see Bruwer). This has little immediate effect on the engine performance, but extended operation leads to a build up of carbon deposits around the injection nozzles and in the combustion chamber. This leads to a decline in engine power and increased smoke in exhaust gases. If left too long, the resultant piston ring seizure and lubricant dilution will lead to engine failure. It can thus be concluded that extended operation on unmodified vegetable oils using injection systems designed for diesel fuel is not possible.

Blending of the vegetable oil with diesel fuel or gasoline appears to be a simple way of eliminating this problem. The resultant saving on diesel fuel is of course less with this method. Preheating of the vegetable oil to improve atomization has been suggested as another possible solution. Upgrading of the oil to an ester is another possibility. This has been reported to give less coking in diesel engines than when operating on diesel fuel, much less smoke production and up to 10% improvement of the efficiency. The maximum power output was reduced by about 10% compared to operation on diesel fuel.

Economic Considerations

For operation on blended or a suitably modified vegetable oil, no changes to a diesel engine are required. It also appears reasonable to assume that the maintenance costs are similar to diesel fuel operation. The cost data collected in Table 11 for compression ignition engines should therefore be applicable.

It is difficult to estimate the costs for a compression ignition engine adapted for operation on straight, unmodified vegetable oil, since such engines are not yet available. It seems reasonable to assume, however, that the modifications will be restricted to the fuel system and that the cost of the engine will not be more than, say, 5 to 10% higher than for diesel fuel, see Table 11.

The fuel cost for operation on vegetable oil or a blended vegetable oil fuel will depend on the price paid for the fuel, and the efficiency of the engine. The efficiency can be assumed to be roughly the same as for operation on diesel fuel.

Environmental Impacts

It appears that operation on blended and suitably modified fuels will not give higher emissions than with the use of diesel fuel.

The emissions of modified compression ignition engines using straight vegetable oil are not known, since such engines are still to be developed. It is reasonable to assume that the emissions can be brought down to the levels of modern compression ignition engines running on diesel fuel.

Development Needs

The use of vegetable oil fuels as a substitute for diesel is still at a developmental stage. So far the technology seems promising and may well be one of the easiest ways of substituting petroleum fuels.

The main emphasis of the present research and development work on vegetable oil fuels is on engine design. Different sizes and models of diesel engines are being tested with vegetable oil, either blended, straight or as an ester. The objectives are to optimize engine performance, study potential operational problems and to discover long-term effects on the engine.

The development of an injection system to allow operation on straight vegetable oil appears as one of the most important research objectives.

Producer Gas

Gasification by partial combustion

A combustible gas, which can be used as fuel for an internal combustion engine can be generated if biomass is burned with less air than is needed to achieve complete combustion. The gas, often called producer gas, consists of the combustible components carbon monoxide and hydrogen and some methane, mixed with carbon dioxide, water vapour and nitrogen. Technically this can be achieved in many ways. For fuelling small engines of up to a few hundred kW the biomass fuel is usually supplied to a vessel, "the gasifier" where a fire is ignited and air supplied at regulated levels. The hot combustion gases are passed through the fuel bed and this leads to generation of the combustible gas.

If the gas contains more than very small amounts of dust and condensible tars, operational problems with the engine or excessive engine wear will be experienced. It is therefore necessary either to design the gasifier to generate a sufficiently clean gas, or to clean the gas by filtering or washing before it is supplied to the engine.

A flow scheme of a power plant with a gasifier operated internal combustion engine is shown below.

The cleaning of the gas is not easy and leads to difficult disposal problems. It is therefore usual to try and generate a gas with a low tar content in order to achieve trouble-free operation of the engine.

Three commonly used designs for small gasifiers are illustrated schematically on the next page.

In each of these the biomass is converted to gas in four stages i.e. drying and pyrolysis of the biomass, combustion of a part of the biomass and reduction of the combustion gases. The difference between the designs is the order of these steps. This is quite important regarding the tar content of the gas.

In the up-draught type, the volatile pyrolysis products i.e. the tars, are carried out with the gas from the gasifier. For this type of gasifier, unless a fuel with low volatile content, like charcoal, is used, the tar content will be too high for engine operation, if the gas is not cleaned.

In the down-draught type the tars have to pass through the hot combustion and reduction zones. If the gasifier is properly designed, this will lead to sufficient combustion and cracking of the tars to make the gas useful as engine fuel without needing further cleaning.

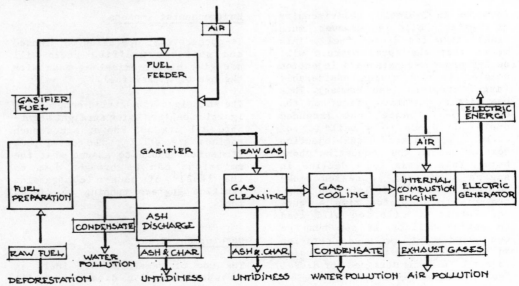

Outline diagram of a producer gas power plant.

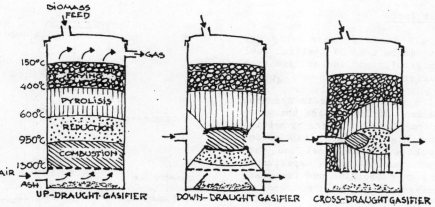

Diagram of Design principles for biomass gasifiers; up-draft, down-draft and cross-draft.

A cross-draught gasifier might be expected to operate like the down-draught type. In practice however this is difficult to achieve. Excessive tar contents in the gas will therefore be obtained unless a fuel such as charcoal is used.

For larger power needs, a gasifier where the fuel particles are kept in suspension by the air/gas flow, i.e. a fluidized bed gasifier, may be used. The Figure illustrates the principle of this type of gasifier. The most important advantages with this gasifier are its compactness and ability to maintain homogeneous gasification conditions. It requires a sophisticated control system and blowers to keep the bed fluidized which makes it relatively expensive for small power needs. Fluidized bed gasifiers will probably not be economic for engines with a power output of less than 300 to 400 kW.

Trouble-free operation of an engine using producer gas as fuel, requires proper matching of the gasifier fuel,

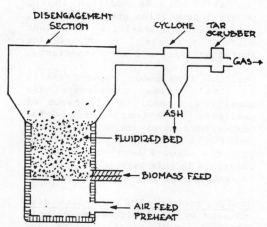

Schematic diagram of fluidized bed gasifier.

the gasifier, the gas cleaning, the engine and the operating conditions. There are several examples of mis-matched systems installed recently in developing countries. The most common mistake seems to be to ignore the importance of the fuel properties and the operating conditions in the gasifier design.

Table 12. Examples of Fuel Specifications

Fuel Property	Down-draught Gasifier		Up-draught	Fluidized Bed
	Wood Type	Charcoal Type	(Lamblon)	(Wärtsilä)
Moisture % Content	Max 25%	Max 18%	Max 50%	Max 60%
Ash Content %	Max 5%	Max 5%	Max 3%	
Bulk Density kg/m^3	Min 150-200	Min 150		
Physical Size	Edge Length 1) 20-80 mm	Min 10 mm	Max Size a	Max 25 mm
	Max Cross-Section Max 60 mm[1]			
	Area 25-30 cm^2			
Volatile Matter	–	Max 2-%	–	–

1) Depending on the size of the gasifier.

Gasifier Fuels

Almost all kinds of reasonably dry biomass can be used as gasifier fuel - but generally not in the same type of gasifier.

Excessive amounts of tar in the gas, the blockage of fuel inside the gasifier, formation of clinkers by melting ash, and high pressure loss caused by blockage in the fuel bed, are the most common problems experienced if the fuel properties and the gasifier design are not well matched.

Examples of fuel specifications for some types of gasifiers are shown in Table 12. In addition to the physical properties specified there, the chemical composition, behaviour during pyrolysis, and ash melting characteristics are also important.

Charcoal from wood, coconut shells or similar types of biomass will usually work well in all types of gasifiers. The temperature in the combustion zone may reach temperatures above 1300°C requiring the use of heat resistant alloys or refractory materials in this part of the gasifier to avoid damage caused by overheating.

Wood, coconut shells, maize cobs and some types of briquettes or pellets made out of agricultural residue have been seen to work well in down-draught gasifiers having a throat below the combustion zone, see Figure. The throat acts to improve the mixing of the gas flow, thereby avoiding cold zones where the vapours can pass without being cracked. Loose and fluffy material like straw or coconut husk will not work well in this type of gasifier. Fuel in the form of fine particles like sawdust cannot be used in down-draught gasifiers because of the high pressure loss. An up-draught gasifier or a fluidized bed gasifier is more suitable for such fuels.

Rice husk and similar materials can be gasified in a type of gasifier developed in China, where the gasification is made in an open vessel, air being sucked in through the opening in the top through which fuel is also supplied.

Gas Cleaning

Before its introduction into the engine the gas must be cleaned to remove dust and tars. Various types of cleaners have been used for this purpose, e.g. cyclones, wet scrubbers, large volumes filled with sawdust, wood wool, pieces of cork or coir fibre. The most efficient system appears to be fibreglass fabric filters or electrostatic filters but

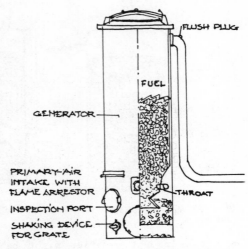

Cross-section of down-draft gasifier for wood and agricultural residues.

such filters are fairly expensive (typically about US$ 50/kW of engine power). Much simpler systems are therefore used for instance in the Philippines and Brazil. However, such simple systems make it necessary to clean the inlet manifold of the engine more often and can be expected to lead to more engine wear, and a shorter engine lifetime than when operating on gasoline or gasoil. Tests made in Sweden on the producer gas operation of tractors, using a fibreglass fabric filter for dust cleaning, show less engine wear than when operating on gasoil.

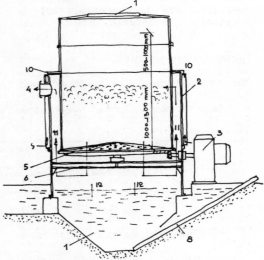

1. Fuel & air inlet
2. Cooling water jacket
3. Furnace grate reduction box
4. Gas outlet
5. Rotary grate
6. Grate support
7. Ash settling pond
8. Ash removing tube
9. Cooking water intake
10. Cooling water effluent
11. Gas
12. Ash

Principle of Chinese rice husk gasifier.

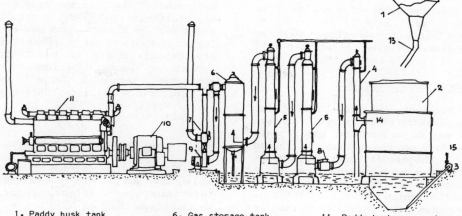

1. Paddy husk tank
2. Paddy husk gas producer
3. Ash pump
4. Ash flashing tube
5. Filter cooler tower
6. Gas storage tank
7. Three-way cock
8. Isolated cock
9. Startin fan
10. A.C. generator
11. Paddy husk gas engine
12. Paddy husk
13. Cooling water
14. Gas
15. Ash removing

Diagram of Chinese rice husk gasifier power station, including gas cleaning system and engine.

In the Chinese type rice husk gasifier system, the gas is cleaned by a series of scrubbers, leading to a potential water pollution problem caused by the presence of tars in the scrubbing water.

Before entering the engine, the gas must also be cooled in order to avoid condensation in the inlet manifold when the gas (which contains some water vapour) is mixed with ambient air.

Cooling of the gas will also lead to an improved engine power output, as it will increase the volumetric energy content of the gas-air mixture supplied to the engine.

Engine Performance

The power output of a reciprocating internal combustion engine depends on the heating value of the fuel-air mixture supplied to the cylinder(s) in each suction stroke and the efficiency of the engine.

The heating value of a producer gas air mixture is less than that of a mixture of gasoline or gasoil with air. Therefore, the power output of an engine will usually be lower for producer gas than for petroleum fuels.

Spark ignition engines can operate on just producer gas. With minimum modification, i.e. replacing the carburettor with a gas/air mixing device and adjustment of the ignition timing, a power loss of 45 to 50% can be expected. Some of this power loss can be avoided by increasing the compression ratio to about 10 to 1 and by reduction of pressure losses in the inlet manifold. By such measures it may be possible to limit the power loss to about 30%.

Compression ignition engines will not work solely on producer gas since the gas will not ignite when compressed. In order to bring about ignition, a certain amount of diesel fuel (or vegetable oil) must be injected into the cylinder. The amount necessary depends on the type of engine. For direct injection engines, 10 to 20% of the diesel oil required for full load operation on diesel oil only, is needed. Pre-chamber engines cannot be operated without disturbances on such small amounts of diesel oil. Since the amount of diesel oil required for ignition is fairly independent of the load, the net saving on diesel oil when compression ignition engines operate on producer gas will depend on the load characteristics. With the practical operation of vehicles in normal use, a saving on diesel oil of between 80 and 70% appears possible according to extensive testing in Sweden.

Energy Efficiency and Fuel Consumption

The efficiency of a producer gas system is determined by the efficiencies of the gas producer and the efficiency of the engine.

The efficiency of gas production depends on the design of the gasifier, the properties of the fuel used and the load. At full load, efficiencies in the range of 60 to 80% have been measured for the gasifier. This means that a system efficiency at full load of about 20% can be obtained for a producer gas plant with a spark ignition engine. With a compression ignition engine in dual fuel operation, a 25% system efficiency can be expected.

This implies a fuel consumption of 1 to 1.5 kg of dry biomass or 0.5 to 0.7 kg of charcoal/kWh for electricity generation.

For the operation of vehicles 1 kg of wood or 0.5 kg of charcoal can be estimated to substitute 0.25 to 0.35 litres of gasoline or gasoil.

Economic Considerations

The cost of equipment needed for the operation of an internal combustion engine on producer gas depends very much on the technology chosen and on the manufacturing situation. There are several commercial companies in Europe, the USA and in developing countries which offer complete producer gas plants or the equipment needed for the conversion of an existing engine to producer gas operation.

Price quotations range between US$ 100 per kW and US$ 1000 per kW for the fuel handling, gasification, gas cleaning and gas cooling equipment. What are believed to be reasonable estimates of the costs for producer gas plants have been collected in Table 13. It should be noted that automatic fuel feeding adds considerably to the specific cost for small plants, typically US$ 800/kW for a 100 kW plant.

Estimates of the lifetime of present day gasification equipment, requirements for operating labour and costs for maintenance of the gasifier system are uncertain due to the limited amount of recent experience. The estimates presented in Table 13 rely on experiences collected during the Second World War.

Hazards and Environmental Impacts

Properly adjusted engines operating on producer gas produce less hazardous emissions than engines operating on petroleum fuels.

The risk of poisoning by the very toxic gas itself (high content of carbon monoxide) constitutes the main problem of producer gas operation. Exposure to high concentrations of carbon monoxide leads to headache, unconsciousness or death, depending on the concentration and exposure time. The immediate effects of carbon monoxide poisoning are well known, and so are the methods used to treat a poisoned person. The experience from the extensive use of producer gas during the Second World War in Sweden and elsewhere, shows that

Table 13. Collection of data for economic assessment of internal combustion engines fitted with gasifier system.

Power output	5	15	60	100		
Fuel	Charcoal	Charcoal	Wood	Wood	Wood	
Fuel feeding system	Manual	Manual	Manual	Manual	Manual	Automatic
Cost of engine only[1] (high compression, spark ignition) USD/kW	540	350	350	165	140	140
Cost for complete gasifier system [1][2] USD/kW	800-400	435-180	550-215	210-75	180-60	980-350
Expected economic lifetime	30 000 hrs or max 15 yrs	30 000 hrs or max 15 yrs	30 000 hrs or max 15 yrs	30 000 hrs or max 15 yrs	30 000 hrs or max 15 yrs	
Lubricant costs USD/Kwh	0.002	0.002	0.002	0.002	0.002	0.002
Maintenance costs Engine USD/kWh	0.020	0.010	0.010	0.005	0.005	0.005
Gasifier system USD/kWh	0.025	0.015	0.018	0.007	0.006	0.030
Operating labour Hour/operating hour	0.35	0.35	0.35	0.50	1.0	0.2
Engine/gasifier system efficiency %	15	20	0.20	0.22	0.23	0.23

1) Installed cost
2) Higher end of range applies for systems manufactured in Europe or USA, lower end applies for systems manufactured in South America or South East Asia.

strict safety regulations and education of operators are very important measures in reducing the risks of acute poisoning. Training and follow-up of producer gas operators in developing countries is therefore necessary, in order to reduce the risk of accidents.

In some cases carbon monoxide poisoning may lead to delayed neurological symptoms including frequent headaches, dizziness, concentration difficulties, incontinence and eyesight disorders. Such delayed neurological effects require specialist treatment. Recovery may take several months or years.

Development Needs

The down-draught gasifier is well proven for engines with a power output of up to about 100 kW and using charcoal, wood blocks or wood chips as fuel. It also appears possible to use this type of gasifier for larger engines, at least up to a few hundred kW. Some types of agricultural residues for instance coconut shells, corn cobs, coir dust (from coconut husk) briquettes, and cotton stalk briquettes have been shown to perform well in this type of gasifier. There are, however, a number of residues with low density and high ash content which cause many problems in down-draught gasifiers. The development of gasifiers for such fuels which can operate with sufficiently low tar content in the gas to allow direct use of the gas as engine fuel, is a very important research objective. Another important research objective is the development of effective gas cleaning equipment which can be made at lower cost than the expensive fibreglass filters.

For engines with a power of above a few hundred kilowatts, up to a few megawatts, fluidized bed gasifiers can probably be used. Development of this technology is being carried out by several commercial companies and the first demonstration plants can be expected to operate within the next two years.

Producer Gas for Liquid Piston Pumps

The liquid piston Humphrey pump, developed in the beginning of this century, was originally designed for operation in gaseous fuel.

Producer gas operation of such pumps is being studied by the Intermediate Technology Development Group in the United Kingdom.

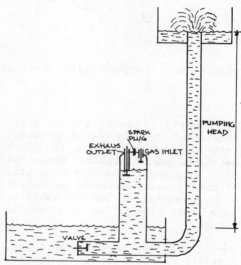

Humphrey pump.

Past and Present Use

The utilization of producer gas generated by gasification of coal or biomass fuel with air to operate internal combustion engines dates back to the end of the 19th century. The interest in this technology has varied with the availability and cost of petroleum fuels and peaked during the Second World War, when several hundred thousand vehicles in Europe were operated on producer gas.

Second World War-type gasifier mounted on automobile.

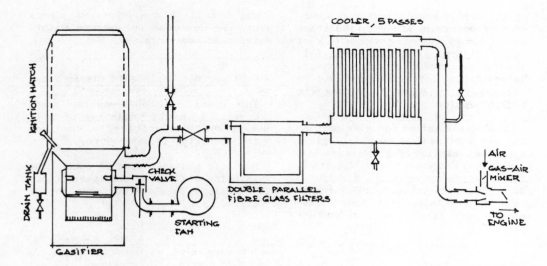

Swedish type Gasifier System for trucks and tractors.

In Sweden, producer gas is still considered the only realistic emergency alternative in case of a new petroleum supply crisis. There are detailed plans for the mass production of the necessary equipment to convert a large number of Swedish vehicles in a short time if such a situation occurs.

During the last decade, there has, however, also been a renewed interest in producer gas as a normal fuel, in particular in developing countries. Industrial production of gasification equipment has been started in the Philippines and in Brazil. Experiments involving the use of producer gas engines for a number of purposes are being carried out in many other countries.

Table 14 shows a list of installations from which operating experiences were reported in a recent conference.

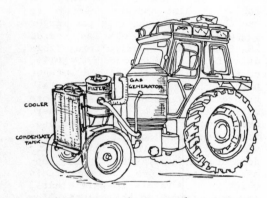

Producer gas unit mounted on tractor.

Producer gas powered fishing boat in the Phillipines.

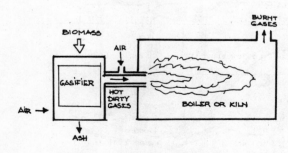

Diagram of direct-heat gasifier system.

Lime boiler in Brazil, fuelled by direct heat gasifier.

Table 14. Overview of gasifier installations for which field experiences were reported at the Second International Producer Gas Conference, Bandung/Jakarta, Indonesia 1985.

Application	Power output kW	Fuel(s)	Operating record	Country operating environment
Water pumping	0.8	Charcoal	1 100 h	Thailand Village school
-"-	About 2 (6 kg/s 17 m)	Charcoal	Report covers only an initial test period of 52 h.	Guatemala 100 household village. Report covers operation with German adviser
-"-	5	Waste wood blocks, - cotton and tuvar stalk	2 000 h	India laboratory
-"-	8		1 250 h	India irrigation
	8	Charcoal	Report covers only initial test period of 87 h	Tanzania test site at fibreboard mill
-"-.	10 (estimated)	Charcoal	Report covers only initial test period of 20 hours	Nicaragua test site at
-"-	24	Charcoal	Report covers only monitoring period of 105 h. Total estimated operating time about 5 000 h.	Philippines state farm
-"-	40	Rubber wood and tamarind wood (blocks)	1 000 h	Indonesia village, irrigation of 45 ha rice fields
Electric power generation	0.8	Charcoal	1 200 h	Thailand village school
-"-	7.5	Sawmill (waste)	1 000 h	India laboratory
-"-	15	Rubber wood (blocks)	1 900 h	Indonesia state owned rubber estate
-"-	25-30	Coconut shell	1 500 h	Sri Lanka coconut desiccating mill
-"-	32	Wood, coconut husk, mixtures of wood and coconut husks	Report covers only initial test period of 142 h.	Seychelles government test site
-"-	40	-"-	Report covers only initial test period of 59 h.	-"-
-"-	40-45	Mixture of coconut shell and husk	500 h	France and Tahiti laboratory
-"-	41	Charcoal	3 000 h	Brazil test institute
-"-	68	Sawmill waste, cotton and tuvar stalk	2 00 h	India mechanical industry
-"-	85	Charcoal	1 190 h	Brazil test institute
Mechanical power generation				
Mill	10-20	Maize cobs	Up to a few hundred h	Tanzania, Ujamaa villages, 4 units
Vehicles				
Landrover	30	Charcoal	5 180 km	Tanzania country-side
VW Pickup	30 (estim.)	Charcoal	3 500 km	Sweden
Five trucks	50-70	Wood chips Wood blocks	50 000- 272 000 km	Sweden commercial traffic
Ten tractors	20-50	Wood chips wood blocks	1 600- 5 600 h	Sweden farm work

Source: Kjellström, Björn (1985)

Bibliography of Vegetable Oils

ASAE (1984) Vegetable Oil Fuels.
Proceedings of Conference on Plant
and Vegetable Oils as Fuels, North
Dakota August 1984.
American Society of Agricultural
Engineers, 2950 Niles Road, St Joseph,
Michigan 49085, U.S.A.

Ascough, W.J. Vegetable Oils as Diesel
Engine Fuel.
Ibid.

Bagne, P. (1982) Bright Prospects
for Plant Oil Fuels.
Article in Alternative Sources of
Energy No. 55, May/June 1982.
107. S. Central Ave. Malaca, MN 56353,
U.S.A.

Bodo, L.B. & Onion, G. (1983) Oxy-
genate Fuels for Diesel Engines: A
Survey of World-Wide Activities.
Biomass, Vol. 3, No. 2, 1983.

Brasilia (1981) Vegetable Oils for
Energy Purposes: Selective Biblio-
graphy.
STI/GID.
Brasil. Ministerio da Industria e
do Commercio Secretaria de Technologia
Industrial.

Bruwer, J.J. et al. Sunflower Seed
Oil as an Extender for Diesel Fuel
in Agricultural Tractors.
Department of Agriculture Technical
Services, Pretoria, South Africa.

Cross, M. (1985) Engines Choke on
a Diet of Vegetable Oil.
Article in New Scientist 4 July 1985.

Cruz, I.E. (1978) Report on Endurance
Tests of a 25/KVA Diesel Engine Gen-
erator Set Fuelled by Crude Coconut
Oil.
Philippine National Oil Company,
Energy Research & Development Centre.

Garrod, Y.B. & Radley, R.W. (1981)
Vegetable Oils as Fuels for Diesel
Engines.
Proceedings of a seminar on Alterna-
tive Fuels for Internal Combustion
Engines, Imperial College, London,
April 1981.
Imperial College, Prince Consort
Road, London, SW7 2BY, U.K.

Vellguth, G. (1983) Performance of
Vegetable Oils and Their Monoesters
as Fuels for Diesel Engines
Article in
Renewable Energy Review Journal
Vol. 7, No. 1, June 1985.
Asian Institute of Technology
P.O. Box 2754
Bangkok. Thailand.

Vellguth, G. (1984) Field Test of
a Di-Diesel Tractor with Methylester
of Rape Oil as Alternative Fuel.
Paper presented at Bioenergy 84.
Elsevier Applied Science Publishers
Ltd., Crown House, Linton Road, Barking, Essex 1G 11 8 JU, U.K.

Bibliography of Producer Gas

Foley, G. & Barnard, G. (1983) Biomass
Gasification in Developing Countries.
Earthscan, 3 Endsleigh St., London,
WC1H ODD ODR, U.K.

Greenpeace (1981) Woodgas Power
Plans. Tractor Conversion Unit.
Greenpeace Experimental Farm, RR1
Denman Island, British Colombia,
Canada VOR 1TO.

Kaupp, A. & Gross, J.R. State of
the Art for Small-scale Gas
Producer/Engine Systems.
University of California, Davis,
March 1981.

Kjellström, B. (1985) Second Inter-
national Producer Gas Conference.
The Producer Gas Round Table, The
Beijer Institute, Box 50005, S-104 05
Stockholm, Sweden.

Kjellström, B. (1984) Producer Gas
as Fuel for Internal Combustion Engines.
The Beijer Institute, Box 50005,
S-104 05 Stockholm, Sweden.

Kjellström, B. (1982) Producer Gas
1982, A Collection of Papers on Producer Gas with Emphasis on Applications in Developing Countries.
The Beijer Institute, Box 50005,
S-104 05 Stockholm, Sweden.

Kjellström, B. (1980) Producer Gas
1980, Local Electricity Generation
from Wood and Agricultural Residues.
The Beijer Institute, Box 50005,
S-104 05 Stockholm, Sweden.

Reed, T.B. (1981) Biomass Gasification
- Principles and Technology.
Noyes Data Corporation, Park Ridge
N.J; U.S.A.

SERI (1979) Generator Gas the Swedish
Experience from 1939-1945.
Solar Energy Research Institute
(SERI), 1536 Cole Boulevard, Golden,
Colorado 80401, U.S.A.

TDRI (1983) Survey of Manufacturers
of Gasifier Power Plant Systems.
G. 180.
Tropical Development and Research
Institute (TDRI). 127 Clerkenwell
Road, London EC1R 5DB, U.K.

13 EXTERNAL COMBUSTION ENGINES AND USE OF BIOMASS FUELS

Steam Engines and Steam Turbines

Working principle

The Figure below shows the main components and the working principle of two types of steam power plants. Common to both types is that water is heated at pressure to generate steam. The steam is piped to a steam piston engine or a steam turbine where it expands and generates mechanical energy in the form of shaft power.

The efficiency of the system is determined by the pressure and temperature difference between the steam leaving the boiler and the steam exhausted from the engine or turbine. In the simplest system shown, saturated steam (i.e. steam at boiling temperature for the pressure prevailing in the boiler) is used. The steam is exhausted at atmospheric pressure, i.e. at a temperature of 100°C. This leads to a cheap plant, but the efficiency will be low, typically 5%. This type of plant also requires a fairly large water supply since water is constantly lost through the steam exhaust of the engine.

In the more sophisticated system, the saturated steam leaving the boiler is further heated by the hot combustion gases. The steam will then become superheated. The engine or turbine is not exhausting into the atmosphere but rather into a closed vessel at vacuum, the condenser, where the remaining heat energy in the steam is cooled, making the steam condense to water. This water is recirculated to the boiler. Such a system can be designed to have a much greater temperature and pressure difference between the steam entering the engine (or turbine) and the steam exhausted. This gives a higher efficiency of typically 10-20%. The efficiency can be improved further still by pre-heating the water before it enters the boiler. Steam extracted from the engine or turbine at some intermediate stage is used for this purpose.

It is also possible to remove the steam from the engine or turbine after it has reached saturation conditions to be superheated again. This latter type of arrangement is only found on large steam power plants having power outputs exceeding 100 MW. The most advanced steam power plants show efficiencies of about 40%.

Possibilities of Using Biomass Fuels

The only components of a steam plant which must be adapted to the fuel used in the boiler are, the combustion equipment and the fuel handling system which transports the fuel from storage to the boiler.

Boilers for solid fuels such as biomass, coal or peat are bigger, somewhat more expensive and require more attentive operation than boilers for liquid fuels. Such boilers have been and are being extensively used. The technology is without any doubt commercially available, even though it might not be quite clear in some cases which type of combustion equipment represents the optimum choice with respect to capital costs, efficiency and operational reliability.

Boilers and combustion equipment for biomass fuels are discussed in a separate chapter, see Chapter 8.

Selection of Steam Data, Cycle Arrangement and Type of Engine

When a steam power plant is designed, there is much scope for choice of

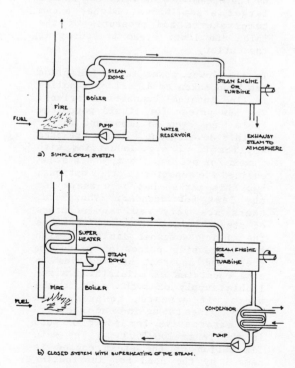

a) SIMPLE OPEN SYSTEM

b) CLOSED SYSTEM WITH SUPERHEATING OF THE STEAM.

Diagram of working principles for steam engine power plants.

steam conditions (i.e. steam pressure and temperature), steam cycle arrangements, and type of engine. The choice made will have a great effect on the energy efficiency of the plant and the costs of its construction. It is generally true that plants designed for a high efficiency will be more expensive to build. The optimum choice of steam data, cycle arrangement and type of engine will depend on the power output needed and the load conditions. It is usually more economic to choose a simple plant with moderate efficiency for uses which demand a small number of annual operating hours and a more sophisticated and expensive plant with high efficiency for plants which operate more or less continuously.

Some of the limitations of the different options will be discussed in this section. Quantitative data on the performance and economy of some specific plant designs will be given in later sections.

The Steam Cycle: Technical Aspects

The choice between an open cycle design, which requires a continuous supply of fresh water to pump into the boiler and the closed cycle design, where water circulates in the system, is, in practice, connected to the choice of steam data and is largely governed by the efficiency requirements. The availability of suitable water for the boiler can also be an important factor in this choice.

The simplest steam power plant would have an open cycle, a maximum pressure of 10 bar and no superheating of the steam. The efficiency of such a plant would be fairly low, hardly above 5%. This type of cycle will often be used for very small power outputs up to about 100 kW. No feed water treatment, except mechanical filtration of the water pumped into the boiler, or, in case of hard water, the adding of softening chemicals to the water, is applied for plants with such steam data. The purpose of water treatment is to prevent deposit build up on heat transfer surfaces in the boiler and to minimize corrosion.

Higher efficiency can be obtained by raising the boiler pressure to 15 bar and superheating the steam somewhat, to say 280°C. This would give an efficiency of up to 7 to 8%. Further superheating to 350°C makes it possible to reach an efficiency of about 10%. The same simple treatment of the boiler feed water is used for pressures of up to 40 bar. Often a semi-closed cycle is used where the exhaust steam is injected into an open reservoir of

water which is supplied with fresh water to compensate for the water evaporated from the surface.

The most advanced steam plants built for power outputs of down to about 200-300 kilowatts operate at a 40 bar boiler pressure, a superheating temperature of 380°C and use a closed cycle with condensation at about 60°C. This produces an efficiency of up to about 18%.

As should be apparent from the examples given, higher pressure and temperature of the steam, generally leads to improved system efficiency. The costs for the boiler and the steam line will, however, increase at the same time and the choice of design will have to be based on a careful economic analysis.

It must therefore be observed that, in order to utilize commercially available components like pipes, valves and fittings most efficiently, the pressure and temperature cannot be selected independently. There are also certain limitations imposed by the choice of engine or turbine as will be further explained below.

Steam Engine or Steam Turbine?

After the water-wheel and the windmill, the steam engine is the oldest shaft power machine. It is, however, limited in power output to some MW and has therefore for large power needs been competed out of the market by the steam turbine which can handle larger steam flows, higher steam temperatures and pressures at the inlet and lower steam pressures at the outlet.

In the lower power range its place has been taken by diesel or gasoline fuelled internal combustion engines. The low prices for liquid petroleum fuels have been an important factor in this development. There are some applications, e.g. industries with a need for process steam, where steam engines are superior to other options, but this market has been small for the last few decades. Therefore, there are very few manufacturers of steam engines around which makes the choice somewhat limited. With the increasing prices of petroleum fuels and apparent difficulties in some countries of maintaining a reliable supply of such fuels to the users, it appears, however, that steam engines can again be an interesting alternative for power outputs in the range 100-1000 kW. The main limitation of the steam engine is caused by the need for lubrication which in practice limits the superheating temperature between 350 and 380°C (compared with a practical economic limit of 350-400°C used for steam turbines of similar power). There is also a flow area limitation at the exhaust, which in practice limits the exhaust pressure to about 0.3 to 1 bar (compared to about 0.2 bar for a steam turbine of similar power). The implication of this is a lesser potential for a high cycle efficiency than can be obtained with steam turbines.

For small power outputs, however, the efficiency of the machine as such can be made higher for a steam engine than for a steam turbine and there is also less economic incentive to choose advanced steam data. The practical efficiency for a steam engine plant may therefore be higher than for a steam turbine plant up to a power output of 500-1000 kW.

Traditional steam engines operate at low speed and are therefore fairly bulky. They cannot be used to drive an electric generator directly, but need a gear to give a suitable shaft speed for the generator. Modern high speed steam engines can be used to drive an electric generator directly. Small steam turbines operate at a speed which is much higher than can be used by an electric generator. These turbines must also employ a gearing system for speed reduction.

The following discussion will be focussed on plants having a power output of less than 200 kW.

Commercially Available Equipment

Boilers

Steam boilers suitable for operation with biomass as fuel are commercially available.

Steam Engines

Low speed steam engines of traditional type, with a power output of up to 160 kW are manufactured in Brazil by MERNACK.

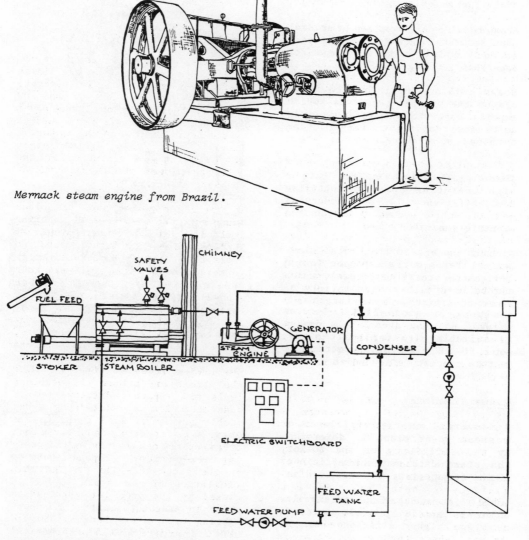

Mernack steam engine from Brazil.

Diagram of Mernack steam power plant system.

MERNACK also make complete units of the locomobile type (steam engine mounted on top of the boiler) with power output of 35 and 160 kW. The engine speed is about 240 rpm. The steam pressure is maximized at 10 and 14 bar respectively and the maximum steam temperature is 350°C.

There are also a number of manufacturers of very small steam engines with power outputs of about 10 kilowatts or less. These engines are usually utilized for the propulsion of small boats.

Advanced high speed engines are manufactured by SPILLING in West Germany. These cover the power range of 10-1800 kW and operate with engine speeds of 750, 1000 or 1500 rpm. Engines may be selected having steam temperatures of up to 380°C and steam pressures of up to 150 bar. These engines are usually applied where there is a simultaneous need for power and heat. For example, there is one such plant operating in Sweden, at Svedjeholmens Saw Mill.

Steam Turbines

Steam turbines are supplied by several manufacturers. However, few of them produce models with an output of less than 200 kW.

Steam turbines in this size range are of the single-stage type, which implies that turbine efficiency is quite low compared to larger steam turbines or high-performance steam engines.

These small steam turbines have a high operating speed to optimize their efficiency. A speed reduction gear is then needed to match the electric generator speed.

Another concept is the "total flow" helical rotor screw expander (using two counter rotating screws), which can be used with a wide variety of steam conditions. It is rugged and can work on steam-fluid mixtures as well as pure steam. This design is suitable also for small sizes, some 50 kW, and development work on this concept is now under way.

Energy Efficiency

The overall energy efficiency of a steam power plant is determined by the efficiency of the boiler, the steam data and the efficiency of the steam engine or steam turbine.

Table 15 shows data for some examples of steam plant designs. As mentioned earlier, higher efficiency can be

Small steam-powered boat.

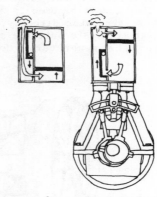

Diagram of reciprocating steam engine.

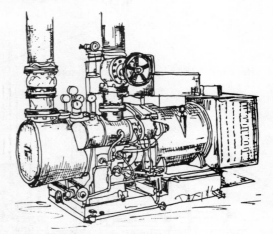

Small scale steam turbine.

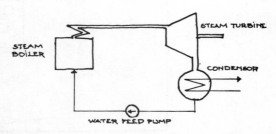

Diagram of steam turbine systems.

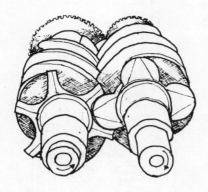

The "total flow" helical rotor screw expander.

obtained at the price of higher equipment cost. It should be observed that the efficiency data related to full power can be somewhat higher, and that the efficiency is usually lower at low loads. The average efficiency will depend on the load characteristics, but will generally be less than the full power value.

Economic Considerations

The capital investment for some types of steam power plants are given in Table 15.

With proper operation and maintenance the life-time of steam power plants can be very long. SPILLING claims that they have delivered engines which have been operating for more than 200 000 hours and are expected to operate for another 30 years.

Maintenance cost estimates work out at about 3% of plant cost per year. Even if the plant utilization is low, these costs will not fall to less than 0.5% of the plant cost.

Operating labour is required mainly for fuel handling and supervision of the boiler and lubricant cost can be estimated at US$ 0.004/kW of electricity.

The cost of feed water treatment will depend on the quality of water available at the site. For plants with a boiler pressure of below 32 bar, it will be the negligible compared to the lubricant cost.

Environmental Impacts

The environmental impacts of main importance are caused by combustion gases from the boiler (see Chapter 8), and by cooling water from the condenser.

The effects of this "thermal pollution" depend very much on the size of the recipient relative to the amount of heat rejected. Even a few degrees centigrade increase in the water temperature of the recipient may have significant effects in the ecological balance and lead to changes in the aquatic fauna and flora.

Thermal pollution of lakes and rivers can be eliminated by the use of air cooled systems. These can be arranged in many ways. The simplest solution is to use a "cooling pond", i.e. a specially made pond or lake from

Table 15. Collection of data for economic assessment of steam engine systems.

Power output kW	5	50	200	200
Characterization	Extremely simple design	Simple design type MERNAK	Simple design type MERNAK	type SPILLING
Fuel feeding system	Manual	Manual	Manual	Automatic
Cost of engine USD/kW$_{el}$	200	180	135	540
Cost for steam boiler and auxiliaries USD/kW	860	1 000	500	1 700
Expected economic life-time	30 000 hours or max 15 years	20 years	20 years	20 years
Lubricants etc USD/kW$_{el}$	0.004	0.004	0.004	0.004
Maintenance etc	2% of plant cost	2% of plant cost	2% of plant cost	2% of plant cost
Operation labour hour/full power hour	0.35	0.35	1.00	0.20
Efficiency fuel - electricity	5	10	10	12

which heat is dissipated by evaporation from the water surface.

Another possibility is to use a "cooling tower", where the cooling water, falling as a spray of droplets, is brought into contact with a rising airstream and dissipates heat by rising evaporation.

Such cooling systems add 5 to 10% to the capital investment depending on the technology chosen.

Research and Development

Steam technology has had a long history of application, has been widely applied and is well documented. New materials, new applications and special requirements resulting from the wish to manufacture the equipment in a developing country, leads, however, to the need for further research and development.

Steam systems which use an organic fluid instead of water as the working media offer the advantage of lower operating temperatures. This is mainly of interest in the use of solar heat for power generation. This concept is being studied for example by Ormat Turbines Ltd. in Israel (see Chapter 17).

Small pumps powered by steam engines are being studied by the Intermediate Technology Development Group in the United Kingdom. A prototype giving 4 litres/s at heights of up to 10 m has been tested. Its fuel requirements are 4.3-7 kg of wood or 11 kg of rice husk per hour. The production cost is estimated at US$ 930 (1982) if manufactured in India.

Pistonless steam pumps are being studied by the Rural Energy Laboratory of the Central Power Research Institute in Bangalore, India. The pump gives 1500 l/h at a 6-7 m head. It requires 6 to 7 kg of wood as fuel per hour. This is equivalent to an efficiency far below 1% but the design is simple and the maintenance cost is low.

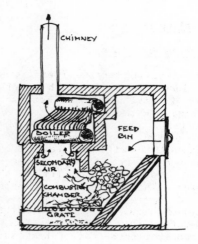

Boiler system for ITDG steam pump.

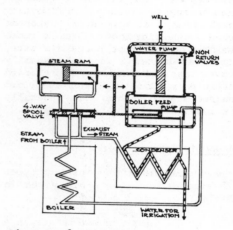

Diagram of ITDG Steam Pump.

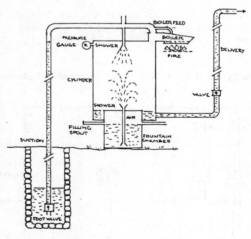

Steam operated pistonless pump.

Trailer mounted steam power system.

Stirling Engines

Working Principle

The Stirling engine operates with a closed system in which a working gas is successively heated and cooled. When the gas is heated it expands. This is used to move a piston which rotates a crankshaft which then will rotate. In this way heat energy is converted to mechanical energy in the form of shaft power. The heat energy can be supplied to the working gas by a solar heater. This, of course, restricts operation to sunny days. The most commonly used method of heat supply is to collect the heat generated when some kind of fuel is burned and transfer this to the working gas via a heat exchanger. Air was used as working gas in the Stirling engines of the 19th century. Advanced designs of today use helium which gives improved performance, but leads to much higher manufacturing requirements since helium may otherwise leak away from the system. The Figure at the bottom of the page illustrates the different phases of the working cycle.

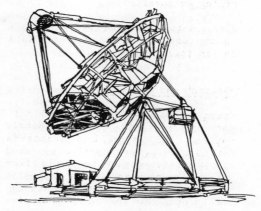

Stirling engine mounted on tracking solar collector.

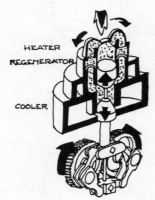

Principle of Philips design Stirling engine with rhombic drive mechanism.

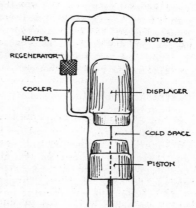

Diagram of single piston Stirling engine.

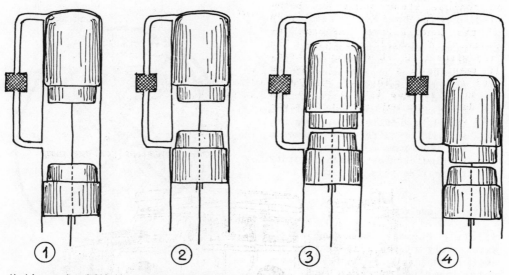

Working cycle of Stirling engine: **1.** Piston at bottom dead centre. Displacer at top dead centre. All gas in cold space. **2.** Displacer remaining at top dead centre. Piston has compressed gas at a lower temperature. **3.** Piston remaining at top dead centre. Displacer has shifted gas through cooler, regenerator and heater into hot space. **4.** Hot gas expanded. Displacer and piston have reached bottom dead centre together. With piston stationary, displacer now forces gas through heater, regenerator and cooler into cold space, thus re-attaining situation **1.**

Possibilities of Using Biomass Fuels

Biomass can be used as fuel for Stirling engines using the four different types of systems illustrated schematically in the Figure.

1   Direct combustion of a liquid fuel derived from biomass (i.e. alcohol or vegetable oil) makes it possible to use the systems already developed for petroleum fuels. Presently available designs can be used and no special modifications are required. The fuel is, of course, more expensive than if biomass is used directly.

2   Indirect combustion of biomass. This means that the biomass is burned in a separate combustor and that the hot combustion gases are used to heat air which is then led in to the combustion chamber of the Stirling engine with its heat exchanger to heat the working gas of the Stirling engine. This system does not require much research and development, but the system will be fairly expensive and the efficiency will be reduced due to the losses caused by the temperature difference in the extra heat exchanger.

3   Gasification of biomass and direct combustion of the gas. Producer gas is generated, similarly to the description in Chapter 12 on internal combustion engines, and this gas is burned in the combustion chamber of the Stirling engine to heat the working gas. This system does not require much research and development either, in particular since the purity of the gas must not be as high as for use in internal combustion engines.

4   Direct combustion of the biomass in the combustion chamber of the Stirling engine means that the combustion chambers developed for liquid or gaseous fuels must be modified. This system is potentially the cheapest and has the highest efficiency. However, it will require some research and development work on the combustion chamber design in order to realize this potential.

State of Development

Stirling engines were used fairly extensively in the 19th century for small power requirements, for instance water pumping to drive fans. They could be used with any fuel available and were safe, durable and comparatively easy to build. However, the efficiencies were low, mainly for thermodynamic reasons, since the maximum temperature of the working

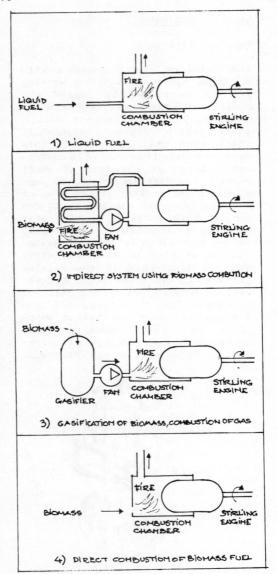

1) LIQUID FUEL

2) INDIRECT SYSTEM USING BIOMASS COMBUSTION

3) GASIFICATION OF BIOMASS, COMBUSTION OF GAS

4) DIRECT COMBUSTION OF BIOMASS FUEL

Different Stirling engine heating and fuel systems.

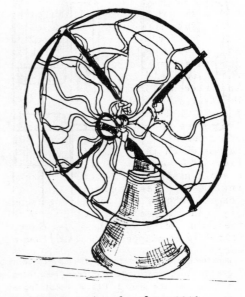

Stirling engine fan from 1918.

gas was limited by the limited temperature resistance of the materials available at that time. Typical efficiencies of such engines with an output of 0.2 to 2 kW range around 2%.

As internal combustion engines became available, the Stirling engine was more or less transferred to the list of technical curiosities. One important reason for the renewed interest in this type of engine was created primarily by the search for automotive power sources giving less air pollution than the internal combustion engine. The Stirling engine allows combustion under more controlled conditions and at atmospheric pressure which is favourable in this respect. More recently, the possibility of using these engines with solid fuels has been a dominant reason for their continued research and development.

The Swedish company United Stirling, which has been active in the development of advanced Stirling engines with helium as the working gas, claims that the commercialization of 10-54 kW engines which operate on liquid fuel or gas is close. Successful tests using all three systems for utilization of solid biomass have been performed. Development of a Stirling engine generator set up using wood chips as fuel is in progress in cooperation with the Department of Energy in the USA. This development is based on direct combustion in a two-step combustor, where the first step generates a hot combustible gas which is finally burned in a cyclone type burner, giving away the heat to the working gas.

Another line of development has been reported on by Beagle, according to whom Sunpower Inc. USA has developed several types of Stirling engines using biomass as fuel. This development includes a free-piston Stirling engine with helium as the working gas and a simple crank linkage engine which uses air as the working gas. The free-piston engine with a 150 W output is suitable for small water pumps (15 litre-metres per second, i.e. more than 10 m^3/h against a 5 m head). It is fitted with a gas burner which means that for use with biomass it should be operated with a small gasifier. The crank linkage engine was designed specifically for manufacture in a developing country (Bangladesh) and for use with fuels including rice husk, sawdust, coir dust or chopped straw. The power output is 4 kW.

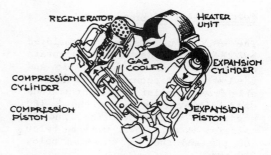

Diagram of modern Stirling engine, the V160 of United Stirling.

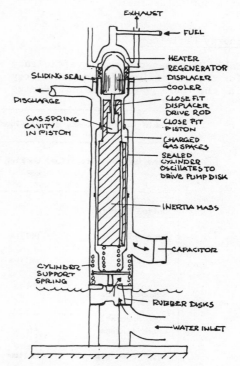

Diagram of free-cylinder Stirling engine with inductance pump.

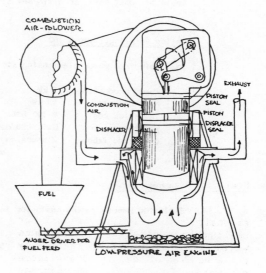

Diagram of biomass fuelled crank linkage Stirling engine.

Environmental Impacts

The continuous combustion employed by these designs allows accurate control of the air/fuel ratio and the low combustion pressure is favourable from the point of view of nitrous oxide generation. For the V 160 engine of United Stirling, the CO emission at full power will be below 100 ppm and the NO_x emission below 200 ppm.

Stirling engines are extremely quiet. For the V160 engine the noise level at 1 m distance has been measured at about 55 decibels.

Energy Efficiency

'Table 16 shows the energy efficiency for the conversion of chemical energy in the fuel to mechanical energy in four types of Stirling engines at full load. At part load the efficiency will be lower. Therefore the average efficiency of an engine operating on a varying load will be less than values given in the table.

Economic Considerations

The cost for the industrial production of biomass fired Stirling engines is difficult to assess, since the types of engines now considered for marketing are only available as prototypes. The lifetime of the equipment and the maintenance costs are, of course, even more difficult to estimate.

The cost data given below in Table 17 should therefore be treated with some caution.

Table 16. Efficiency at full load of some types of Stirling engines.

Type and identification	Working gas	Max power kW	Total efficiency %
United Stirling			
Type 4-95	Helium	20	24.5
Type V160	Helium	9.4	20.5
Sunpower Inc.			
Type "Dacca" crank linkage	Air	4	5.7
Type "Free piston"	Helium	0.15	About 10

Compiled from different sources.

Table 17. Cost data for different Stirling engines.

United Stirling, type V 160

Small electric power plant using straw as fuel. Reject heat recovery.

Power production	8 kW
Heat recovery	23 kW
Installed cost	8 100 USD (1 013 USD/kW)
Maintenance costs	0.10 USD/operating hour
Operating labour	2 hours/day for fuel handling and ash removal

Sunpower Inc.

Estimated manufacturing costs for industrial production:

Dacca[1]	4 kW	750 USD (190 USD/kW)
Free piston[2]	0.15 kW	50-100 USD (330-660 USD/kW)

1) Engine only. Combustor for biomass probably not included.
2) Engine - pump package. Combustor for gas.

Compiled from different sources.

Bibliography

Beale, W.T. (1981) Stirling Engines for Developing Countries.
Sunpower Inc. 6 Byard St. Athens OH 45701, U.S.A.

Bernås, M. (1984) Introduction to the Reciprocating Steam Engine Technique.
BEVI International, S-384 00 Blomstermala, Sweden.

Copley, D. (1983) Hot Air Engine, Power from Waste.
Rev. David Copley, St. Peters Methodist Church, High Park Street, Liverpool 8. U.K.

CPRI, Technical Note on Steam Operated Pistonless Pump.
Central Power Research Institute, (CPRI), Post Box 1242, Bangalore, India.

Edström, S.E. (1973) Helical Lobe Machines for Heating, Refrigeration and Air Conditioning.
ASHRAE Journal August 1973.

Hislop, D. (1985) Small Steam Systems for the Third World.
Article in Appropriate Technology Vol. 12, No. 2. September 1985.
I.T. Publications Ltd., 9 King Street, London WC2E 8HW, U.K.

Hislop, D. (1980) Survey of Demand for Small Steam Power Generating Equipment for Manufacture and Use in Rural Areas of Developing Countries.
Intermediate Technology Development Group (ITDG) Ltd., 9 King Street, London WC 2E 8HW, U.K.

Hislop, D.W. et al. (1982) A Small Steam Pump for Irrigation in the Third World.
Intermediate Technology Development Group, 9 King Street, London, WC 2E 8 HW, U.K.

Hislop, D. & Joseph, S. (1984) Small Steam Systems for The Third World.
Intermediate Technology Development Group (ITDG), 9 King Street, London, WC 2E 8HW, U.K.

Reading University (1984) Small Engines and Their Fuels in Developing Countries.
Proceedings of the Conference at Reading University, September 1984.
The Energy Group, Engineering Department, University of Reading.

Peter Rogers et al. (1980) State-of-the-art Review of Biomass Fueled Heat Engines.
Meta Systems Inc. 10 Holworthy St. Cambridge Mass. 02138, U.S.A.

United Stirling (1983) Some facts about United Stirling, and V160
Box 856, 201 80 Malmö, Sweden.

14 COMPARISON & SELECTION OF BIOMASS ENGINES

Factors Influencing the Choice

In the preceding chapters of this section different technologies for the generation of shaft power using biomass as the primary energy source have been discussed. It should be realized that it is not possible to identify any of these as the general best choice, without considering in detail the application, the operating conditions and the infrastructure of the community where the equipment will be used. In this chapter the factors which will influence the choice of technology will be briefly discussed.

It should be apparent from the technical descriptions that each of the technologies has its own characteristics. The best choice depends on how well these conform with the requirements of the application considered. It is believed that each technology will find its own niche in the future energy systems of the developing countries. The choice of the most suitable technology for the generation of shaft power must always start with a careful consideration of the requirements of the particular application considered, and any other circumstances which may influence the choice.

Economic factors will, in almost every case be of great importance in the choice of technology. The characteristics of the load are of great importance to the economic evaluation. Therefore, even for the same type of application, for instance, electricity generation, the most economic technology will often be different depending on the load characteristics.

Operational reliability can be a very important factor for some applications, for instance the electric power supply of a hospital, or compressors for a cold store. Any required degree of reliability can be obtained with all the technologies discussed earlier, by use of back-up systems or in some cases energy storage. The implication of this is that the systems which show higher failure probabilities and longer down-times require more expensive solutions to give the overall reliability required. The costs to obtain the required reliability will therefore have to be considered in the economic evaluation.

Fuel versatility has implications for the operational reliability in cases where interruptions of the normal fuel supply are a possibility.

Fuel versatility may also in some cases give possibilities of switching between fuels in order to use whichever fuel is cheapest at that moment. However, fuel versatility often leads to a more expensive plant. The advantages of fuel versatility must therefore be evaluated in the economic analysis.

For industrial installations and some types of agricultural applications where the residues from the operation are used as fuel (e.g. saw-mills, coconut processing plants, maize mills, rice mills), fuel versatility is seldom a strong requirement. However, for road vehicles moving over great areas where it might be difficult to find a homogeneous fuel and for agricultural applications where available fuel may vary between seasons (e.g. pumps), fuel versatility may be of great importance.

Convenience can partly be translated into economic terms to the extent that it can be expressed as amount of labour required from the operators. It must be accepted, however, that assessment of convenience will mainly depend on subjective judgement. In fact, the convenience considerations may often be a major reason why a technology other than the most economic is chosen.

Convenience will depend on such things as possibility for automatic start-up, possibility for unattended operation, as well as the frequency of inconveniences caused by actions such as adjustments to the operation, refuelling, cleaning of filters, removal of residues and the like. The time required for starting the engine can also be considered a convenience factor.

Automatic start and unattended operation will not be an important requirement for those applications where the operator is present during operation and has adequate time available to carry out the tasks required. This would be the case in most mobile and industrial applications. Automatic start is probably only necessary for emergency generation of electric power. Unattended operation will seldom be a necessary requirement, but on the other hand the need for a special operator may, however, in some cases lead to such high costs that economic viability is impaired.

Weight and space requirements will
be an important factor in most mobile
applications and will also be important for equipment which must
be transported to remote sites.

Production of useful reject heat
can be important in governing the
choice of technology in some industrial and agricultural applications.
It is obvious, for instance, that
a steam powered plant will show great
advantages if there is a need for
steam as well as mechanical power
in the process concerned.

Retrofitting of existing equipment
can be an important consideration
in those cases where a steam-engine
or a diesel engine is already utilized
for generation of shaft power. An
economic evaluation of back-fitting
as an alternative to installation
of a new technology is obviously
necessary, but such things as needs
for re-training operators, lack of
experience and possible deficiencies in the infrastructure if a new
type of technology is introduced,
must also be considered.

Environmental Impacts and Operational
Hazards should not be neglected when
the choice of technology is made.
For example, noise can be an important
factor in the choice of technology
for an electric power plant located
close to a village, whereas it may
not be so important for some industrial applications.

Solid and liquid residues should
be handled and disposed of in such
a way that minimum harm is caused
to the environment. This will obviously have implications for the
economic evaluation.

Operational hazards must be kept
at a level which is considered acceptable by the operators and the society
in which the equipment is used.
For some applications, like public
transport, it may be considered particularly important that the risk
of accidents is low.

Intangibles like prejudices, subjective assessment of acceptable
extra costs for improved convenience,
national independence as regards
fuel supply and manufacturing of
equipment and acceptability of certain
environmental impacts and operational
hazards may be as important to the
choice as many of the factors listed
above.

Economic Evaluation

There are several economic factors
which need to be considered before
the most suitable technology is
chosen.

The overall economy, estimated for
a certain period of time (often the
lifetime of the equipment) is ideally
the most important criteria. In
practice, however, cash-flow considerations and assessments of the
predictability of the economy may
be as important in making the choice.
In particular for small industries,
artisans, farmers and private persons
with limited economic resources,
these other factors may in fact be
more important than the overall economy.

For the national economy as a whole
it is important that the influence
of these other factors is minimized
so that in general the choices are
based on the most favourable overall
economy.

There are several methods of evaluating the overall economy. For the
purpose of illustrating the relative
importance of different factors for
some applications it is sufficient
to use a simple comparison of annual
costs where the capital costs are
calculated as constant annuities.

The information and procedure required
for such a comparison will be discussed below. An example to illustrate how the comparison can be made
and to highlight some important differences between the technologies
presented earlier in this chapter
will also be given.

The capital investment must be estimated for each type of plant considered. It is essential that the
comparisons are made for complete
installations, ready for commerical
operation and that the plants are
designed to meet the same criteria
regarding operational reliability
and environmental impacts. If this
is not done, it will be difficult
to use the economic evaluation for
objective decisions.

The annual capital cost can be calculated by the annuity method (which
assumes that a constant cost each
year is allocated to cover depreciation and interest) if the interest
rate and the economic lifetime of
the equipment is known. The latter
must be estimated, based on available
experience with similar equipment.

Other fixed charges like insurance
and taxes must be established. They
are often proportional to the capital
investment.

The annual fuel costs can be calculated from the fuel price, the
average overall efficiency of the
plant and the required annual energy
output.

It should be observed that the efficiency depends on the load, which

means that the average efficiency will depend on the load characteristics. The same type of plant producing the same annual energy output, may show different average efficiencies, depending on whether the operation is mainly at full load for relatively few operating hours, or at low load for a larger number of operating hours.

The load characteristics are often summarized by estimation of "the load factor" or "the equivalent number of full power hours". The latter is calculated as the number of hours a plant needs to operate each year at full power in order to generate the total annual energy output.

The annual cost for operating labour depends on the wages of the operators and the number of work-hours required for the operation.

The annual cost for maintenance and supplies should include costs for maintenance labour, spare parts and supplies, such as lubricants and the like.

The annual cost for disposal of residues depends on the amounts and character of the residues generated, and the cost for their handling and disposal.

The total annual cost is obviously calculated as the sum of the annual costs discussed. The average cost per unit of energy can be calculated by dividing the total annual cost by the output of energy generated annually.

Economic Comparison between Different Types of Small Electric Power Plants

In order to illustrate the importance of different factors, a comparison has been made of six possible solutions for generating electricity in a small power plant with a maximum power output of 200 kW.

The options considered are:

1 Diesel engine

   a) Using diesel fuel at US$ 320/m^3 (US$ 8.5 GJ)

   b) Using a dual fuel system with 20% diesel oil and 80% ethanol at US$ 170 per cubic metre (US$ 8/GJ)

2 Producer gas engine

   a) Using wood at US$ 25 per ton (US$ 1.7/GJ)

   b) Using charcoal at US$ 120 per ton (US$ 4.1/GJ)

3 Steam engine

   Using wood at US$ 15/ton (US$ 1/GJ)

4 Stirling engine

   Using wood at US$ 15 per ton (US$ 1/GJ)

The price assumed for diesel oil is approximately that valid in Europe including taxes and charges. The local price may be considerably higher in some developing countries. The price of ethanol estimated for Brazilian conditions (see Chapter 11), has been assumed. Cost estimates made for other countries indicate prices which are up to three times higher.

The cost of wood will vary between different locations and may be up to twice as high as the value assumed here. Alternatively it could also be also be much less, for example when using sawmill residues. The cost difference between wood used for producer gas and steam production reflects the additional fuel preparation required for the gasifier fuel.

The cost of charcoal depends on the location, the cost of wood and the manufacturing technology. Variations by a factor of two up and down are possible.

The cost of operating labour will be assumed of US$ 4/h. This may easily differ by a factor of four up or down between different countries.

Finally the real interest rate i.e. excluding inflation is assumed to be 10%. If this is compared to the interest rate actually charged by the banks, inflation must be deducted from the bank rates.

Comparisons will be made for two types of load conditions characterized by 500 and 3000 equivalent full power hours respectively. A load factor of 50% will be assumed which implies that the average load is 50% of the peak load.

The performance characteristics of the different technologies have been given in earlier tables 11, 13, 15. The results of the comparisons are presented in the Figure on next page in diagrammatic form.

The comparison between ethanol and diesel fuel is simple since the capital costs and the operation and maintenance costs are assumed to be the same in both cases. The outcome of the comparison is therefore entirely dependent on the fuel cost assumed.

Disregarding the ethanol option, it appears from this comparison that diesel engines presently work out as the most economic solution for small power needs i.e. below about 50 kilowatts and particularly for short annual operating times.

For Stirling engines, sufficient cost data for an economic comparison are only available for a 10 kW output. The economy does not appear to be better than for producer gas engines. It should then be observed

that the low wood cost assumed for Stirling engines implies that no special fuel preparation such as chipping or cutting into small pieces is required.

A closer comparison between the estimated costs for diesel engines and producer gas engines shows that the high costs for operational labour make a very important contribution to the high estimated cost for producer gas engines. In order to illustrate the effects of a lower labour

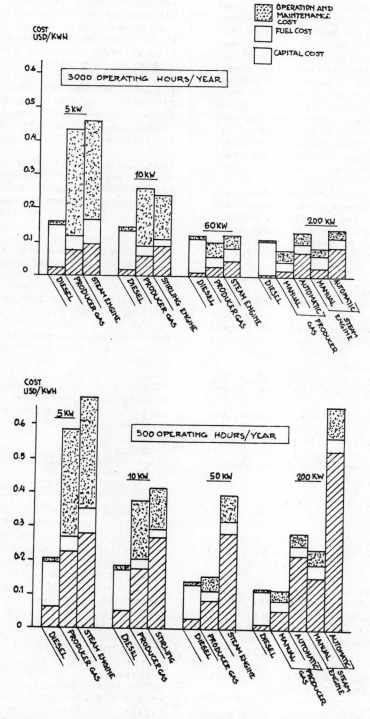

Required price increase for diesel fuel in order to make producer gas economic in different applications.

cost or a situation where for some reason the time required for operation is not considered to cost anything, a comparison with diesel engines has also been made assuming zero labour cost. The results of these estimates are presented in the Figure below which illustrates the change in diesel fuel price required to make the cost for producer gas operation and diesel fuel operation equal. The diagram shows that with a labour rate of US$ 4/h producer gas will be economic for present European diesel fuel prices, for a power output above about 80 kW for 500 annual full power hours and above about 30 kW for 3000 hours. Automatic feeding will not be economic at present oil prices, but an increase of the oil price by a factor of 1.5 would make it economic for powers above 100 kW. Use of charcoal as fuel will only be economic for more than 3000 annual operating hours or higher oil prices. An increase of the oil price by a factor of 1.5 would make charcoal more economic than diesel fuel over a certain range of powers and load conditions. Wood will still be more economic of course, but charcoal has certain advantages as regards convenience in operation (less risk for operational disturbances due to tar in the gas which can be caused by much low load operation, moist fuel or wrong size of fuel pieces).

If the cost of operational labour is neglected, producer gas will be economic even with present oil prices for much lower power outputs – even at 5 kW for 3000 annual full power hours if wood is used as fuel.

Steam engines should be considered as an alternative to internal combustion engines in the power range above about 200 kW if biomass is the preferred fuel. This is particularly true if there is a simultaneous need for steam and shaft power as is the case for some industrial applications. The comparison between producer gas systems and steam systems indicates that producer gas is consistently more economic in the power range studied. However, in longer operating times, i.e. in the range of 3000 hours per year, the cost difference is not very large and is probably within the uncertainties of the estimates.

It is possible, for instance, that if high requirements of operational reliability are necessary, a producer gas power plant may require at least a diesel engine as back-up. This will reduce the small cost difference shown for 3000 operating hours between the producer gas plant and the steam plant.

It can be concluded, therefore, that diesel engines can be expected to be the most economic solution for the present European diesel fuel prices and for power needs below 10 kW.

Producer gas may compete with diesel engines in this power range if the diesel fuel prices are twice as high as the present European price or if the labour cost is much less than US$ 4 per hour.

Producer gas appears as the most economic alternative for powers in

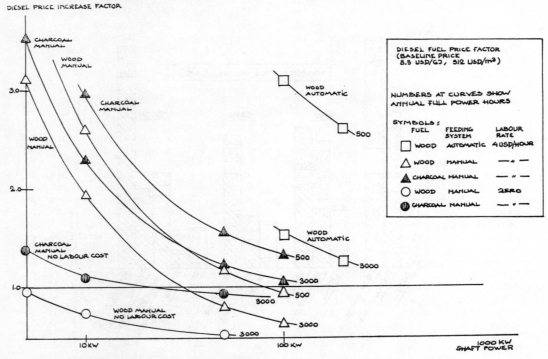

Comparison of estimated costs for shaft power generation using different systems, for 5, 10, 50 and 200 kW respectively.

the range 50-200 kW in particular for annual operating hours in the range of 1000-3000 hours. For longer annual operating times, steam engines may be competitive, even at power levels of 50 kW.

It should be apparent from this discussion that the best choice of system depends largely on the specific situation and that this must be carefully defined before the choice is made.

It should also be apparent from the Figure above comparing estimated costs that even if producer gas or steam systems are more economic in certain ranges of power and operating hours, the differences with respect to the costs for diesel power generation are very small. Since diesel systems have a much lower capital investment, are more convenient and do not require a re-organization of the fuel supply, training of operators and other measures made necessary by the change of technology, it is quite understandable why the market interest for alternative power systems is presently low.

In those locations where fuel prices are higher and the wages lower, the general economic situation is often so difficult that it is not possible to make the necessary investments to acquire the most economic biomass powered energy source.

Future Strategies for Developing Countries

The primary motive for countries investigating the possibilities of increased use of biomass fuels from wood or agricultural crops, is to reduce the costs of their petroleum imports. Countries with a food and land "surplus" have the possibility to develop large biomass-based energy programmes. Brazil, Thailand, the Philippines and the Sudan belong to this group of countries.

Improved national independence in the energy supply, may also be a motive for using such fuels. Increased used of biomass fuels may, however, involve very difficult problems of land use competition. This is particularly the case if the readily available technology of ethanol production from agricultural products containing starch and sugar is employed. It is therefore of great importance to develop the technologies for alcohol production from cellulose materials, together with other technologies for the direct use of cellulose material e.g. producer gas and steam power.

For countries which are net importers of both agricultural products and energy, the critical question is how to make the best use of their agricultural land. This question cannot be determined by strict economic criteria. Consideration has to be given to the fact that an increase in land used for energy production could affect the food supply for a large number of low-income people.

As has been shown in this Chapter, there are indeed several possibilities for the use of biomass to substitute petroleum fuels. The extent to which each technology can be used or should be used to substitute petroleum fuels in the most profitable way will depend very much on the specific situation in each country.

It is important to recognize that the widespread introduction of a biomass energy system will not be successful unless there is an integrated programme for the introduction covering:

- the fuel supply

- the education of operators and maintenance personnel

- the organization of the supply of customer service and spare parts

- financial support to the pioneer users

- research and development based on feedback from the users.

The operation and maintenance of solid biomass engines such as producer gas engines and steam engines require somewhat more insight in to the process and considerably more discipline than the operation and maintenance of an engine operating on liquid fuel. The fact that almost 1 million vehicles were operated on producer gas during the second World War indicates, however, that the level of education required to operate a gasifier system is not extraordinary. Still, the demands on the operator may be a major obstacle to the rapid introduction of technologies for shaft power generation from solid biomass in some developing countries. This problem is particularly important for small power needs. The situation is slightly different as regards larger plants in the power range above a few hundred kilowatts. The number of such plants in each country will be fairly small, the annual operating times will be longer and a small number of people

will be involved in the operation
and maintenance of the plants.

Possibilities for the local manufacture of equipment may be an important
consideration in the choice of technology in many countries. It is
therefore essential that field tests
are carried out with equipment not
only adapted to the user requirements
in developing countries, but also
adapted to the local manufacturing
capabilities.

Because of this, it is recommended
that countries interested in substituting petroleum fuels by biomass
fuel for shaft power generation,
initiate field testing of as many
of the possible technologies as can
reasonably be assumed to be of interest for petroleum substitution.
It must be understood that rapid
introduction of any biomass-based
technology for substitution of diesel
fuel or gasoline will require a strong
commitment and a carefully designed
programme for the introduction. This
will require a reasonable amount
of previous experience within the
country, which cannot be obtained
without field testing.

Prior to or parallel with the field
tests, careful studies of the biomass
supply and possible utilization should
be carried out with the aim of

-   quantifying the amount of biomass
    of different types available for
    energy purposes, particularly
    considering possible food/fuel
    competition.

-   determining the optimum use of
    this fuel resource for different
    types of applications.

It is probably possible with a well
planned and managed approach to petroleum substitution with biomass
fuels to obtain a good basis for
judgement of the possibilities within
a period of 2-3 years, and to obtain
significant results within less than
10 years.

Section III
SOLAR ENERGY

15 SOLAR WATER HEATING

Introduction

The most common conception of the use to which solar energy can be put centres on the heating of water in solar collectors. It is indeed true that solar water heaters are the most common form of solar technology now in use around the world with such heaters installed in over two million homes in Japan, in 600 000 homes in Israel and in well over 30 000 homes in the USA.

In addition, solar water heaters are in regular use throughout northern Australia, where fuels are expensive, and in Greece and Cyprus, both of which are making major efforts to switch from electric to solar water heating in order to reduce their dependence on oil from which the bulk of their electricity is generated.

The most common type of solar water heater incorporates a 2 m^2 flat-plate solar collector and a 200 litre storage tank. The price of this heater ranges from US$ 50-500. In 10 years, which is assumed to be the life time of the solar water heater, a corresponding conventional heating system would consume about US$ 1500 worth of oil.

In the more prosperous regions of the world, the market for solar waters heaters is growing as home owners, hoteliers and local authorities realize their fuel-saving and hence cost-cutting benefits. In the poorer nations these benefits are also recognized to some degree but the solar water heater market has yet to emerge with any great impact since few sectors of these countries' populations can either afford solar systems or can see the need for hot water.

Typical solar water heating installation.

In most developing countries, the raw materials for solar water heater manufacture are unavailable and imported systems, or system components for assembly in these countries, often attract heavy import duties thereby making solar water heaters even more expensive than in the wealthier nations. The problem is compounded by the fact that the earning power of the majority of people in the developing world is a mere fraction of that in the developed world which leads, in effect, to a double block on the widespread dissemination of solar water heating systems in developing countries.

In addition to the problem of cost, only the wealthier sections of developing country communities, at present, use hot water which is mainly heated by electric water heaters. The poorer sections have generally never used hot water in any quantity and hence show little interest in investing in any form of water heating system, despite the fact that washing in hot water reduces the incidence of disease and infection.

Nevertheless, the United Nations Conference on New and Renewable Sources of Energy in 1981 stated that: "In developing countries, although the energy currently used for water heating is in the order of 5-8% of total domestic energy consumed, the availability of reliable solar water heaters at affordable costs could reduce pressure on commercial energy, i.e. electricity produced from oil".

Solar Water Heater Sizing

The average consumption of domestic hot water varies with culture. In the USA an average of 100 litres per capita per day is used whilst in Israel and New Zealand this figure is about half. In Mauritius and Zambia the average consumption of hot water by those few to which it is available is normally 25 litres per capita per day. A lifestyle which does not include a hot shower or bath each day and an automatic washing machine in the USA requires 38-45 litres per capita per day. A temperature of 60°C is usually adequate for all household uses. Water conservation devices can reduce consumption significantly, but on the other hand installation of hot water often increases water consumption so a 150 litres tank is often sufficient for most families in developing countries.

Total solar radiation varies from country to country. Generally speaking, countries between latitudes 15° and 35° have the highest levels of total solar radiation. The next most favourable region is the equatorial belt between the 15° latitudes, but here humidity is higher and clouds more frequent. The third best region is between the 35° and 45° latitudes, but here the potential is limited by daily and seasonal variations. A radiation intensity of 0.7 kWh/m^2 with ten hours of daylight is equivalent to an energy potential of 7 kWh/day. Some places have a high potential of about 8 kWh/m^2/day like Israel, Australia, Egypt and Peru. Others may have about 3-4 kWh/m^2/day. This varies from place to place and with season. e.g. Nandi, Fiji 4.7; Bet Dagan, Israel 5.4; Nairobi, Kenya 5.7 etc. Solar collectors often have an efficiency of about 50%. The use of selective coatings increases this figure to between 55 and 56%. In the best designs some 10-15% of energy is lost due to optical reflection and about 15-20% due to re-radiation. Apart from this there are heat losses in the piping system and storage tank as well.

If the inlet water temperature is 25°C and 150 litres are to be heated to 60°C, the energy required will be 6.1 kWh.

If the average solar intensity is 5 kW/m^2 and the solar water heating system has an overall efficiency of 50%, then this means that a collector area of 2.44 m^2 is needed. Normal solar water heaters have 2-4 m^2 collectors and a tank of 150-250 litres.

Solar Water Heater Technologies

An extensive range of solar water heater technologies has been developed throughout the world to suit the individual climatic and economic conditions prevailing in different areas.

Design, materials, system efficiency, expected lifetime, price, etc. do vary a great deal. Integrated systems, where the tank and the collector are integrated into the same unit are the cheapest and simplest designs. This concept is of interest mainly to the domestic sector. The simplest concept is a black plastic bag sold at a price range from US$ 3-25.

Plastic bag solar water heater used for shower.

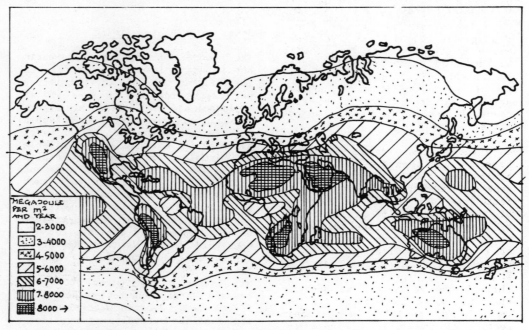

Average annual solar irradiation.

Simple, do-it-yourself systems using the same simple approach are black painted trays, filled with water, and covered with plastic or if pressurized water is available, a black garden hose in which water is heated. Designs like this can be very cheap and cost anything from US$ 25-250.

A more advanced form of the integrated solar water heater consists of a black water tank inside an insulated box which has a glass wall to admit sunlight.

The water is heated during the day and has to be used in the afternoon or evenings as it will cool off during the night. The design can be improved by insulating the glass wall at night-time like in the "bread- box" design. This makes it possible to also have hot water in the morning. Integrated systems have also been mass produced in commercial enterprises. Japan is estimated to have about 2 million solar water heaters installed and about 80% of these are integrated black tube systems. These systems cope well with the social habit of taking a hot bath in the evening. The concept is based on a number (about 5-6) of black polyethylene plastic pipes serving as collector and storage tank. The pipes are usually placed in a steel box (sometimes insulated) and covered with polycarbonate glazing.

The solar water heater systems mentioned above are all of the integrated collector/water storage type whereas the thermosiphon water heater discussed below consists of the water storage tank separated from the collector.

Separated systems, comprise a separate collector (absorber) that is connected to a hot water storage tank via a piping system. The storage tank, being a separate unit, can easily be insulated to store the hot water over night. This concept is of interest to the domestic sector, hotels, hospitals, schools, offices, army camps, workshops and certain industries.

In a thermosiphon system the tank´s base is positioned above the top of the collector and water from the tank is circulated through the collector and back to the tank by means of the natural circulation caused by heating water in the collector. Cold water fed to the bottom inlet of the collector from the base of the tank heats up and, as it does so, becomes lighter thus rising through the collector and back to the tank. It is admitted to the tank through an inlet in the upper section of the tank and, once returned, forms a layer of hot water

Simple plastic hose solar water heater.

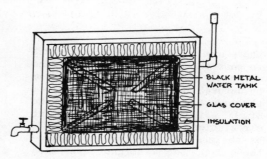

BLACK METAL WATER TANK

GLAS COVER

INSULATION

Simple integrated solar water heater.

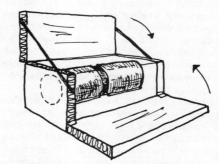

"Breadbox" solar water heater.

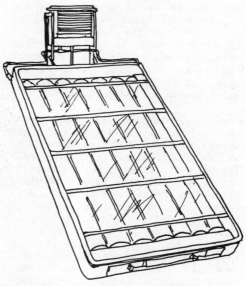

Integrated Japanese solar water heater.

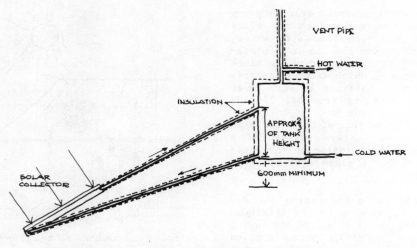

Diagram of thermo-siphon water heating system.

above the cold water which has yet to pass through the collector. Once water in the tank has been heated by natural thermosiphon circulation, through the collector, it can be drawn off for use. For efficient circulation, the tank's base should be at least 30 cm above the collector's top, but a height separation of 60 cm is often preferable in order to avoid reverse flow from the tank to the collector which can occur during cold nights when the collector is not functioning. In situations where the storage tank cannot be sited above the collectors, it is necessary to employ a water pump to force water through the collectors.

Centrifugal pumps have proved satisfactory in such systems and, although the inclusion of a pump adds to the collector's cost, the additional expense can be recovered through the system's greater efficiency. The pump can be controlled by a thermostat which activates the pump when the temperature of water at the top of the header tank exceeds that at the bottom of the tank, and shuts it off when there is little appreciable difference between the upper and lower layers of water in the tank — that is, when water is not being heated in the collector.

Experience has also shown that, where the water is calcareous, of high oxygen content, dirty or chemically contaminated, clogging of the collector and corrosion may be a problem. In such cases piping of wider dimensions (more than 15 mm in diameter) may be required throughout the system.

The Israelis have concentrated on separated, pressurized thermosiphonic systems. The Israeli concept has the advantage of being self-regulating, independent of electricity and requiring little maintenance. Most

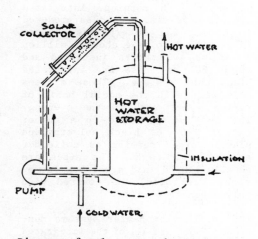

Diagram of solar water heater system with pump.

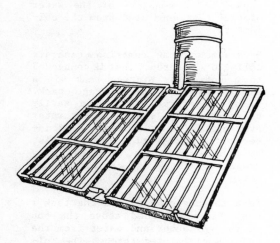

Separated solar water heater from Israel.

systems are made of steel tanks, with copper piping, a glass cover plate and a copper or aluminium absorber plate. The price of a system with a collector area of 2 m^2 and a storage tank of 120-150 litres is about US$ 500-600. The collector accounts for about 50% of the price and installation for about 25-30%.

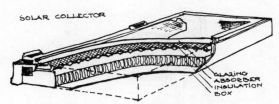

SOLAR COLLECTOR

GLAZING
ABSORBER
INSULATION
BOX

Principle of solar collector.

Future Developments

The general principles of solar water heating are now well-known and understood but several obstacles still block the path towards the widespread use of such systems in the developing world. Most obstacles can be traced back to three general constraints - high cost and insufficient lifetime and reliability of the systems. In view of the large amount of research being conducted into this field of solar technology, improved materials in the fabrication of collectors and simpler systems could solve these problems.

Efforts are therefore being made to develop new, lower cost materials and combinations of materials capable of withstanding conditions usually associated with developing countries such as high heat and humidity, and water of high sediment and corrosive element content. Lightweight reinforced concrete is being tried in the construction of both collector and storage tanks, as is plastic and, in some cases, rubber. Although these materials offer cost advantages over steel and copper (currently the commonest materials in use), the need to employ sophisticated and expensive machinery in the complex fabrication of systems using these materials detracts from their attractiveness in the developing world. In addition, repairs to plastic and

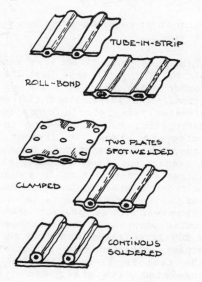

TUBE-IN-STRIP

ROLL-BOND

TWO PLATES
SPOT WELDED

CLAMPED

CONTINOUS
SOLDERED

Principles of different absorber types.

rubber require special tools and techniques often unavailable in developing countries.

Until systems are available which use cheap materials to produce reliable and efficient solar water systems, and until those systems are matched to local needs, until installers are trained to install systems to maximize the effectiveness of systems available and until factors inhibiting costreduction from manufacturers and governments are substantially reduced or negated altogether, the use of solar water heaters cannot be expected to grow significantly in the developing countries in the near future.

Production of Solar Water Heaters

The number of solar water heaters installed so far in the developing world is very small in comparison with the developed countries. Production is possible and is taking place in the developing world but, in most cases, many of the components needed in solar water heater fabrication are unavailable locally and have to be imported.

In some developing countries, however, the small market for solar water heater industry is expanding. In Papua New Guinea, for example, about 8000 solar water heating systems have now been installed, about 90% of them in the domestic sector. Virtually all systems in use are of the thermosiphon type with 250-300 litres water tanks and a collector area of about 4 m^2. These are mostly imported from Australia. System costs in 1981/82 averaged about US$ 1050 - a price which produces significant savings since electricity and diesel prices are high at 15-20 US¢/kWh and 35 ¢/l.

In Nepal, five manufacturers are presently making solar water heaters with the largest, Balayu Yantra Shala, also engaged in system installation. Total annual solar water heater production is 250-300 including both thermosiphon and flat tank integrated systems. Thermosiphon systems, with separated tank and collectors, cost in the region of US$ 900 while the flat tank (integrated) system costs about US$ 600. All materials and components used in the systems are imported from India.

In Zimbabwe the solar water heater market has been tapped over the last 15 years by four main companies, two of which are now well-established, despite a small market resulting from low electricity prices (4.8 US¢/kWh). All heaters sold in the country are locally-produced although most materials and components for fabrication are imported. About 500 integrated systems costing US$ 700-800 have been sold in addition to about 1500 domestic-sized thermosiphon systems at a cost of US$ 2700-3200 each. About 100 large-scale systems for hotels and hospitals have also been installed.

Kenya has five main solar water heater fabrication and installation companies which have installed approximately 4200 m^2 of collectors to date. Originally, complete systems were imported from Australia, Israel and Germany but import restrictions and high import duties have forced the companies - particularly Total Oil (formerly Instrumentation Ltd) which claims to have captured 80% of the market - to manufacture systems locally using both imported and locally-produced components. Most Kenyan systems for domestic use consist of three collectors and a 150 litre water tank in the thermosiphon mode costing about US$ 1190 per system. The price includes a 40% importation duty and 15% sales tax. The high price has reduced the domestic market for solar water heaters whereas commercial and institutional customers account for about 85% of the market since installations in these sectors have associated tax benefits.

Simple integrated solar water heater from Zimbabwe.

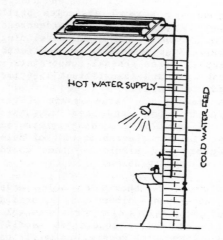

Roof mounted integrated solar water heater from Zimbabwe.

Do-it-yourself Solar Water Heaters

Home-built solar water heaters, (thermosiphon design) can cost about US$ 50 for the materials only. For a single family residence a collector area of 1.5-3.0 m^2 and a storage tank of 150-200 litres has been found to give satisfactory results. Brace Research Institute in Quebec, Canada, has developed and published plans for a solar water heater. The unit has been specially designed to incorporate low-cost materials generally available, even in relatively remote parts of the world.

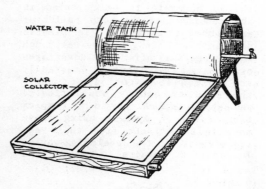

Solar water heating system from Kenya.

The life of the equipment is in the order five years with negligible maintenance, after which time it will probably be necessary to spend about US$ 10 to replace the oil-drum-based hot water tank and overhaul the installation. The absorber is constructed from a corrugated galvanized steel sheet of standard roofing material and a sheet of 22 gauge flat galvanized steel which are rivetted and soldered together.

Another type of solar water heater has been built at the Molefi Secondary School in Botswana. The design described below was built at a cost of US$ 50. It consists of a number of pipes (made of plastic) through which water flows. The pipes are in contact with an "absorber", in

this case corrugated iron painted black. The sun's rays heat the absorber which in turn heats up the pipes. The water pipes (12 mm in diameter) are laid in every second corrugation and soldered on to the corrugated iron. To prevent heat loss, the whole system is insulated. A transparent cover (usually glass) allows solar radiation in and helps to trap the heat inside the insulated box. A rubber strip acts as an air-tight washer to prevent the heat inside escaping. The tank used is a 200 litre galvanized drum and the piping connecting the collector to the tank is 40 mm black PVC pipe insulated with glass wool. This system has been operating in Botswana for two years and has given no problems.

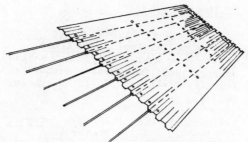

Solar collector integrated in metal sheet roofing.

One concept could be to use the metal sheet roofing as the absorber in a solar water heating system. In Sweden aluminium roof sheets with integrated copper piping for solar water heating is commercially available.

There are several manuals available on how to build a do-it-yourself solar water heater besides the one from Brace Research Institute. Most of the designs can be built with recycled materials for less than US$ 50 or with all new materials for about US$ 500. There is a great variety of possibilities so it should be possible to find a solution for every location and the materials available in that location. (Some manuals or books describing this are listed in the references).

Economy

The economics of solar water heaters depend on initial cost (which depends on technical concept, size, manufacturing process, duties, taxes and price competition), maintenance cost, expected lifetime, whether there is an existing water heating system or if backup is needed etc.

There are several different ways in which the economics of such systems can be assessed. These are; by annualized cost, by pay-back period or by internal rate of return (IRR - the expected return on capital over the expected economic lifetime). An investor who is reluctant to take risks may be more concerned about recouping his money as fast as possible than obtaining the best profitability over the entire project lifetime. He would then pay greater attention to the payback period method than to internal rate of return.

Integrated systems cost US$ 300-400 in Japan, but almost double this in developing countries e.g. US$ 600-800 in Nepal and Zimbabwe. Thermosiphon systems that cost US$ 500-600 in Israel, cost US$ 1000-2000 in developing countries such as Kenya and Papua New Guinea.

With heavy importation tariffs levied on imported systems and components, the price of systems available in the developing world can be up to five times the cost of systems in countries like Israel and Japan where home-produced solar water heaters have high market penetration. Recouping the initial investment laid out on a solar water heating system in terms of reduced expenditure on other fuels and power, differs from place to place depending upon the initial system cost, the cost of competing fuel and power supplies, insolation levels, collector and system efficiencies and interest rates on borrowed finance for the system. In Papua New Guinea, which has extremely high fuel and electricity costs, pay-back is estimated at two to four years, in Zimbabwe the period is about six years and in Kenya it is between three years (against standard electricity tariffs) and eight years (against off-peak electricity tariffs).

The combination of high cost, lack of full demonstration and commercialization and the questionable reliability of newly developed systems has been mainly responsible for the poor market for solar water heaters in developing countries.

Dissemination Factors

The dissemination of solar water heaters in the developing world is affected by a great number of inter-related factors of which the price of the system and of other energy sources is the most critical. In terms of prices of competing energy sources, the cost of electricity is often the most crucial since most water heaters currently available in the developing world are electrically powered. As a result, in countries like Bangladesh, Bolivia and Zambia, where electricity from centralized sources costs 3 US¢/kWh solar water heating cannot compete, whereas the 15 US¢/kWh cost of electricity in Papua New Guinea, Liberia and Fiji makes solar water heaters economically attractive.

However, even in countries with high electricity costs, if the initial cost of the solar water heater is also high, few individuals will be able to raise the initial capital even if the installation can show a high rate of return on investment. If a potential customer decides to invest in a solar water heating system, he will expect value for money in that the system should work effectively for many years. At present, simple, cheap systems are less efficient and have shorter lives than more expensive, more sophisticated systems - a situation which automatically reduces the attractiveness of such systems. Equally important is the quality of installation work and the ease of maintenance of the system. Higher priced systems, such as those in Israel, are virtually maintenance-free while the cheaper Japanese systems require regular servicing. In the developing countries which have yet to acquire the experience with solar water heating systems that the developed countries possess, installers often install systems badly since they lack the specialist knowledge resulting from training and experience. The resulting poor performance hinders efforts by manufacturers to persuade potential customers to buy systems.

Since the high price of solar water heaters is the major drawback to their widespread dissemination, some developing countries have introduced financial incentives aimed at alleviating the financial burden on the potential user. In Papua New Guinea, for example, following a total ban on the installation of electric water heaters in all new buildings, the government has introduced a tax benefit scheme allowing a 100% tax depreciation allowance for the year of expenditure and installation of solar equipment. In Mauritius, all import duties on imported solar water heating systems and components of those systems have been abolished and in the Philippines, investment in a solar water heating system has been made tax deductible. Manufacturers of systems also receive favourable tax treatment.

In Fiji, a tax allowance scheme has been introduced allowing the installer to claim 40% of the cost of installing the system as a tax deductible item. A further 20% allowance is applicable if the installed system is manufactured locally. In other countries certain financial factors inhibit the introduction of solar water heaters. In Sri Lanka, for example, kerosene and electricity prices are subsidized while Zambia maintains low electricity tariffs. In Kenya cheap off-peak electricity (used mainly for water heating) does little to encourage hot water users to switch to solar water heaters. In addition, heavy import duties together with a 15% sales tax on installed systems add further to the already high cost of solar water heaters in that country.

Finally, it should be remembered that the vast majority of people, especially in rural areas, have never used or expect to use significant quantities of hot water and will therefore be extremely unlikely to invest sizeable amounts of scarce money in any form of water heating systems let alone an expensive solar system.

Bibliography

Andersson, B. (1984) Solar Water Heating in Developing Countries. Study Prepared for SIDA. Swedish International Development Authority, Industry Division, 104 25 Stockholm, Sweden.

Brainbridge, D.A. (1981) The Integral Passive Solar Water Heater Book. The Passive Solar Institute, P.O. 722, Davis, CA 95617, U.S.A.

Chinnery, D.N.W. (1971) Solar Water Heating in South Africa. 248 Bulletin 44. National Building Research Institute, CSIR, P.O. Box 395, Pretoria, South Africa.

Werkgroup Ontwikkelingstech (WOT) 1982. Flat Plate Solar Collector, a theoretical and practical Construction. Vrijhof 152, Postbus 217, 7500 AE Enschede, The Netherlands.

Wasserthal, W. (1981) Solar Water Heater, Mode in Tanzania. Arusha Appropriate Technology Project, P.O. Box 764, Arusha, Tanzania.

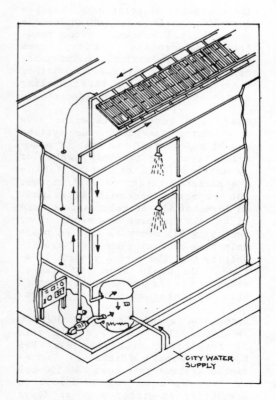

Diagram of solar water heating system for larger houses.

16 PHOTOVOLTAICS

Introduction

Photovoltaic (PV) cells have been developed over the last 100 years in order to directly convert sunlight into electricity by transforming photons of light into electricity. Cells capable of performing this function are connected together and packaged into modules capable of generating several watts of electricity, and modules are connected together to generate larger quantities of electricity using only the sun as the fuel source.

In theory, therefore, they appear to offer unparalleled potential for those developing countries which have both good average insolation rates and large numbers of people in rural and remote areas which are not connected to public power grids.

Electrical energy in such areas is generally supplied by diesel generators which are expensive to run and often cannot operate due to the inavailability or high cost of fuel in the area. PVs are capable of fully replacing diesel generators and also of bringing electricity to areas without them. As a result, PVs could power a wide variety of electrical appliances ranging from lighting to refrigerators and water pumps.

In general, PVs have proved themselves to be largely reliable and durable under extreme conditions, but a number of important refinements, mainly to the complete system as opposed to the modules themselves, are still needed before PV systems will be totally appropriate for use in developing countries. However, although system reliability is a cause of concern, a larger worry concerns the high cost of PVs.

Even though PV module and system costs have declined by a factor of 10 over the last 10 years, PVs can still only compete economically with conventional electricity generating systems in the most remote areas to which the transportation of fuel represents a major operating cost.

Nevertheless, in a study by Rosenblum in 1982, for the US National Aeronautics and Space Administration, it was stated that "for the present, the cost of PV systems in regions of moderate to good insolation is equal or less than that of diesel generation systems for applications with load requirements of about 10 kWh/day or less. Within this range of load requirements lie many important rural applications of imme-

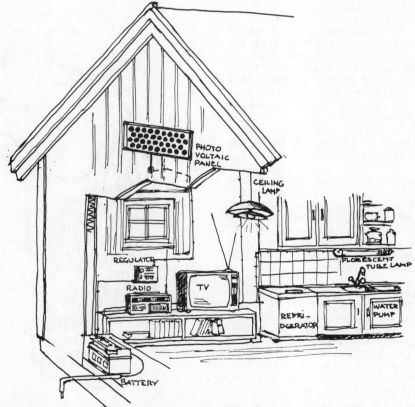

Special low-power appliances are now developed for use in photovoltaic systems, e.g. for isolated weekend cottages in developed countries.

diate relevance to developing countries such as water pumping for domestic and agricultural use, refrigeration and lighting."

The same study states that: "Over the next few years it may be anticipated that PV energy systems will be cost-effective for powering load requirements of about 100 kWh/day - suitable for small rural industry and village applications. Finally, within 20 years, if cost projections are born out, PVs will become the least expensive and most reliable source of energy for most decentralized electrical power applications in the developing world".

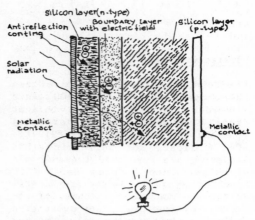

Principle of silicon photovoltaic. cell.

Photovoltaic Technologies

Most PV cells generating electricity in modules and arrays today, consist of silicon wafers sawn from crystal blocks and connected together in the modules. Typically, each cell has a diameter of about 10 cm and can generate approximately one watt of electricity so, when 30 cells are connected together in a module, the module is capable of generating 30 W of electricity at peak sunlight conditions. By connecting together many modules into arrays, it is possible to generate many kW or even MW from PV generators.

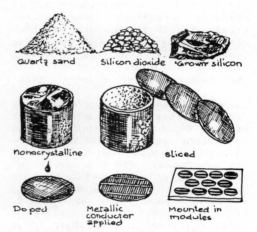

Diagram of production scheme for silicon mono-crystalline cells.

Photovoltaic modules are rated in peak Watts (Wp). This is a reference value of the maximum power output from the module when operating at a cell temperature of 25°C under a solar irradiance of 1000 W/m^2 and is a higher power output than is achieved on average in the field. Cells in good sites produce four to six watthours per day per peak watt. In such sites it takes an array of about two m^2 to produce one kWh a day, so sizeable arrays are needed to produce even modest amounts of power.

In the developing world, it is usually impractical, and indeed extremely expensive, to generate more than a few kW from PV generators since the cost of generating one unit of electricity from large capacity diesel generators is far less than from small capacity generators.

As a result, most uses for PV arrays at present in the developing world are for powering low demand electrical appliances and the market "niche" is currently restricted to a number of specific applications. Such appliances include electric fencing, battery chargers, lighting for hospitals and small communities, refrigerators - especially for vaccines in remote health centres, water pumps - for both potable and irrigation

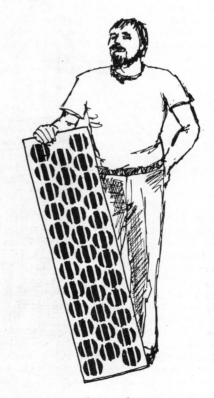

Solar module.

water, remote telecommunications transceivers and relay stations, and cathodic protection equipment installed to protect pipelines and bridges in remote locations.

PV arrays for powering electric fences consist of a single module containing about 10 PV cells, a battery and regulator and cost in the region of US$ 300. A simple PV-powered battery charger capable of recharging a number of rechargeable batteries at one time costs about US$ 50 including the PV array, a converter, an accumulator and battery holders.

Moving up the rated capacity scale, a five module PV array together with five storage batteries, 15 lamps, a floodlight for operations and a small refrigerator for storing vaccines used in medical centres costs in the region of US$ 5000.

Experiments, looking into the Future

In Indonesia, the West German Research and Technology Ministry provided DM 50 million to part fund the installation of large PV arrays in three locations. The first is a 5.5 kW PV generator at Picon which pumps water to irrigate rice fields during the dry season. The 648 module system is linked to a battery storage bank and powers pumps to lift 100 m^3 of water an hour to irrigate a 16 ha area. At the village of Cituis, a 25 kW PV generator comprising 2730 modules and a battery storage bank has been installed to power an ice-making plant, a salt water desalination plant, a weather station and a data recording unit. In one week in June 1983, the ice-making plant produced 1368 kg of crushed ice for use in fish preservation and the plant's average capacity is estimated at 1 kg of ice per kWh. During the same week in June, the reverse osmosis desalination plant produced 19 m^3 of drinking water having originally drawn the water from a 45 m deep well. Average desalination rate is estimated at 50 litres of water per kWh.

At the third Indonesian site on Sumba Island, three PV powered water pumps have been installed to pump water for drinking. The first PV generator generates 5.7 kW to lift water a height of 30 m all year round to supply drinking water to about 5000 people. The second system uses a 3.6 kW array to pump 20 m^3 a day from a 70 m deep well to supply drinking water to 4000 people and the third system uses a similar array to the second plant to deliver 30 m^3 a day from a 45 m deep well for 2000 people.

In Saudi Arabia, a 350 kW generator

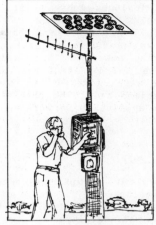

Solar powered telephone.

Solar electric fencing.

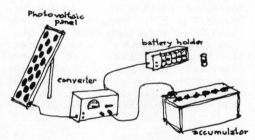

Solar battery charger.

1 Photovoltaics
2 Batteries, Controls
3 Ice Making Plant
4 Desalination Plant
5 Rain Collection
6 Well Pump
7 Water Storages
8 Workshop
9 Radio Equipment
10 Weather Station
11 Field Station
12 Spare Parts

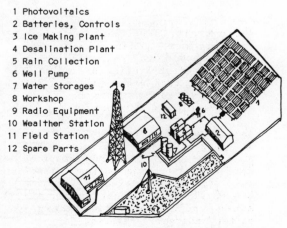

Diagram of photovoltaic system of coastal village in Sumba Island with 800 inhabitans. The majority are fishermen.

which includes lenses over the PV modules to concentrate sunlight on the cells, has been in operation for two years in a remote region supplying power to three villages. Despite initial teething problems during which time 48% of the cell connections had to be repaired, the system appears to be working satisfactorily.

Solar Pumps

Photovoltaic pumping systems have now developed to the stage of being commercially available. There are several well-developed systems available, although there is still scope for improvement. The United Nations Development Programme (UNDP) project GLO/80/003, (See Chapters 17, 25 and 26) with the World Bank, has looked closely at the technology and evaluated its economy. (Sir William Halcrow and partners in association with Intermediate Technology Power Ltd have made this study.)

The pumping systems are designed for small-scale use. Typical applications include supplying 60 m^3 of water per day, through lifts of 2 to 10 m which is suitable for irrigation of land holdings with areas of up to 2 hectares, or supplying typically 20 m^3 of water per day through lifts of 10 to 30 m that is suitable for meeting the water needs of small villages having populations of up to 1500. Drinking water is more valuable than that for irrigation and it is thus economical to obtain it from greater depths. The hydraulic power output requirement of the above pumping systems lies in the range 100 W to 800 W. Photovoltaic systems are par-

ticularly suited to regions where diesel costs are high and wind speeds low, provided that conditions are suitable with a favourable solar irradiation regime and a steady demand for water all the year round. At the moment, in many places it is already economical to pump water through the smaller lifts specified in the depth ranges above. As costs fall to the levels expected in 5 years, solar systems will be cheaper than most other alternatives (except in the case of very low cost diesel systems or an electrical grid) over the head range of 2-20 m and subsequently it can be expected that the ranges over which water can be pumped economically will expand.

The UNDP study looked at three different concepts in detail:

1   Pumping 60 m3/day through a static head of 2 m for irrigation.

2   Pumping 60 m3/day through a static head of 7 m for human water supply and/or irrigation.

3   Pumping 20 m3/day through a static head of 20 m mainly for human water supply applications.

A photovoltaic pumping system can be relatively simple, comprising a flatplate photovoltaic array with associated support structure, a direct current (DC) electric motor directly coupled to a water pump, and pipework from source to delivery point. More complex systems incorporating power conditioning items such as storage batteries or a DC-AC inverter for an alternating current (AC) motor may be appropriate for particular applications.

Photovoltaic water pumping system.

The UNDP project consisted of four stages from July 1981 to February 1984. The first stage looked at the state of the art and concluded that most systems were poorly developed, poorly matched and that considerable improvements could be made. The second stage studied the technology in principle and produced a list of ideal specifications. The third stage stated the performance of the improved solarpumps and predicted prospects for the future. The fourth stage took the form of a hand book on solar water pumping. It can thus be seen that solar-pumps are now well investigated and that a good background exists for further use of the technology.

With respect to the pump applications discussed above, the UNDP recommended the most suitable models available in each category (see above).

1) One of the UNDP tested systems in this category that looks promising is the Solar Electric International (Malta) whose tender price in 1982 was US$ 6000. It uses a solar power PV-array made in the USA with a nominal output of 351 Wp. The motor is an AEG (West Germany) brushless, rated at 360 W and the pump is KSB (West Germany) Aaquasol 100 litres floating centrifugal pump. This is a reliable system and is easy to install. The projected specific capital cost is high at US$ unit 2.5/kJ.d for a lift of 2 m but this could probably be reduced by better matching of the array and the pump-set. The maximum sub-system efficiency was 33% over a lift of 2 m (but reached a peak of 44% for a lift of 1.5 m) with a motor efficiency of 67-72% and a pump efficiency of 63% This design pumped 63 m^3 of water as tested and therefore performed favourably with respect to the UNDP guidelines of 60 m^3/day. The projected capital cost is US$ 4415 (with PV power costing US$ 5 per Wp). Further improvement might include cheaper PVarrays, (85% efficient) motors and axial (propeller) pumps with an efficiency of about 72%. This could mean 1987 prices of US$ 1150 per unit and a specific capital cost of US$ 0.96/kJ.d. unit.

2) Another promising-looking pump is the AEG (West Germany). Its tender price in 1982 was US$ 10 150. It uses AEG PV-array with a nominal output of 614 Wp. The motor is an Engel (West Germany) and the pump is a Loewe (West Germany) submerged centrifugal pump on floating unit. With relatively minor improvements, this easy to operate system could be very good, with a specific capital cost of US$ 0.9/kJ.d. The sub-system efficiency is also good, with a peak of 57%, at the design lift of 7 m with little fall-off across the range of between 5 and 10 m. The motor efficiency is 75-82% and the pump efficiency is about 67%. Local manufacture of the sub-system would be feasible. The projected capital cost is US$ 4614 (with PV-power costing US$/5 W. It pumped 55 m^3 when tested, which with minor improvements could reach the UNDP target.

Other improvements might include cheaper PV-arrays, motors of 85-90% efficiency and centrifugal pumps (self-priming or operating submerged) with an efficiency of 72-75%. This could mean 1987 prices of US$ 3370 per unit and a specific capital cost of US$ 0.8 kJ.d.

Further improvements might include progressing cavity pumps (rotary motion positive displacement pumps), cheaper PV-arrays and more efficient motors and pumps.

This could mean 1987 prices of US$ 4030 and a specific capital cost of US$ 1.01/kJ.d.

3) For a typical village water supply, needing 40 litres per capita/day with a population of 500-1500, a peak flow rate of 1 to 3 litres per second is needed, which at a head of 20 m means a hydraulic power requirement around 200 W.

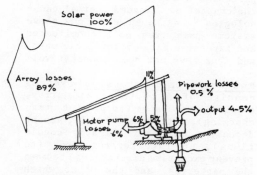

Flow diagram of photovoltaic pump system.

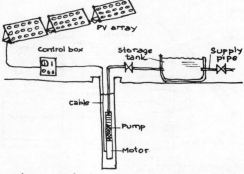

Diagram of photovoltaic pump system.

One of the UNDP-tested systems in this category that looks promising is the Grundfos (Denmark) whose tender price in 1982 was US$ 13 360. It uses Arco solar PV-arrays with a nominal output of 840 Wp. The motor is a Grundfos, 3-phase AC motor with inverter, and the pump a Grundfos multistage centrifugal pump. The complete system was easy to install and performed very well. The current cost of the system is relatively low since the pumpset is based on a standard mass-produced AC unit supplied by a variable frequency inverter. Sub-system efficiency was 40% for the static head of 20 m. The daily overall systems efficiency of the entire solar pumping system was the best of all systems tested. Inverter efficiency was 96%, motor efficiency 68-73%, and the pump efficiency was 50-62%. The projected capital cost was US$ 7812 and the projected specific capital cost was US$ 1.2/kJ.d. The system pumped 24 m^3/d. when tested.

Active Solar Cooling Technologies

Some cooling needs in developing countries such as the cooling of vaccines or electronic equipment in data and telecommunication units, are very important and cannot be allowed to fail if there is a power cut. Active solar cooling systems fall into two categories, those driven by photovoltaic electricity discussed here and those driven by thermal processes which will be covered in the chapter on advanced thermal technologies - although until now these systems have been more complicated and expensive than photovoltaic ones, and thus have not been competitive.

Photovoltaic/compression systems consist of a photovoltaic array, a voltage regulator (to prevent overcharging), batteries (12 volt, amount depending on local climatic conditions), an inverter/regulator (to prevent excessive discharging) between the batteries, and the motor which drives the compressor that cools the refrigerator.

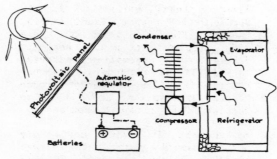

Diagram of photovoltaic compressor refrigerator.

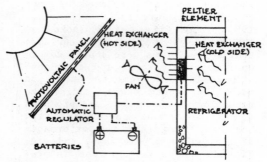

Diagram of photovoltaic compressor refrigerator.

Photovoltaic/thermoelectric systems consist of a solar panel, a regulator, batteries and the thermoelectric (peltier) element. When actively connected the element has a cold side (that cools the refrigeration box) and a hot side that is cooled by a fan. The system is reliable as there is only one moving part (the fan) and is often used to cool electronic equipment. Although this system is used in refrigerators in remote Norwegian houses in temperate climates, the system cannot meet the loads specified by WHO, for freezing ice-packs in tropical climates due to the higher ambient temperature.

Among the active systems, photovoltaic/ compression has been chosen by the WHO to be used in the "Cold Chain for Storing and Transport of Vaccines", since these systems can meet the criteria set up by WHO. The WHO specifies that a refrigerator should be able to maintain internal temperatures below +8°C in an external temperature of +43°C during the day and +32°C during the night. It should be able to produce 1 kg of ice every 24 hours, the temperature should never drop below -3°C and the capacity should be 30-40 litres with top opening and a hold-over time of 6 hours when there is a power cut.

The WHO checks the refrigerators that are available, then laboratory tests their performance. After the laboratory test the manufacturers often make some adjustments and then the refrigerators are put on field testing so that the most suitable design is found. By spring 1983 models from the Solar Power Corporation (Adler Barbour), Polar Products and Western Solar (all from USA) and Electrolux-Kreft (from Luxemburg) had passed laboratory tests and had undergone field testing, although systems from about six other manufacturers were under consideration. These are: B.P. Solar Systems Ltd. U.K. with the LEC refrigerator; Solovolt International U.S.A. with the Marvel refrigerator; Leroy-Somer, France with the Leroy-Somer refrigerator; Areo Solar U.S.A. with the Sawafugi re-

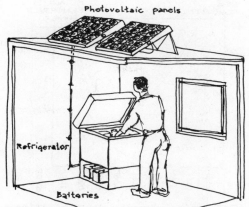

Photovoltaic refrigerators are used for vaccine preservation in remote health stations.

frigerator; Solarex Australia with the Solarex refrigerator; Solar Force France with the Frigesol refrigerator, and there are also other companies using the same refrigerators. Field testing is a prerequisite for all new products. They must be fully proven before they are put into widespread use.

Solar cooling remains an expensive technology. Current purchase prices are today (1985) between 3500-6000 US$ per unit. Bearing in mind that spares and repairs should be added to the cost of solar equipment and that the ten year total costs of purchasing and running kerosene refrigerators is only US$ 2500, it is hard to see an economic justification today, but the aim is to get the price of a PV refrigerator to about US$ 1500 by mass production.

In many areas, an effective cold chain cannot be built using equipment which is dependent on kerosene or gas. Both these fuels are scarce and kerosene has been found by WHO surveys in nine countries to be of unusable quality for refrigerators. In these and other cases there will be stronger arguments than economics to consider.

In spite of sharp differences of opinion today, there is little doubt that photovoltaic and solid absorption cooling devices do work effectively and are capable of going a long way to solving the needs of the vaccine cold chain at the local level. However, a number of problem areas exist in which a joint effort by manufacturers and international agencies is clearly needed.

Future Developments

The principle of converting sunlight to electricity is well-known and research is now concentrating on cheaper photovoltaic cells and system efficiency. Most researchers are concentrating on developing thin film cells made from non-crystalline silicon or from other, more efficient sunlight-to-electricity conversion materials such as gallium arsenide or cadmium sulphide. However, since silicon is the world's second most abundant element after oxygen, most researchers in the PV field expect future PV systems to contain silicon cells rather than the more efficient yet less abundant and more expensive elements.

A great deal of work is taking place on developing systems of producing the ultra-pure silicon, needed in the production of silicon PV cells, more cheaply, and researchers are looking not only at monocrystalline silicon but also polycrystalline and amorphous silicon. In addition, a number of research teams have now turned their attention to reducing the cost and increasing the reliability and efficiency of the complete PV system, e.g. by using bi-facial cells and light focussing systems.

Most people in the field expect PV system costs to decline sharply in the coming years - a price reduction which is essential if the full potential of PV's is to be realized.

Production and Economics of Photovoltaics

The PV cell, module and array production require sophisticated and expensive equipment plus a high level of technical ability. Since most developing countries possess neither the funds nor the expertise to produce their own PVs, only a few developing countries - including India and Brazil - have local production capabilities. As a result, PV systems used in most areas of the developing world are imported and so attract associated high import duties, consequently significantly raising the price of PVs in these areas.

PV cell manufacturing companies are found mainly in the USA, Japan, Germany and France, and cells from companies in these countries are packaged into modules in a number of other countries, mainly in Europe, but also in India, China and Brazil. Soon, it is thought, Saudi Arabia, Pakistan, Tunisia, Egypt and Kenya will also acquire module assembly facilities. By assembling the modules locally, PV costs in developing countries will be cut, not only through reduced importation duties but also through lower transportation costs.

At present, PV module costs vary between manufacturers but the average price is currently around US$ 10/W of electricity generated from the modules. Considering that the price has declined from about US$ 2000/W in the late 1950s to about US$ 200/W in the late 1960s, to US$ 20/W in the late 1970s and to US$ 10/W in the beginning of the 80s. Today (1985) prices are about 7 $//Wp and it is reasonable to assume that prices will continue to tumble and finally even reach a figure that is competitive not only with larger scale remote power plants fuelled with diesel but also, eventually with grid electricity in some areas. However, that point is still some way off especially when one takes into account that as PV cell and module prices decline, the cost of ancillary equipment - including array frames, power conditioning equipment, cabling and other connecting equipment will rise proportionately against cell and module costs in the calculation of overall cost per watt.

It is generally estimated that such ancillary equipment today costs in the region of US$ 0.35 per watt (excluding electrical load regulation costs of US$ 0.1 per watt) and believes that the cost will decline only marginally in the coming years.

Nevertheless, the future market for PVs is still potentially extremely large. It has been estimated that the market size and break-even cost of PVs against conventional power plant for the following applications are as follows:

	Annual Potential Market (MW)	Break-even Systems Price (US$/W)
Communications	110	11-29
Cathodic protection	55	13-53
Low-lift water pumping	85	3.7-4.5
Medium-lift water pumping	210	4.25-9.5
Remote village power	35	4.0-12.4

In early 1982, calculations showed that water pumped with a PV pump cost between US 6 ¢ per m^3 for irrigation water rising to 20 US¢m^3 for drinking water. The cheapest pumps available (including PV modules, submersible pump and regulator) at present cost in the region of US$ 15-20 per watt with about 80% of the cost of the pump taken up in the cost of the PV modules. It is calculated that PV pumps will be truly competitive with diesel pumps when the cost of total PV pumping systems declines to below US$ 5/W.

Many in the PV industry feel that such prices are not as far off as some believe, especially if the scal-

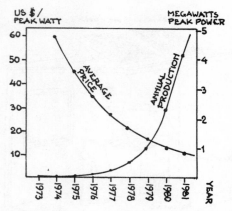

Development of price and production of photovoltaic modules in the U.S.

ing up and automating of PV production plants continues and if companies begin large-scale production of the cheaper, non-monocrystalline silicon cells in the near future. There are signs that such production is starting in several places, which strengthens the general belief that the historical PV cost decline rate will continue.

Once again, though, unless a sizeable market for PV systems develops, sufficient investment for large-scale cheap PV system production is unlikely to become available and, until that investment is available, companies in the field will find it difficult to increase production and reduce costs. This cost/production vicious circle is experienced with most new technologies and it is hoped that, unlike some other promising technologies which have not developed due to a shortage of investment capital, PVs will eventually break out and develop to reach their full potential.

Dissemination Factors

It has been recently calculated that there are a total of 20 MW of PV arrays installed throughout the world generating from a few watts to one megawatt of electricity and that the installed capacity is doubling annually. In 1984 the worldwide production of PVs were 28 MW. However, by far the greatest proportion of installed capacity is in the USA and Europe with relatively little of the capacity being in the developing world despite the competitive nature of PVs in remote areas with small electricity demands.

The reason for this anomaly lies mainly with the cost of PV systems and the lack of encouragement given by governments of developing countries to potential users of these systems. In most cases, heavy importation duties levied on complete or partial

PV systems raises the price of these already expensive items to the point at which they become totally uncompetitive with e.g. diesel generators, and only those with duty-free status or the most wealthy sections of the community can afford them. Aid projects are therefore the largest customers for PV systems in the developing world where PVs are being used for water pumping, telecommunications, health centre and village electrification projects as outlined in the Technology section of this chapter.

PVs will only become widely disseminated in the developing world if system prices decline substantially, if governments reduce or eliminate import duties on PV systems, if local manufacture and assembly of the cells, modules and systems is started, and if the potential users of such systems are educated and made aware of the capabilities of the systems available. At present, only a very minor portion of populations in the developing world are aware of the existence and potential of PV systems.

Conclusions

PV systems have the unique ability to provide power for local use in remote areas where power lines do not exist and which have irregular supplies of expensive fuel. They can bring power to places which would not otherwise have power and can enhance the quality of life in such places while remaining environmentally benign. They can generate power for water pumps in areas short of this essential commodity, they can power medical health centres which previously did not have adequate or reliable power supplies and they can bring lighting and telecommunications to villages which are largely cut off from the outside world and where activity stops when the sun goes down.

Unfortunately, the high cost of PV systems prohibits their use in places without sufficient capital to afford the initial high cost of installing the systems and it is precisely these which could benefit from their installation.

As costs decline, it is reasonable to assume that more remote regions in the developing world will install PV systems but, unless a great deal of funding is forthcoming from the aid agencies and governments of developing countries, it is unlikely that PVs will play any significant role in the energy scenarios of developing countries in the near future, and that their use for some time to come will be restricted to only a few specific applications.

Bibliography

Aarse, A.T. (1984) Power From The Sun - Light at Night.
Botswana Technology Centre, P/Bag 0082, Gaborone, Botswana.

Aylward, M.B. & McNelis, B. (1984) Small-scale Solar Pumping Systems - Present Status and Future Prospects.
Article in International Journal of Ambient Energy Vol. S No. 3, July 1984.
Ambient Press Limited, Hornby, Lancaster, LA2 8LB, U.K.

Derrick, A. & McNelis, M. (1983) A Technical and Economical Review of Small-scale Photovoltaic Refrigeration Systems for Medical Use.
Paper submitted to ISES Solar World Congress, Perth.

Florin, C. (1983) Photovoltaics - International Competition for the Sun (article).
Article in Environment Vol. 25 No. 3 April 1983.
4000 Albemarle Street N.W., Washington D.C. 20016, U.S.A.

Garner, I.F. (1980) A Sunny Future.
Environmental Studies, Newcastle Upon Tyne Polytechnic, Newcastle Upon Tyne, U.K.

Halcrow, Sir W. & Partners & Intermediate Technology Power Ltd. (1983) Small-scale Solar-Powered Pumping Systems: The Technology, its Economics and Advancement.
UNDP Project GLD/80/003 executed by the World Bank.
UNDP, 1, United Nations Plaza, New York. N.Y., 100 17 U.S.A.

Halcrow, Sir W. & Partners & Intermediate Technology Power Ltd. (1984) Handbook on Solar Water Pumping.
UNDP Project BLO/80/003 executed by the World Bank.
UNDP, 1, United Nations Plaza, New York, N.Y. 100 17, U.S.A.

Howes, M. (1982) The Potential for Small-scale Solar-Powered Irrigation in Pakistan.
The Institute of Development Studies (IDS), University of Sussex, Brighton BN1 9RE, U.K.

Lovejoy, D. (1984) Cost Trends for Photovoltaic Systems for Developing Countries.
Article in Natural Resources Forum, United Nations, New York, N.Y. 10017 U.S.A.

Luons, R.A. (1985) Making and Selling Solar Cells.
Article in Alternative Sources of Energy No. 73, May/June 1985.
107 S Central Avenue, Milaca, M.N. 56353 U.S.A.

McNelis, B. (1982) Solar Energy for Developing Countries. Refrigeration & Water Pumping.
Proceedings from a U.K.-ISES conference, London, January 1982.
U.K. Section of the International Solar Energy Society (U.K.-ISES). 19 Albemarle Street, London, W1X 3HA, U.K.

Palz, W. & Fittipaldi, F. (1983) Proceedings of the 5th International Photovoltaic Solar Energy Conference, held at Kavouri (Athens).
Kluewer Academic Publishers, P.O. Box 322, NL-3300 Att. Dordrecht, The Netherlands.

Rosenblum, L. (1982) Practical Aspects of Photovoltaic Technology Applications and Cost.
NASA CR-168025. Lewis Research Centre, Cleveland, Ohio, 44135, U.S.A.

Sørensen, B. (1980) Renewable Energy.
Academic Press, London, U.K.

WHO (1981) Specifications for Photovoltaic Refrigerators.
EPI/9.81/CC.
World Health Organization (WHO) 1211, Geneva 27, Switzerland.

WHO (1981). Solar Refrigerators for Vaccine Storage and Icemaking.
EPI/CCIS/81.5.
World Health Organization (WHO). 1211 Geneva 27, Switerland.

WHO (1985) Solar Powered Refrigerators for Vaccine Storage and Icepack Freezing.
Status Summary June 1985.
EPI/CCIS/85.4 Corr. 1.
World Health Organization (WHO) 1211, Geneva 27, Switzerland.

17 ADVANCED SOLAR HEATING SYSTEM

Introduction

While simple, low-efficiency solar thermal systems are suited, in the main, to domestic uses, a breed of more advanced, higher-efficiency systems has been developed to cater more for industrial and commercial applications. Such systems, when installed in areas of moderate to good insolation, are capable of raising high temperatures for pre-heating boiler feedwater, for direct water and air heating in industrial processes and for steam production. However, despite the existence of such systems, some of which are now reaching an advanced state of development, it is not expected that advanced solar thermal systems will play a major role in industrial and commercial energy production in the near future as their economics cannot yet be shown to be competitive with conventional energy generating technologies.

Although the high cost of available systems is the single most important barrier to the widespread dissemination of advanced solar thermal systems in both the developed and developing worlds, an additional barrier – that of the reliability of the system – also acts strongly against their acceptance by industry. Many managers of industrial plant are sceptical that solar systems can fulfil the requirements of industry since such systems depend heavily on climatic factors. If the sun fails to shine for a number of days at a time, the managers say the system cannot hope to be able to provide the quantities of energy required on a regular and uninterrupted basis, especially for those plants working on a 24-hour cycle. In addition, managers are far from convinced that solar thermal systems can always supply the temperatures and pressures required, and that expensive back-up systems will always be required.

If these objections can be overcome, it is believed in many circles that industry, especially in the developing world (where, in general, solar radiation levels are high and regular) could take maximum advantage of the available systems. This could significantly reduce oil, gas, electricity and fuel-wood consumption so as to help to alleviate the burden of costly fuel and energy imports on developing world countries and consequently to contribute towards these countries´ development.

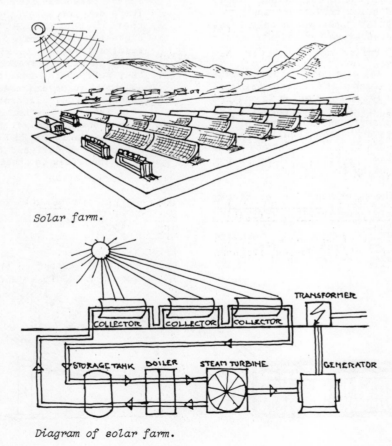

Solar farm.

Diagram of solar farm.

Classification of advanced solar collectors.

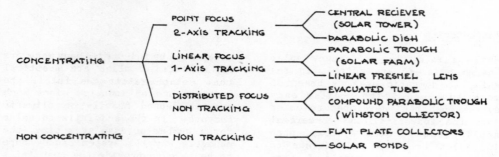

High-efficiency Solar Collectors

The function of a solar process heat system is to collect energy from the sun, transfer it to the desired location, and convert it into a usable form. The major components needed to carry out these functions are 1) solar energy collectors, 2) heat transfer fluids, 3) piping or duct work, 4) pumps, and 5) controls. The collectors are the most expensive and most critical determinant of the system's performance.

Collectors can be classified by their optical properties. The two major categories are 1) non-concentrating and 2) concentrating collectors.

1 Non-concentrating collectors fall into two types,
a) flat-plate collectors which are the most common and widely used solar collectors and
b) solar ponds which only exist as prototypes for industrial applications.

a) A flat plate collector consists of a black plate or absorber placed in a low box and which is covered by a sheet of transparent glass or plastic and insulated at the sides and bottom. Sunlight passes through the clear sheet and heats up the absorber, heat being retained by the insulation.

b) The solar pond is a relatively deep (2-5 m) pond filled with brine of a controlled concentration. The purpose of the dissolved salt is to provide an artificial density gradient which acts as a barrier to natural convection in the pond to allow localized heat build up. (This concept will be more extensively discussed later in this chapter).

2 Concentrating collectors come in a much wider variety of types. These can be divided into three groups according to their tracking characteristics:

a) point focus/two axis tracking
b) linear focus/single axis tracking and
c) distributed focus/fixed non-tracking collectors.

a) The most common point focus/two axis tracking concepts are the central receiver (solar tower) and the parabolic dish collector. These designs offer the highest thermal performance (i.e. the highest temperatures). The central receiver concept consists of a number of two-axis tracking mirrors that reflect the sunlight on to a central absorber (receiver). The parabolic dish collector consists of a dish shaped reflective surface with a small receiver suspended above in the focus, and the whole structure is designed to rotate in 2 dimensions to track the sun.

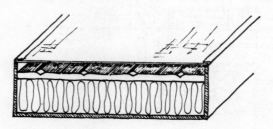

Principle of flat-plate solar collector.

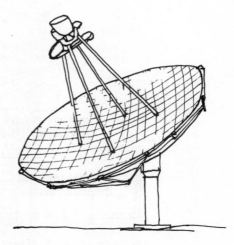

Point-focus tracking solar collector.

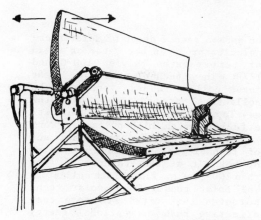

Linear focus, parabolic trough tracking solar collector.

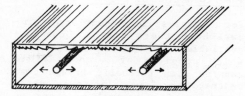

Linear focus Fresnel lens tracking solar collector.

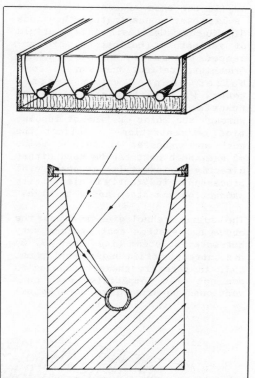

Non-tracking compound parabolic trough solar collector (Winston collector).

b) Linear focus collectors only require single axis tracking which can be either north-south or east-west. The most common concepts are the parabolic trough and the linear Fresnel lens. These systems use a reflective parabolic surface or a linear Fresnel lens to concentrate the sunshine on to a linear absorber. The tracking can involve the whole structure, just the absorber, or just the concentrator. Both the parabolic trough and the linear Fresnel lens collectors are now commercially available.

c) The most common non-tracking concentrating solar collectors are the evacuated tube collector and the compound parabolic trough collector (the Winston collector). Both these collectors have a less concentrated focus point than the other types and thus feature relatively larger absorbers. The evacuated tube collector consists of concentric glass tubes with a vacuum between the inner collecting surface and the outer transmitting surface. The vacuum eliminates a lot of heat loss and the collectors can work efficiently at higher temperatures than flatplate collectors. The tubes are mounted in a frame with a gap between adjacent tubes. The back of the frame consists of reflectors that give some degree of light concentration. The Winston collector is a version of the parabolic trough collector which does not track the sun. It consists of numerous compound parabolic surfaces that focus sunlight in a diffuse manner down to a large absorber that is situated at the bottom of the compound parabolic trough. Because the parabola is compound and the absorber fairly large, light will always be reflected onto the absorber.

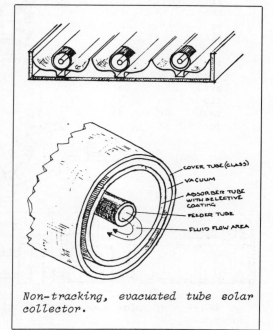

Non-tracking, evacuated tube solar collector.

The Production of Process Heat

For a large-scale production of process-heat there are two concepts available, namely "the solar farm" which consists of a whole field covered with parabolic trough concentrators or "the solar tower" which consists of a central receiver on a tower and a whole field of tracking mirrors.

Temperatures at the point of focus, in the case of a "solar farm" can reach several hundred degrees Celsius or, in the case of central receiver "solar tower" concentrators, thousands of degrees Celsius, since a field of reflectors (heliostats) are arranged separately on sun-tracking frames to reflect the sun on to a boiler mounted on a central tower. With both systems, a heat transfer fluid or gas is passed through the point or line of insolation concentration to collect the heat and transfer it to the point of use. Such heat can be used either directly in industrial or commercial processes or indirectly in electricity production via steam and a turbine.

The solar technologies such as the above two systems that produce very hot water or steam (i.e. over 177°C) are currently still under development and, in general, these technologies are not cost competitive with conventional power sources such as oil or gas. A second group of solar technologies produces hot water or steam at temperatures between 100°C and 177°C. These technologies (e.g. linear focus collectors and evacuated tube collectors) are generally past the development stage and are often cost-effective with high-priced conventional alternatives. The final set of technologies produces hot water or hot air at temperatures of less than 100°C, e.g. flat-plate collectors and solar ponds. These solar technologies that often are more cost effective than conventional energy sources will also probably be utilized to meet lower temperature needs (i.e. hot water, hot air, or low-pressure steam) in the near-term.

In order to further the distribution of these technologies development must proceed in order to reduce the cost and improve the efficiency of each collector type. However, a good collector is not enough, the whole system must be well designed, carefully constructed and proven to function, so that the break-downs of the system can be reduced or elminated.

One of the technolgies that theoretically could be cheaper than others is the solar pond. However, this design is still difficult to assess because very few have actually been built.

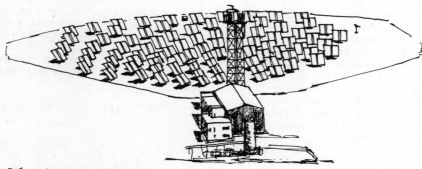

Solar tower system.

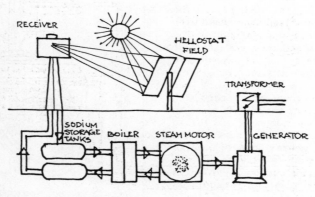

Diagram of solar tower power plant.

Solar Ponds

Solar ponds are large brine ponds approximately 1-2 m deep, in which vertical gradients of salt concentrations are maintained so that the most concentrated and hence the most dense solutions, are found at the bottom of the pond. The pond acts as a solar collector and a heat store. Heat is generated at the bottom of the pond by the pond's black-coloured bottom absorbing incident radiation, transmitted through the water. Unlike fresh water ponds, where heated water expands, becomes less dense and rises to the surface (a process known as convection), in solar ponds the process of convection is inhibited. In spite of the temperature increase in the water at the bottom, its specific gravity remains greater than that at the top if a sufficiently large vertical concentration gradient is maintained. Early experiments gave temperatures of up to 96°C at the bottom of the pond, although a practical working temperature appears to be about 90°C. Thus the solar pond is a potential source of low-temperature heat. The energy cost is so low that conversion to electricity by means of low temperature turbines becomes practical, despite the low thermodynamic efficiency of converting this heat to mechanical or electrical energy.

The concept of solar ponds was first investigated in Israel in 1958. In the early 1960s a considerable amount of work was done, mostly in Israel, but the work was discontinued in 1966 concluding that solar ponds were not economically competitive. But the rise in oil prices in the early 1970s changed that and work was restarted in 1975. A number of solar ponds were constructed, and a 6 kW unit was demonstrated in the late 1970s. A 150 kW solar-pond-electricity- generator was built at Ein Bokek on the Dead Sea in early 1982 by Ormat Turbines Ltd. The ponds were 1500 m^2 and 7000 m^2 in size respectively. A 5 MW^2, 250 000 m^2 solar pond has now been built at Beit Arava, also on the Dead Sea, and there are plans for a similar plant at Salton Sea in California (also built by Ormat Ltd). Ponds have also been built in the USA, Australia, Portugal and India.

The electricity cost is estimated to be in the range of US¢ 13.5/ kWh(e) (1980 estimate) for large plants and US¢20/kWh(e) (1980) for small plants of around 500 kW. It has been determined from Israel's experience that for units of say 50-150 kW, the solar pond would often compete with diesel installations of equivalent capacity. This can be quite encouraging for developing countries where there is often a need to provide small amounts of power in remote areas.

The economics of solar ponds have yet to be fully evaluated since, for the present, only a small number of demonstration ponds exists. Without full operating and performance data taken over an extensive period, any calculation as to the economic attractiveness of such ponds is difficult. Probably the simplest, least expensive use of solar ponds is for thermal applications. In developing countries there is a need for low-temperature energy especially in agriculture and agro-industries. At the Ohio Agricultural Research and Development Center in Wooster, Ohio, a solar pond is used to heat a greenhouse, and researchers at Ohio State University in Miamisburg are looking at solar ponds in order to supply agricultural and industrial process heat.

Climatic limitations restrict the construction of ponds to between

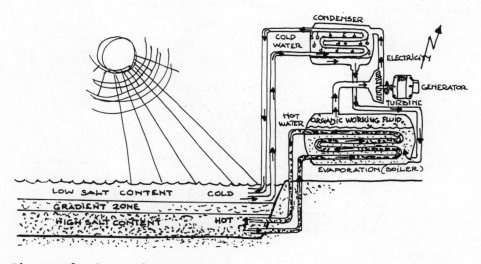

Diagram of solar pond power plant.

40° north and 40° south and, since most developing countries lie within this belt and since pond construction techniques can incorporate much local capability and materials, the use of this concept is expected to increase substantially in the developing world over the coming years.

Solar Pond Technology

The vertical configuration of a salt-gradient solar pond normally consists of three zones. Adjacent to the surface there is a homogenous convective zone that serves as a buffer zone between environmental fluctuations at the surface and conductive heat transport from the layer below. This is the upper convective zone (UCZ). At the bottom of the pond there is another convecting zone, the lower convecting zone or LCZ. This is the layer with the highest salt concentration and where the high temperatures are built up. For given salinites and temperatures in the upper and lower convective zones, there exists a stable intermediate gradient zone. This zone keeps the two convective zones apart and gives the solar pond its unique thermal performance. This intermediate zone provides excellent insulation for the storage layer, while simultaneously transmitting the solar radiation. To maintain a solar pond in this non-equilibrium stationary state, it is necessary to replace the amount of salt that is transported by molecular diffusion from the LCZ to the UCZ. This means that salt must be added to the LCZ, and fresh water to the UCZ whilst brine is removed. The brine can be recycled, divided into water and salt (by solar distillation) and returned to the pond. The major heat loss occurs from the surface of the solar pond. This heat loss can be prevented by spreading a plastic grid over the pond's surface to prevent disturbance by the wind. Disturbed water tends to lose heat faster than when calm.

Due to the excessively high salt concentration of the LCZ, a plastic liner or impermeable soil must be used to prevent infiltration into nearby ground water or soil. The liner is a factor that increases the cost of a solar pond. A site where the soil is naturally impermeable, such as the base of a natural pond or lake, or can be made impermeable by compaction or other means, will allow considerably lower power costs.

The optical transmission properties and related collection efficiency vary greatly depending on salt-concentration, the quantity of suspended dust or other particles, surface impurities like leaves or debris, biological material like bacteria and algae, and the type of salt. It becomes obvious that much higher efficiencies and storage can be achieved through the utilization of refined or pure salt whenever possible, as this maximizes optical transmission.

The solar pond is an effective collector of diffuse, as well as direct radiation, and will gather useful heat even on cloudy or overcast days. Under ideal conditions, the pond's absorption efficiency can reach 50% of incoming solar radiation, although actual efficiencies average about 20% due to heat losses. Once the lower layers of the pond reach over 60°C the heat generated can be drawn off through a heat exchanger and used to drive a low-temperature organic Rankine cycle (ORC) turbine. This harnesses the pressure differentials created when a low boiling point organic fluid or gas is boiled by heat from the pond via a heat exchanger and cooled by a condenser to drive a turbine to generate electricity. The conversion efficiency of an organic Rankine cycle turbine driving an electric generator is 5-8% (which means 1-3% from insolation to electricity output).

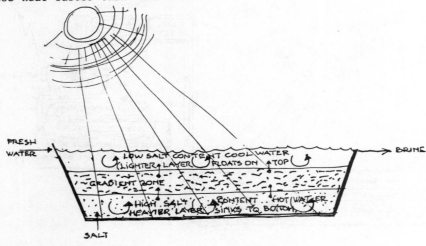

Principle of solar pond.

A solar pond has a considerable amount of built-in storage so that almost no variation in yields is detected between night and day and indeed over a period of several sunless days. However, there might be a 1:4 differential in output between summer and winter depending on the variation in solar radiation.

Organic-rankine-cycle

Ordinary rankine-cycles use water as the working fluid, and need operating temperatures of 370°C-600°C. In order to be able to utilize lower temperatures other working fluids with boiling points lower than that of water is needed. The introduction of organic working fluids promises to expand the range of the rankine applications. Small, decentralized units - particularly those responding to uneven power demands - work better under low temperatures, on the order of 65°C-370°C. Small decentralized steam systems use a group of organic compounds including the chlorobenzenes, fluorinol, isobuthane, freon, and toluene. None of these substances could be called benign. Freon has been implicated in the depletion of atmospheric ozone, while all the others are highly toxic and some are extremely flammable. They do however, operate efficiently within the low temperature range most appropriate for small power systems. Organic fluids have the additional advantage of remaining in the vapour state when water condenses. This property leads to longer turbine life, since the impact of tiny condensation droplets on turbine blades is the major cause of their wear.

The efficiency of organic rankine cycle (ORC) engines, falling between 5-20%, are not spectacular either, and they do not appear to depend on scale.

The costs of ORC units in the 5-100 kW range are still high - more than 3000 $ US/KW (1980 $) installed. But interest in them is increasing.

One of the most promising thermal sources is the solar pond, but other energy sources might be solar collectors, geothermal steam, combustion of agricultural and forest wastes, and industrial waste heat.

Ormat industries in Israel are presently drawing energy from a solar pond at the Dead Sea (300 kW of electricity). They produce ORC in the power range 200 W-5000 kW and can utilize heat sources between 80-200 C. They have different applications like electricity generation and water pumping with low temperature heat sources.

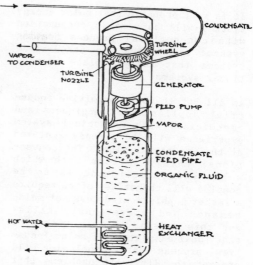

Organic Rankine cycle generator, of the ORMAT-type.

The French Company Sofretes produced ORC that used a screw engine instead of a turbine for water pumping (1-30 kW), and solar collectors were used as a heat source. There were many problems with these solar-pumps and they were finally abandoned because of the photovoltaic competition. Another possible use of ORC is to drive compressors. ORC will probably be most useful when large amounts of waste heat is freely available. And it could in such cases produce cheap electricity.

The system consists of four major components: an expander, a boiler, a feedpump, and a condenser. The boiler will be the most expensive part of the installation (20-30% of total cost). As an expander a screw engine will probably be preferred at smaller sizes and a turbine at bigger. But it must be concluded that it is a new technology with little experience and that small installations are complicated and expensive.

Solar Energy for Industrial Application

Since approximately 40% of a developing country's energy consumed in the industrial sector is consumed in the form of process heat at temperatures below 170°C, it can be considered that advanced solar systems could potentially play an important role in energy production in industry. This includes agro-industries (e.g. sugar plants), foodstuff industries (for vegetables, fruit, fish, milk, breweries, etc), paper and pulp production, textile manufacture and chemical industries.

It is most probable that solar energy systems will be used in conjunction with conventional process heating systems rather than as stand-alone systems that fully replace conventional systems.

In all but the most primitive industrial applications, engineers and technicians with mechanical system experience will be already employed at the industrial site. The conventional energy systems with which solar energy equipment is to be coupled will themselves often require a rather sophisticated level of maintenance and operational skills. For this reason, it is not thought that local adaptation of solar industrial process heat technologies will present the difficult problems that face most sophisticated technologies supplying energy to agricultural or rural poor sectors of developing countries.

The most probable path to obtaining solar industrial process heat in developing countries in the near to mid-term future will be to import total systems and/or components while continuing to develop indigenous knowledge and manufacturing capabilities.

There are many industrial applications which use temperatures below 100°C and such temperatures can be adequately raised using flat-plate collectors. However, until local industry is made aware of the capabilities of available advanced solar thermal systems through demonstration, and until engineers are trained to maintain these systems, it is unlikely that many such systems will be installed in the developing world in the near future.

In addition, and more importantly, initial installation costs are still

SOME TYPICAL PROCESS HEAT APPLICATIONS

Name, location	Application	Type of collectors	Collector area m²	Field output temperature °C	Process temp.	Status	Financing
Coca Cola Queanbeyan Australia	Can Warming	flat plate	94		60°C	In operation since 1976	governmental
Swan Hill Laundry Victoria, Australia	Laundry	parabolic trough 1-axis	126	steam water	86°C oil savings; 535 l/week	In operation;	governmental
SHARP Electronics Nara, Japan	Semiconductor prod.process	flat plate	1184		25°C water	In operation since 3/1980	governmental and private
Oyama Factory Showa Aluminium Co. Tochigi, Japan	Aluminium Washing	flat plate	144	80°C	water	In operation since 1975	governmental and private
Hankabe Aichu, Japan	Dyeing	cascaded flat plate tubular vacuum	440 538	135°C	steam	under construction	governmental and private
RAM, Lactaria Castelana (Alcorcon, Spain)	Pre-Sterilisation of Milk	parabolic trough 1-axis	580	180–240°C	159°C (steam)	In operation since 6/1980	governmental and private
Carcesa (Merida, Spain)	Sterilisation of tinned meat	parabolic trough 1-axis	1120	180–240°C	162°C (steam)	In operation since 1981	governmental and private
Ucal (Aguas de Moura, Portugal)	Sterilisation and Pasteurisation of milk	parabolic trough 1-axis	1150	280°C	200°C (steam)	under construction	governmental and private
Anic (Pistioci, Italy)	Process steam for chemical plant	parabolic trough 2-axes	1728	280°C	190°C (steam)	under construction	governmental and private
BIBS & Co. (Macon, Georgia, USA)	Process steam for textiles	parabolic trough 1-axis	10,000		260–315°C	In operation	private
Lone Star (San Antonio, Texas, USA)	Process steam brewery	parabolic trough 1-axis	880	274°C	177°C (steam)	In operation	governmental and private
Westpoint Pepperell (Fairfax, Alab, USA)	Textile drying	parabolic trough 1-axis	697	195°C	160°C (steam)	In operation since 12/1978	governmental and private
Home Laundry (Pasadena, Calif. USA)	Laundry	parabolic trough 1-axis	604	190–210°C	170°C (steam)	In operation since 2/1980	governmental and private
Caterpillar factory, (San Leandro, Calif. USA)	Washing of parts	parabolic trough 1-axis	4000	114°C	steam	In operation since 2/1983	governmental and private

Source: Friedrich, F.J. Solar World Congress 1983, Perth.

too high at present for most industries to consider switching from existing heat generating equipment to solar systems and, until prices drop quite substantially, few industries will consider replacing oil, electricity, gas and wood-burning boilers with solar systems.

Production Of Advanced Solar Thermal Systems

The fact that the systems described here are of the advanced variety of solar thermal technologies implies that advanced design, construction work, sophisticated equipment and know-how, are required in the fabrication of such systems. This is essentially true with point and linear focussing concentrators, central receivers, advanced flat-plate systems and evacuated tubes, but is not truly accurate in the case of solar ponds.

In addition, few advanced systems are in use in developing countries and, without the benefit of experience from demonstrations of the technologies, it is unlikely that these countries will, in the near future, acquire either the technical know-how or an awareness of systems available.

The situation is different with solar ponds which uses relatively simple components compared with the other advanced solar thermal systems. Construction of the pond requires little sophisticated engineering knowledge, although manufacture of the low-temperature Rankine cycle turbine does require a certain level of design and manufacturing capability not always available in the developing world. As a result, for the foreseeable future, solar ponds may be constructed in developing countries once their performance has been proved but it is likely that turbine and generator sets matched to the pond's capabilities will have to be imported.

Economic Aspects

The economic viability of solar industrial process heat systems has to be viewed from two perspectives. First, the costs of a specific system have to be compared to conventional energy alternatives to determine if the technology is cost-effective (micro-economics). If enough comparisons yield favourable results, then it is important to examine the broader impacts of the technology on the entire internal economic system and balance of trade (macro-economics). We will begin with a review of the micro-economic issues.

Solar industrial process heat systems do not produce a unique form of energy. Their application will be to replace conventional energy sources (primarily gas and oil) in existing industrial facilities, or to be placed in new facilities when they are built. In either case, the relevant comparison is with conventional energy sources.

The economics of industrial application of solar energy is extremely site-specific, and it is therefore extremely difficult to reach any general conclusions about the economic feasibility of solar industrial process heat systems. There are thousands of variations possible in the way the systems are engineered, the type of process to which they are applied, and the type of technologies and components used. The costs of a solar process heat system include capital equipment such as collectors, storage, controls and piping/ductwork. The cost of routine maintenance and the cost of replacing major components during the system's operational life must also be included. Using typical costs of collectors and other components currently available, the majority of solar industrial process heat projects studied have not yet been cost effective.

A recent analysis reviews the economics of ten solar industrial process heat installations in specific locations. Most of the case studies occurred in the United States. They were generally not cost effective for several reasons. First, the demand for energy in many industrial processes is large and is not well matched to hours of peak solar energy. As a result, most early solar process heat designs mix conventional and solar energy equipment together to supply the needed heat. Under this arrangement, the benefits of the solar energy systems are limited to savings in conventional fuels. The capital cost of the conventional system is not avoided by installing the solar energy system and cannot be included as a benefit. Therefore, the price and escalation rates in conventional fuels become prime determinants of the benefits of the solar energy system.

Industrial solar heating technology is new and breakdowns frequently occur. In addition, experience shows that repairs, spare part supplies and maintenance on new and complicated technologies is always a particularly difficult problem in developing countries. The problem of maintenance is further not only economic, but also a question of the physical availability of skilled and trained man power to supervise and run new technologies.

Let us now turn to the macro-economic aspects of solar industrial process heat technologies. Solar industrial process heat technologies could have an impact on the economy of a developing nation if sufficient production investments were made to make low-cost solar equipment locally available. Lower costs would stimulate increased demand for the equipment which would lead to further production investments. Such reinforcing events could enhance the overall economic prospects for the country.

In theory, the macro-economic impacts of the technology could be significant because of the central role of energy in development and in trade balances. In practice, though, the small industrial sector of most developing countries, coupled to the need for imports of advanced solar technologies and the general lack of skilled manpower will most probably mean that the contribution of industrial solar energy on national energy balances will be negligible for the foreseeable future. Only under very special conditions with respect to climate, heat load and type of industry will it be possible for solar energy to compete with conventional energy in industrial applications.

Advanced-solar-cooling Methods

Solar cooling can be divided into passive and active methods, and the active methods can be powered by photovoltaics or thermodynamic solar collectors (passive methods are described in a separate chapter). Up until now thermodynamic methods have not been very successful although they do work. Most of them use an absorption cycle (solid or liquid), and a lot of development work is under way. The WHO has tested some of the systems available and concluded that for the time being these systems cannot compete with the photovoltaic designs.

Solid Absorption Refrigerator

Function: During the day, ammonia gas is driven out of a solar collector (filled with calcium chloride and cement). The gas is condensed by cooling in a water tank to liquid ammonia and passes into the evaporator. At night, as the collector cools and the pressure falls, the liquid ammonia passes into the evaporator and boils at a temperature of -12°C. The frozen surface of the evaporator can make about 9 kgs of ice over 8 hours. The ammonia gas then returns to the collector and is reabsorbed into the calcium chloride ready for the next day. The ice can be removed for use, or left to keep the refrigerator cool in times when there is no sun.

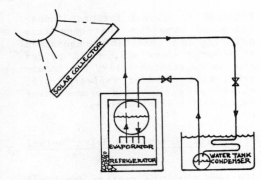

Solid Absorption/Calcium Chloride Refrigerator.

Another concept is using a mineral called zeolite in the collector and water vapour as the gas.

The most advanced projects however, are the prototype designed by the Technical University of Denmark, tested at the Institute for Energy Research, in Khartoum, Sudan, and the zeolite refrigerator produced by the Zeopower Company in Natick, Massachussetts, U.S.A.

Liquid Absorption Refrigerator

Function: Ammonia gas is boiled out of a solution of water and liquid ammonia in a solar collector during the day. The gas is condensed to liquid ammonia and is stored until the collector cools at night and the pressure in the system falls. The liquid ammonia then passes into the evaporator and boils at -12°C. The frozen surface of the evaporator can make about 5 kgs of ice in 8 hours. The ammonia gas then returns to the collector where it is reabsorbed into a solution with water ready for the next day. The ice can be removed for use or left to cool the refrigerator in times of no sun.

The three most important projects are based in: the Centre for Appropriate Technology (CAT), Delft University, The Netherlands; the Asian Institute of Technology (AIT), Thailand; and the Centre for Industrial Innovation (CII), University of Strathclyde, United Kingdom.

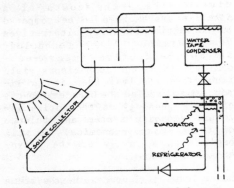

Liquid Absorbtion/Ammonia Refrigerator.

Advanced-Solar-Pumping Methods

UNDP Project GLO/80/003 (see Chapters 16, 25 and 26) run by the World Bank, looked at small-scale solar-powered pumping systems. It concluded that several photovoltaic solar-pump systems were commercially ready and available, while no solar thermodynamic pumps are yet being commercially produced and recent installations are few in number.

The study concluded that thermodynamic solar water pumping should be considered as a technically viable and potentially economic competitive option in the near future, but reductions in the cost of photovoltaics below US$ 5/Wp (which will soon be achieved) may result in PV-systems costing less than thermodynamic designs.

Working solar thermodynamic power systems were demonstrated as long ago as the 1870s. In Egypt a solar irrigation system using parabolic trough solar collectors was installed in 1919. Sofretes (France) installed a number of 1 kW solar pumps in different countries in Africa in the late 1970s. The pumps were powered by flat-plate solar collectors, coupled via a heat exchanger, to a reciprocating Organic Rankine Cycle (ORC) engine operating at about 80°C. The majority of these are now abandoned.

The main problem with the solar thermal pump system, and one which is also generally associated with most advanced solar thermal technologies, is the system's complexity. Regular maintenance to keep all system parts, especially the moving parts, functional is necessary and in regions remote from engineering skills, mechanical failures are common. Unless local maintenance capabilities are available, the installation of advanced solar thermal systems is not considered appropriate.

Of the 50 organizations having known activities involving solar thermodynamic pumps, only seven commercial manufacturers appear to be maintaining a serious development of small-scale designs and four are involved in the development of simpler systems which might be better suited for local manufacture in developing countries.

Dornier System GmbH (FRG), in association with BHEL (India), are testing a 300 W hydraulic power pump using 25 m² of direct evaporating solar collectors to run an ORC engine.

Wrede Ky (Finland), has tested a 300 W ORC with a parabolic trough solar collector and a condenser "after cooler" to ensure a high operating

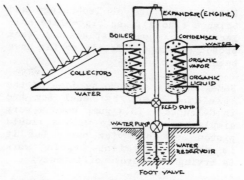

Advanced solar thermodynamic pump system "Sofretes".

temperature difference (high Carnot efficiency). It uses a double-acting reciprocating expander and a double-acting water pump.

Solar Pump Corporation (USA) are developing a 100 W prototype of a flat-plate collector and an ORC system.

Jyoti Limited (India), are developing a 500 W reciprocating steam engine, powered by a glass strip reflector parabolic trough solar collector.

Foster Miller Incorporated (USA), are developing a direct acting Rankine cycle engine pump. An organic fluid, "refrigerant 11" or similar, is evaporated in a solar collector and admitted into the expansion cylinder where it acts on the piston.

Cranfield Institute of Technology have developed a multivane expander as part of an ORC which uses evacuated glass tube solar collectors. It might be more appropriate for larger systems.

More simple solar pump designs are made by:

J Vanek (USA). A parabolic trough collector attached to a steam pump (based on the Savery pump designed in 1968).

The Technical University of Denmark uses flat-plate collectors with planar reflectors to power a simple steam pump. Field trials are being made in Tanzania.

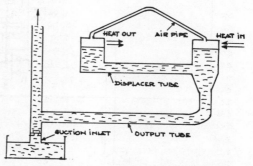

Simple solar thermodynamic pump system, "The Fluidyne Pump".

Birla Institute of Technology and Science (India), use flat-plate collectors to power displacement tank prototype pumps. Trials so far have been disappointing.

Some work is also being done on the "Fluidyne pump" by Metal Box Ltd in Calcutta. This pump uses a very simple stirling cycle with a liquid piston and no moving parts but it is very inefficient. Ongoing work is trying to improve efficiency.

Solar-thermodynamical Pump Development

Results of mathematical modelling indicate that higher temperature solar collectors result in lower system costs for a given output. Single axis tracking solar collectors, such as parabolic trough or linear Fresnel lens with Rankine cycle engines, are shown to have potentially lower system cost than flat-plate systems, and are, in some cases, nearly as well developed. They are also less influenced by environmental parameters such as heat sink temperature and air temperature. One of the major barriers to the use of parabolic trough collectors has been the requirement for a simple low-cost tracking system. Thermohydraulic tracking systems which require no electrical power appear to be an advance towards fulfilling this requirement. Parabolic trough solar collectors are used by Wrede Ky (Finland) and Vanek (USA). The use of evacuated compound parabolic collectors appears to be a potentially cost-effective approach for manufacturing a collector array, as no tracking is required, although vulnerability to breakage and poor potential for local manufacture are disadvantages. These devices are now commercially available. Evacuated tubular solar collectors are used by Cranfield (UK).

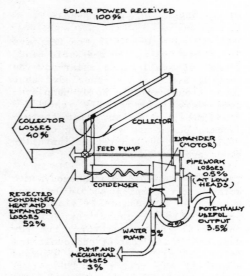

Flow diagram for thermodynamic solar pump.

The least-cost approach appears to be point focussing by 2-axis tracking solar collectors combined with Stirling or high pressure Rankine engines. Except for manually tracked low-concentration collectors, these systems are, however, furthest from commercial development.

The strong potential for local manufacture of some of the simplest thermodynamic systems is an attractive advantage, but in the absence of comprehensive performance data it is not yet clear whether locally-produced simple systems will be more cost-effective than conventional thermodynamic or photovoltaic systems. There is a risk that simple designs may result in low efficiency and hence in large collector areas, high overall system cost and poor reliability and life expectancy.

The capital costs of thermodynamic systems, if produced in quantity are believed to be competitive with current projected capital costs for

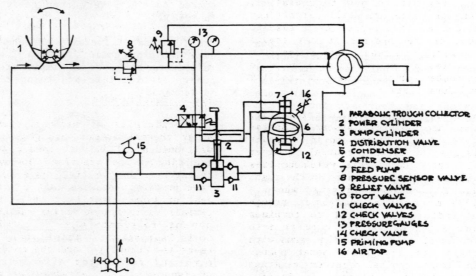

Principle of Wrede Ky solar pump system.

photovoltaic systems. However, no proven thermal system is commercially available and the investment required for sufficient development is likely to be large.

It is likely that solar thermodynamic pumping will be more cost-effective in large scale systems than in small-scale systems. As the size of the system is increased, the fixed cost proportion of any installation becomes less significant. In addition, development costs for a small system may well be similar to those for a larger system. A larger system may also justify the cost of employing a local person to maintain and operate the system. Solar ponds may be a cost-effective option in large-scale systems, but do not appear to be suitable for small-scale water pumping: at present their costs are unattractive and they require a large area of land. Reliability of thermodynamic pumping systems remains a cause for concern. Satisfactory field trials are required to build up confidence in the thermodynamic approach.

No reliable solar thermal pump is currently commercially available and progress has been slow. Current activity is still at the prototype stage, and performance and cost data on solar thermodynamic pumps are limited.

Dissemination

Advanced solar thermal systems are far from being widely disseminated throughout developed countries, let alone the developing nations. Prime reasons for the limited spread of this technology hinge on the high cost of the available systems and reticence by industry and commerce to change from tried and trusted "old faithful" heat and power generating systems to new technologies, many of which are still largely regarded as prototypes. As a result, most of the systems in use around the world have been funded by research agencies as experimental models to be tested for efficiency, reliability and durability.

However, in the USA in particular, there are commercial systems in operation. At the larger end of the scale is the 10 Megawatt central receiver at Barstow, California, which supplies electricity to consumers in the area, and at Coolidge, Arizona, where a 2000 m^2 area of parabolic trough collectors is linked to turbines generating 22 000 kWh a month into the local grid, supplying farms with power for irrigation water pumps. In Europe, central receivers generate electricity at Adrano in Sicily, at Thermis in France and near Almeria

in Spain. At the Spanish site, a 500 kW rated array of parabolic concentrators has also been installed to compare the power generating efficiencies of a central receiver with trough concentrators. A similar concept has been built at Nio in Japan where a 1 MW central receiver and a 1 MW array of parabolic concentrators have been installed and are being compared for operating efficiency.

In the medium range of heat and power generation, three systems installed in the USA stand out as examples of interest. One is installed at Macon, Georgia, the second is at Greensboro, North Carolina, and the third is operating at Louisville, Georgia. All installed by the Luz International Company, the systems use parabolic concentrators to supply process heat to local agro-industry plants. The power users buy the energy, rather than the system, from a consortium of investors set up by Luz as a solar power utility.

Luz has also installed similar systems at Sha'ar Hanegev in Israel where a 500 m^2 array of concentrators supplies heat to a potato processing plant and a 2500 m^2 area of concentrators generates process heat for a textile mill. Middle Eastern countries are showing interest in this area of technology. Kuwait, for example, has installed 100 kW of parabolic trough collectors at Sulaibkhat which generate 80 kWh of electricity daily.

Solar systems installed in the developed world can be partly experimental since technical back-up facilities exist to service, modify and repair the systems. In the developing world, on the other hand, such back-up facilities are not so commonly in evidence. As a result, it would be unwise for a consumer in remote areas (where solar systems are presently most cost-effective) to install an advanced system. This is particularly true of systems incorporating tracking mechanisms which require regular service.

Future Developments

The principles of high-efficiency solar thermal systems are generally well-understood and, in some cases, are being put into practice. However, researchers in the field have found from prototype systems that a great deal of improvement in terms of materials used, system design and component fabrication is possible. Such improvements should help to raise the durability and the conversion and reflection efficiencies of advanced systems, and should eventually lead to price reductions.

In addition, a great deal of thought is currently going into system design in order to reduce the number of moving parts in advanced solar systems and also to improve working fluid heat transfer efficiencies. It is well-known that many demonstrations of early advanced solar systems failed due to the lack of mechanical maintenance required to keep the system operational, especially in remote, hostile environments.

In almost every case with advanced solar thermal systems, the key development issue centres on improving efficiency and system reliability and, at the same time, reducing cost in order that developed systems can begin to compete with conventional power generating and heating installations in both the developed and developing worlds.

Conclusions

Advanced solar thermal systems have the theoretical potential to displace significant amounts of high-cost, imported oil and other fuels in developing countries, especially in the industrial and commercial sectors.

Such systems, either already commercially-available or under development, are capable of generating either heat or electricity or both, and can do so in locations remote from central areas where industry has tended to concentrate in the past, due to the availability of fuel supplies. The use of solar systems could help alleviate pressure on central locations and could bring a manufacturing base to areas which remain under-developed for want of local industry. However, systems suitable for this sector incorporate highly sophisticated components which are expensive, require regular maintenance and are still largely in the early days of development. As a result, only a handful of these systems have been installed in the developing world and very few are planned. This leads to a lack of mass- production (and hence no decline in system prices), a lack of demonstration through which to educate potential users of the systems' capabilities, and a lack of training in the maintenance and possible future fabrication of these systems.

So, in spite of the fact that there is a large and growing demand for industrial process heat in developing countries, and that there are a large number of current solar technologies that have the technical capability to meet a part of that demand, development of the technologies has to proceed further in the developed countries before they can be considered for widespread use in the developing world.

Bibliography

Costello, D. & Lawrence, K. (1983) Prospects for Solar Industrial Process Heat Systems in Developing Countries. 7730 Sangre de Cristo Road, Littleton Colorado 80127, U.S.A.

Derrick, A. & Kenna, J.P. & Gillett, W.B. Small-scale Solar-thermodynamic Water Pumping Systems: a review of development. Paper in: Proceedings ISES Solar World Congress Perth, Pergamon Press (1984). Pergamon Press Ltd., Headington Hill Hall, Oxford OX3 OBW, U.K.

Dunstan, R.J. (1983) Prospects for Solar Electricity Generation in Australia. An Economic Overview. Solar Energy Research Institute of Western Australia, G.P.O. Box R1283, Perth, Western Australia.

Friedrich, F.J. (1983) Thermal Applications of Solar Energy in Industry. Paper in: Proceedings ISES Solar World Congress Perth, Pergamon Press (1984). Pergamon Press Ltd., Headington Hill Hall, Oxford OX3 OBW, U.K.

Halcrow, W. Partners & Intermediate Technology Power Ltd. (1984) Small-Scale Solar-Powered Pumping Systems: The Technology, its Economics and Advancement. UNDP Project GLO/80/003. Executed by the World Bank. UNDP. 1, United Nations Plaza, New York, N.Y. 100 17, U.S.A.

ISES (1981) Solar Thermal Power. Special issue of Sun World Vol. 5 No. 3 & Vol. 5 No. 4. Magazine of the International Solar Energy Society (ISES). Pergamon Press Ltd., Headington Hill Hall, Oxford, OX3 OBW, U.K.

Kutscher, C.F. et al. (1981) Design Approaches for Solar Industrial Process Heat Systems. Solar Energy Research Institute (SERI), 1617 Cole Boulevard, Golden, Colorado 80401. U.S.A.

Meade, W.R. (1982) The Use of Solar Ponds for Energy Production in Developing Countries. Senior Theses, Brown University, Providence, R.I. 02912, U.S.A.

Ormat Turbines Ltd. (1981) Brochure Material. New Industrial Area, Yavne, Israel.

Tabor, H (1979) Solar Ponds. The Scientific Research Foundation, Jerusalem, Israel.

WHO (1981) Solar Refrigerators for Vaccine Storage and Ice Making. EPI/CCIS/81.5. World Health Organization (WHO), 1211 Geneva 27, Switzerland.

18 SOLAR WATER DISTILLATION

Introduction

Although water is a basic necessity for man´s survival, the United Nations estimates that over 2000 million people in the world still do not have regular access to safe and sufficient supplies of drinking water.

In an effort to increase the availability of fresh water in remote, arid regions a wide variety of technologies have been developed, among them solar distillation systems which have been proved to work effectively in distilling saline or polluted water without requiring power generated by using commercial fuels. However, even though solar distillation systems have been proved to work effectively, they are only economically acceptable under a limited range of conditions. Competing technologies such as flash distillation, vapour compression, electrodialysis and reverse osmosis systems, all tied to commercial fuel power supplies, have been shown to be more economic than solar stills for distilling large quantities of water, despite rising fuel prices.

As a general rule, solar distillation systems can be considered only for areas where:

- saline or polluted water reserves are available but fresh water resources are in short supply

- solar radiation levels are high

- cheap and reliable commercial energy supplies do not exist.

Solar distillation is, at present, only economically attractive in areas where local fresh water requirements do not exceed 200 000 litres/day and where the cost of importing water by vehicle or long-distance pipeline is exceptionally high.

Distilled water production from a solar still is calculated to cost in the order of US$ 0.8/m^3. In view of this price using currently available systems, it is clear that solar-distilled water can only be a cost-effective and practicable solution for drinking water supplies and not for agricultural purposes.

The high capital investment required for the installation of solar distillation systems is generally regarded as the primary reason why such systems are not in more wide-scale use today. The user should also be aware of possible deleterious effects of the still on the environment especially concerning the disposal of brine following salt water distillation.

Solar Distillation Technologies

Simple solar distillation technologies are well-established. In Las Salinas, northern Chile a large basin-type solar still was built in 1872 and worked effectively for 40 years. In Australia and Greece stills built in the 1960s continue to operate satisfactorily and in the developing world, India, Pakistan and China all have a number of solar stills working effectively.

Basin-type solar water still.

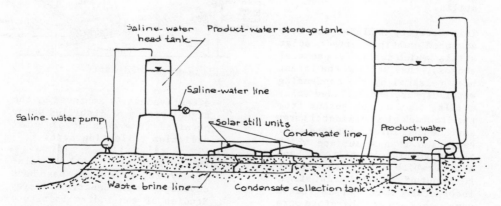

Diagram of basin-type solar water still.

Solar distillation systems can generally be divided into four categories:

1   Simple, single basin stills. These are shallow depressions in the ground (in which the bed is blackened to enhance solar absorption) with a transparent cover placed over the depression. Using the greenhouse effect, salt water (or polluted water) is evaporated by incoming radiation and the resulting condensation forming on the inner surface of the cover is collected. (See Figure).

The single basin solar still, of which many different designs exist, is the cheapest and most proven of all the solar distillation systems currently available. To maximize the still's efficiency, the cover should be sloped at an angle great enough to allow the easy passage of water towards the collector, but not be so steep as to reflect useful solar radiation. If possible, the still should be made air-tight, the basin water-tight and a thin layer of insulating material used to prevent contact between the ground and the water. Shallower water depths down to 2.5 cm speed up condensed water production as does a black dye added to the water. In addition the daily yield of fresh water will be increased significantly if the water is pre-heated.

Single basin solar stills cost US\$ 15-20/m^2 and have a life expectancy of about 20 years if properly designed, constructed and maintained. Typically, such a still has an efficiency of 30% corresponding to a daily distillate of two to five 1/m^2 depending on solar radiation levels and design.

2   Advanced solar stills are designed to obtain higher yields of distilled water per m^2 of collecting area than simple, single basin stills.

This category includes the more advanced multiple effect solar stills (which distill via several stages), that utilize the latent heat released by the condensing water vapour, and inclined solar stills, that do not suffer from the drawback of a horizontal water surface that intercepts less solar radiation than a surface which is appropriately tilted. Such stills, which have yet to progress fully out of the development stage, incorporate more sophisticated components and are therefore more expensive than simpler still types. Nevertheless, their cost-

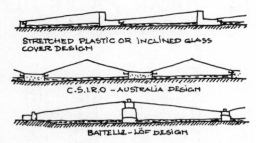

Different basin still designs.

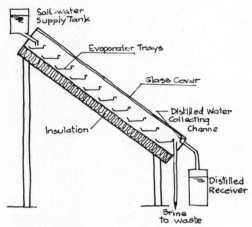

Inclined solar still.

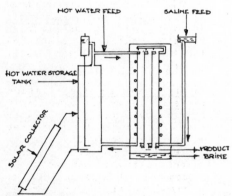

Diffusion solar still.

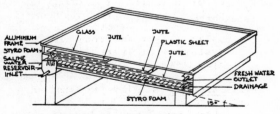

Multiple-Wick-solar-still.

effectiveness is improved by the higher yields of distilled water obtained from each m^2 of solar radiation collecting area.

An inclined wick still of relatively robust design has been developed at the Centre for Energy Studies of the Indian Institute of Technology. (See Figure). This multiple wick solar still consists basically of a unit of two aluminium frames enclosing a sheet of glass which covers several interspersed layers of jute cloth and plastic sheets. The overlapping layers of cloth, dyed black to absorb radiation and separated by sheets of black plastic, have their upper ends dipped in a reservoir of saline water. The capillary action of the cloth sucks up the saline water, which then flows down the length of the cloth which is exposed to the sun's rays. These exposed ends provide a continually wetted surface from which the water evaporates. A constant water level is maintained in the reservoir and a trough along the lower wall collects the distillate. A copper tube provides an outlet for excess saline water.

This solar still is light and easily portable. The daily yield is significantly higher than for basin type stills and it can be built for about half the cost of basin stills having the same area. On a typical cold, sunny day in Delhi, India, the daily distillate output was 2.5 litres/m^2 corresponding to an efficiency of 34%. Other advantages are that it can be orientated to any angle in order to receive maximum solar radiation and salt forming on the blackened cloth can easily be brushed off.

3   <u>Solar-assisted distillation systems</u>: These are essentially hybrid stills, using the combination of commercial fuel and solar energy to power the distillation process. Such systems are cost effective and are frequently able to compete with larger-scale stills powered solely by commercial fuels. Continuing development of these stills, which are largely in the research stage, is expected to produce systems capable of filling the gap between the economic small production capacity solar stills and conventional systems with large production capacities.

4   <u>Emergency stills</u>: One kind of emergency still is made from an air-inflated plastic bag for use in emergency life rafts. As many as 200 000 of these were used by the US Navy during the second World War.

The "earth water" still is made by covering holes dug in the ground with transparent plastic covers and placing a beaker inside the hole to collect condensed moisture. This may be a useful emergency source of fresh water in arid regions.

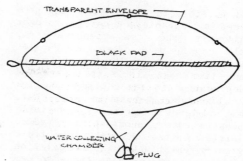

Life raft type solar still.

Economics of Solar Stills

Costs of the different types of solar stills vary widely depending on the level of sophistication of design and construction, with estimates ranging from US$ 15-50/m^2 of solar radiation collecting area. However, higher cost systems are usually found to require lower maintenance and operation costs. Average fresh water production from solar stills is in the order of 2-4 litres/m^2 and day although highly sophisticated, highly capital-intensive systems can be made to produce much more, up to 100 litres/m^2/day.

The cost of distilled water production is a factor of the productivity of the still in relation to its capital cost, and amortization and interest rates on capital borrowed for the still's construction. By using these parameters, it has been calculated that, on average, solar stills built in developing countries are capable of producing water at a cost of US$ 1-2/m^3 - a cost which is far from being competitive with natural fresh water supplies at about US$ 0.01-0.05/m^3 or with distilled water from large plants powered by commercial fuels producing water at an average price of US$ 0.25/m^3. However, solar stills are competitive with commercial fuel stills in regions where fresh water demands of local communities and industry do not exceed 100 000-200 000 litres/day. Stills powered by commercial fuels are heavily dependent on economies of scale which result in higher-priced distilled water from small-capacity plants.

Cost estimates for solar stills can be calculated from experience. In Somalia a UNIDO/UNICEF solar distillation plant with a 2000 m^2 evaporating surface costs in the region of US$ 15/m^2 (1967) breaking down as US$ 23 000 for materials and US$ 4600 for labour. Production (including collected rainfall) is 5000-6000 litres/day of fresh water.

Solar stills will inevitably become more attractive as their costs decline. As a result, the use of as many locally-available materials as possible in the construction process is desirable, and careful thought given to the most expensive item, the cover, will often produce important savings. The cover is the one item most liable to damage and degradation so the designer and installer should bear in mind that, although most currently-available plastic covers substantially reduce initial costs, the material offers greater resistivity to solar radiation than glass and it also tends to deteriorate.

Dissemination Factors

In contrast with many other solar energy technologies, solar stills are already used throughout the world and data collected from their operating performance have been used in determining construction and water production costs as well as in proposing improved and more advanced system designs.

Important factors in assessing the appropriateness of a solar still to a particular site include estimated costs of building and operating the still at a specific location, climatic conditions affecting the operation and the actual demand for fresh water in the region.

In general it can be assumed that solar stills can be cost-effective in areas receiving less than 250 mm of rain a year, where solar radiation is plentiful and where reliable supplies of commercial energy do not exist, and where demand is limited.

From the evidence available it is clear that even though the technologies of solar stills are well-known and a number of them are in operation around the world, the cost of water from such devices precludes their use for anything other than producing drinking water for humans and animals.

Cost of the proposed system is usually regarded as the single biggest obstacle to the widespread dissemination of solar stills since they have been shown to be uncompetitive with most conventional distillation processes except for small-scale water production. This situation is unlikely to change in the foreseeable future.

Finally, in justifying a decision to install a solar still, one must take into account the actual need for such a system. Expenditure on a solar still in a remote region of the developing world where population densities are extremely low is difficult to justify if areas with higher population densities are competing for scarce financial resources.

Bibliography

Headley, O.S.T.C. & Morris, J.B. (1980) Solar Stills for Production of Distilled Water
Article in Ambient Energy, Vol. 1 No. 4 October 1980.
Ambient Press Ltd., Hornby, Lancaster LA2 8LB. U.K.

Malik, M.A.S. et al. (1982) Solar Distillation.
Pergamon Press, Ltd., Headington Hill Hall, Oxford, OX3 OBH, U.K.

Seufert, C. (1978) Survey of Solar Distillation/Desalination Devices for Small Quantities.
German Appropriate Technology Exchange (GATE) GTZ, Dag Hammarskjöld-Weg 1, D-6236 Eschborn 1, Federal Republic of Germany.

United Nations (1970) Solar Distillation - as a Means of Meeting Small-scale Water Demands.
Sales No. E.70.II.B.1., United Nations Publications, New York, U.S.A.

19 SOLAR DRIERS

Introduction

Developing countries suffer heavy losses of food in the post-harvest period. Direct grain losses are mainly caused by rodents, birds, spillage and contamination, and are often as high as 10-15% of the crop.

This resulting high level of crop wastage is of critical importance in the economies of countries largely dependent on agricultural produce, and such large food losses further aggravate economic development problems already being experienced by most developing countries.

Most crops require special treatment to prevent rapid decomposition and the growth of fungi. Drying of crops after harvesting is an especially important process in the preservation of agricultural produce. Yet it is at this stage where much of the crop deterioration takes place.

At present most crops produced in the more remote areas of developing countries are dried using the open-air, sun-drying method.

Traditionally, sun-drying is carried out by spreading produce out on the ground and exposing it to the sun during the day and covering it at night to protect it from rain, dust and other damaging elements.

Open-air drying is appropriate for some cases but brings major disadvantages in others. For example, product quality can only be optimized when drying temperatures are controlled. Also the produce cannot easily be protected from the elements and scavenging animals, especially when it needs to be spread over large areas in order to dry. Because of the lack of control over the drying process using the traditional method, the quality of the produce suffers and part of it is lost through the effects of fungal proliferation, infestation by parasites, reduced germination (caused by too high temperatures), and rodent infestation. Moreover, continued wetting and drying in variable weather conditions may cause cracking in kernels, which reduces the quality of the crop.

The Drying Process

The drying of organic produce such as vegetables, fruits, coffee, grain, fish, cocoa, tobacco and timber, has long been a regular and important feature of village life in rural areas of developing countries. Most of the drying process is carried out using open-air drying techniques and the drying of foodstuffs and other products is one of the oldest uses to which solar energy has been put.

The open air sun-drying process is, therefore, a very hit-and-miss affair. Since there is little control over drying rates, crops can be either over-dried or under-dried, leading to the development of fungi and bacteria in under-dried crops and grain case hardening and splitting (spilling the contents) in over-dried crops.

In order to adequately dry a material, it is first necessary to know the initial moisture content of the material to be dried and the desired moisture content of the final product. Drying rates are controlled by the rate at which heat is applied to the product, the rate at which the product's internal moisture is released from its surface and the rate at which moist air is removed from the area surrounding the product. The drying rate is controlled by varying the heated air's temperature and humidity and, since insolation levels can vary widely, solar driers must be carefully designed if steady drying conditions are to be achieved.

Even warm, dry air can only absorb so much moisture at one time and, unless the air is frequently replaced, it will become saturated and the drying process will slow considerably or stop altogether. As a result, best drying is achieved when the air mass moves constantly and slowly over the product being subjected to the drying process.

In general, solar driers dry crops more slowly than conventional oil-fired hot air systems since the drying process is carried out using lower temperatures and high air flow rates whereas conventional systems use high temperatures and low air flow rates. However, rapid drying rates are often inappropriate for crops such as rice which is dried to best effect when subjected to the slower drying rate usually obtained with solar driers.

To maintain optimum drying conditions in a drier, some method of measuring the temperatures of the ambient air, the air at the inlet, the air at its outlet and the temperatures of the drying material, should be included. In the same way, a means of measuring the moisture content of the material, both before and

219

after the drying process has taken place, needs to be incorporated, together with a method of measuring the relative humidities of the air entering and leaving the drying chamber. Controlled drying in specially-designed driers produces much better quality products with greatly reduced losses.

After harvesting, grains which usually have a moisture content of 20-30% in hot, humid climates, should have this reduced to 13% within two to three days if the product is to be prevented from deteriorating. This "safe drying time" increases moderately in lower temperature climates and considerably in areas of low relative humidity. Fungal development is greatest at 93-99% humidity but drops to being virtually absent when relative humidity falls below 65%. Optimum drying temperatures are between 29° and 40°C and, if the grain is to be used as seed, drying temperatures should not exceed 43°C. The drying process virtually ceases at temperatures below 0°C. Grain losses due to fungal development and attacks by parasites during storage, usually regarded as being between 10% and 30% under normal conditions, can be reduced to about 10% if the grain is dried quickly.

Conventional Hot-air Driers

In most current commercial crop driers heat for the drying process is generated by the combustion of oil, gas or wood or by electrical means (see Figure). Such drying methods are inappropriate to small-scale farmers in the developing world as fuels are usually only available to farmers in limited quantity and at prohibitively high cost. Likewise, conventional heaters are often not available or are imported and sold at high prices making them inappropriate for remote areas. Since exhaust gases from the combustion of most fuels are dirty, a heat ex-

changer is often necessary in order to avoid contamination of the produce, again adding to total cost.

Solar Drier Technologies

Although many solar drier models are now available, it is still recognized that further research is needed to optimize designs, to increase the lifetime of materials used in their construction, and to match individual driers to different crops and climates.

The wide variety of solar driers developed over the years can be generally divided into:

1 ## Natural open-air driers

These are very simple constructions. The material to be dried is placed outdoors on a tray, a rack or the floor, and is dried by ambient sunshine and wind. Some of these driers can have a fixed or movable roof to protect the crop against rain.

2 ## Direct solar driers

In these driers the material to be dried is placed in a transparent enclosure of glass or plastic. The sun heats the material to be dried and the enclosure causes a heat build up due to the "green house effect". Sometimes ventilation through the enclosure is introduced via vents or chimneys.

3 ## Indirect solar driers

In these driers the sun does not act directly on the material to be dried, thus making them useful in the preparation of those crops whose vitamin content can be severely diminished by the action of sunlight (e.g. vitamin A in carrots). Air is heated in a

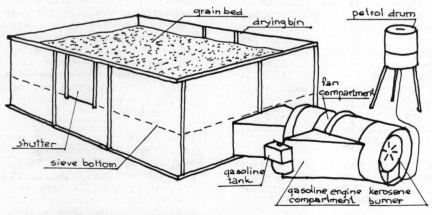

Conventional oil-fuelled batch drier.

special solar collector and is then ducted through to the drying chamber where the material to be dried is situated, and released. A chimney is usually introduced in order to improve air flow.

4 Mixed mode driers

In these driers the combined action of solar radiation incident on the material to be dried and the air preheated in a solar collector provide the heat required for the drying operation.

5 Hybrid systems

These are solar driers in which another energy source such as a fossil or biomass fuel or electricity is used to supplement solar energy in the drying process, e.g. for additional air heating during cloudy period or for increased ventilation using fans.

Those five basic types of driers will now be discussed in more detail below, using examples of typical designs operating in various countries.

Natural Driers (1)

Open-air coffee drying is employed worldwide mainly on small-scale holdings since it is generally regarded as the cheapest method of drying coffee beans. During the day the beans are exposed to the sun whereas at night, and during rainy periods, the trays containing the beans are covered over. In this method, no insulation is used and all construction materials are locally-available.

Normally, freshly harvested beans have a moisture content of 50-52%, plus some superficial water added during washing. The desired figure for this is in the region of 11%. In open-air drying about 24 kg of beans are spread over each m^2 area. When dried to a moisture content of 11% this weight is reduced to about 12,5 kg. Trays used in the drying process

usually measure 1.2 m long by 0.5 m wide and 15 cm deep. Coffee is spread in the trays to a depth of about 3 cm.

Crib driers are often used for drying cassava. One drying system in use in Colombia consists of a chamber made of wood, angle iron and wire mesh capable of containing 0.5 m^3

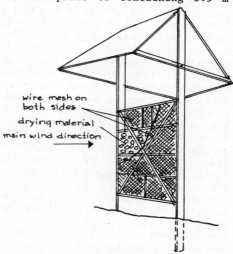

Crib drier for cassava.

(or 300 kg) of cassava cubes. The drier is loaded from the top and unloaded from a wooden door at the bottom. A wooden roof is used to protect the crop from rain. The unit's price is calculated at about US$ 60, including labour costs, and the drying process takes about three days. However, the drier should only be used when the relative humidity is below 60% and the wind speeds exceed 4 m/s. Delicate materials like fruit should not be dried in this type of unit since the pressures at the bottom of the chamber are too high for good quality dried products.

Direct-Solar-Driers (2)

A solar tent drier consists of a rack on which the product to be dried is placed, and a transparent plastic tent which is placed over the rack. This simple drier does not

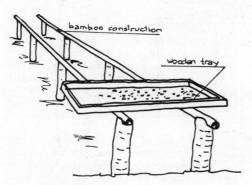

Open air tray drier for coffee.

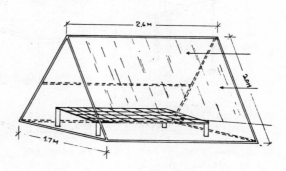

Solar tent drier.

decrease drying time significantly, but protects the crop from rain, birds and rodents. In Gambia, plastic tent driers are used to dry fish. With this drier design the main challenge is to make them both cheap and durable.

Another interesting concept is the see-saw drier developed by the Institute for Tropical Agricultural Products (ITIPAT) in the Ivory Coast to dry coffee and cocoa beans. The see-saw drier consists of a rectangular wooden frame divided lengthwise into parallel channels of equal width, and crosswise by means of retaining bars. The bottom of the drier frame is made of black painted bamboo matting and receives the material to be dried. The frame is covered by a film of transparent polyvinyl chloride (PVC). This simple drier is suitable for small-scale drying operations and is easy to operate. During operation the drying tray is moved along an east-west plane. This permits the material to be dried to face the sun more directly during both the morning and the afternoon. This increases the effectiveness of the unit and leads to a more evenly dried product. By this means, under good conditions the drying time can be reduced by two days, but when the climate conditions are unfavourable for drying this difference is less significant. The water content of cocoa at the end of the drying period is within 7-8% which is impossible to reach with ordinary open-air drying methods. It is absolutely essential to stir the cocoa at least once a day. The cost of a drier with the overall dimensions of 5 m by 1.5 by 8 cm is about US$ 12.5.

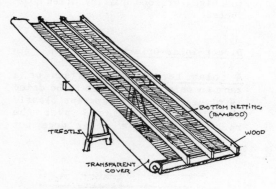

See saw drier.

The Swedish University of Agricultural Sciences in Lund, Sweden, has developed a solar chamber drier for grain in Nicaragua which incorporates a drying chamber on poles raising the system 0.5 m off the ground so as to prevent rodent attacks. Air is allowed to permeate the grain from below through a steel mesh mounted

on a rectangular wooden frame. To increase the chamber's internal temperature, and to prevent rain from entering the chamber, the structure is covered with a transparent PVC foil. The drier also incorporates a chimney positioned so as to enhance the air pressure difference within the chamber, and a black PVC sheet facing the sun which acts as a solar energy absorber heating air which rises up the solar chimney. This set up achieves rapid drying speeds, reducing the grain moisture to 17-18% which can be further reduced to about 10% by frequently stirring the grain in the drying chamber. Experiments have shown that it is possible to reduce grain moisture contents to 10% using this system.

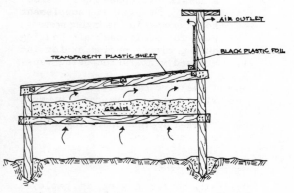

Solar grain drier.

The grain layer in the drying chamber should be 5-10 cm thick and the chimney at least 1.5 m high to promote optimum ventilation. The chamber floor should be at least 10% air-permeable and the chamber and chimney should be made to be as air-tight as possible. The cost of material for the drier is calculated at about US$ 35/m². A polyester film costing about US$ 17/m², was found to be the most suitable material for the transparent covering. The steel mesh costs in the region of US$ 8/m². However, by using locally-available materials in the construction, overall costs can be reduced to US$ 20-25/m².

Indirect Solar Driers (3)

The indirect solar drier concept is flexible in that it can be incorporated into any crop storage building e.g. a bin or a barn. A wall or roof can be converted into a solar collector and can heat air to be ducted through the product to be dried.

Alternatively, it is possible to connect a separate solar collector to the crop storage building. This solar collector can be permanent or, like the inflatable "air mattress" solar collector, only be installed during the time of drying.

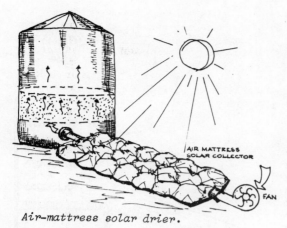

Air-mattress solar drier.

One <u>solar drying barn</u> in South Dakota, USA, is aligned longitudinally on an east-west axis, with the solar collector on the south facing side of the roof. The roof surface is painted black to absorb the solar energy. A transparent plastic material is supported 8 cm above the roof on a frame. The air drawn by a fan, enters the top of the roof and moves through the collector roof down the south wall into the outside air duct. From here, the fan pushes the warmed air into the inside air duct and through the grain by way of a perforated floor.

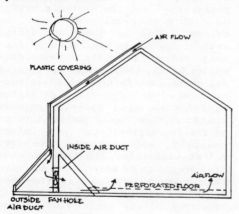

Solar drying barn.

Mixed-mode Driers (4)

<u>The solar rice drier:</u> If rice is multi-cropped, the second or third crop has to be harvested during the wet season when the grain has a high moisture content. The harvested paddy cannot be stored with a moisture content of over 14% if quality deterioration and losses are to be minimized, and the excess moisture must be removed by some drying process.

Even during the other seasons, the initial moisture content of rice may be between 16-25% at harvest time.

Patio drying, mat drying and field drying in the sun have been practised for centuries. These methods can be used if the original moisture content of the paddy is not too high, if there is enough sunshine, if it does not rain and if there is sufficient labour to manage the drying process. Successful traditional drying of rice harvested in the wet season, however, seems to be difficult. Losses with traditional sun drying due to rodents, birds, insects, micro-organisms, spillage and contamination are about 17%. To reduce and even eliminate such considerable losses, the Asian Institute of Technology in Bangkok, Thailand, has produced a solar rice drier. It has been designed so that it can be made by the farmer himself with locally-obtainable materials at a low cost. The paddy can be dried in one or two days and is protected from rain, insects and birds. Moreover, its milling quality and germination is superior to normally dried rice.

The solar rice drier comprises of three important parts. The solar collector is the heart of the drier. It consists of a collector, a matt black substance (e.g. burnt rice husk) spread on the ground with a clear, low-cost plastic cover. The collector covers an area, (approxi-

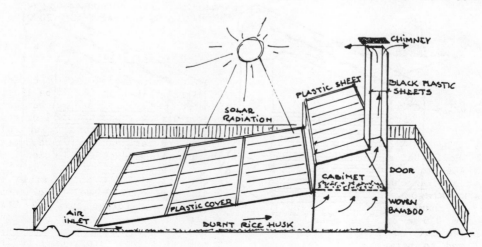

Solar rice drier.

mately 30 m² which is about three times that of the drying cabinet's floor area. The clear cover slopes up from a 10 cm air inlet towards the drying cabinet in order to direct the heated air through the grain.

The drying cabinet is positioned at about waist height. Its floor must be made of a perforated material (woven bamboo) which allows the air to pass through easily, but prevents the grain from falling through. The sides of the cabinet should be strong enough to resist the grain pressure and a door allows the loading and unloading of the crop at the back. To hold one ton of rice, the drying cabinet area needs to be 11.5 m² with a rice bed depth of 15 cm.

The ventilation chimney consists of a light, strong frame covered with matt black plastic and a cover above the chimney keeps out the rain. The chimney material must have high durability and be weather resistant. The base of the chimney is at the highest part of the roof. Chimney height is 2 m with a cross sectional area of 0.5 m². For a one ton capacity drier two chimneys are sufficient. Initial moisture content of the produce is around 22% and final moisture content is about 11%. The drying period is two days and the temperature of warm air should not exceed 40°C to prevent the rice cracking. Stirring is necessary at least once or twice during the day time to obtain uniform drying.

Fruit and vegetables can also be dried in chamber-type solar driers. These work in both direct and indirect modes and consist of a solar collector to heat the air, the chamber itself containing racks on which the produce is dried, and a chimney to induce natural ventilation by convection. In this system, air, pre-heated by the collector located at the bottom of the drier,

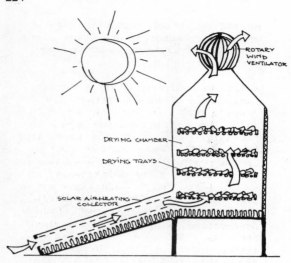

Chamber type solar vegetable drier.

is admitted at the base of the chamber and rises through the crop. The moisture-laden air is released through vents in the rear wall or through the chimney. The chamber itself can be made of a black material to prevent solar radiation reaching the product directly, or of transparent material to enhance the direct drying process.

Solar timber driers: sawn-wood has a high quality only when it is properly dried and the open-air drying of timber is usually slow and unreliable. Although speed and drying reliability can be improved considerably using commercially-available drying kilns, these are sophisticated and expensive. Solar driers are normally able to halve the drying time experienced with open-air methods and can improve timber quality through controlled drying techniques. Unfortunately, solar driers have very modest capacities but even so they have proved to be more cost-effective than conventional fossil-fuelled driers in many cases. In developing countries, their small size is not such a big disadvantage. Few timber

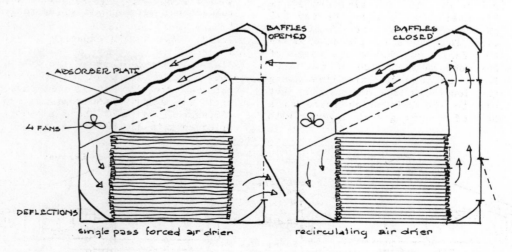

Solar timber drying kiln.

ducers in these countries require drying facilities of any great size since local requirements for seasoned timber are small.

Solar timber drying kilns usually consist of a wooden frame with a sloping roof facing the sun with an angle of inclination equal to the latitude in which the drier is located. The wall facing the sun is made of black-painted plywood in order to enhance solar radiation absorption, and all other faces consist of a double layer of transparent, usually polyethylene, sheeting. Between the roof and the timber stack there is a blackened metal heat absorber which is about 60 cm below the roof and parallel to it. Fans are placed in the upper rear part of the drier. These blow the air over and under the black absorber sheet, and thus warm the air as much as possible. Baffles force the air flow to circulate through the lumber pile. Vents to control air flow into and out of the drier are installed in the walls. Two modes of ventilation are possible. In the early stages fresh air is passed through the drier, but in the last days of seasoning, air is recirculated inside the drier to increase the temperature. The rate of drying depends, to a large extent, on the control of air circulating in the drier.

Drying of wood from green to 12% moisture content takes 28-32 days. The design makes use of large quantities of fairly low-grade (up to 60°C) heat to permit an adequate drying rate without reducing the quality of the timber.

Solar timber driers are built in different parts of the world with a wide variety of capacities. In Uganda, for example, a drier with a capacity of 19 m3 capacity has been built using 4 mm PVC sheet for the roof and 2 mm sheet for the walls. The kiln contains six 50 cm diameter fans powered by two 1.5 kW electric motors for air circulation. Four vents are placed opposite the fans just below the eaves of the roof on each side of the kiln. Solar collectors extending along the length of the kiln supply hot air to the kiln which rests on a bed of packed gravel covered with

tarmac. This type of drier costs in the region of US$ 4000. Life expectancy is about 15 years except for the plastic sheeting which needs to be replaced every two years.

Hybrid Driers (5)

Hybrid driers combine solar energy absorption systems with heat from combustion - a combination most appropriate to areas of low-to-moderate insolation, where solar power cannot be used continuously or when the scale of operation and precise drying requirements make solar drying usable only as a complementary source of energy. In any hybrid driers, however, dried product quality can often be affected by the gases and smoke emitted from the burning of the combustible substance used to raise heat, and a heat exchanger needs to be incorporated into the drier's design.

Hybrid driers are now coming into wide-scale use in the developing world where conventional energy sources are either scarce or expensive and where solar systems alone have insufficient capacity or heat production capabilities. In Thailand, for example, a fuel wood/solar drier has been developed for curing tobacco. The unit consists of a brick-built curing barn of 3.5 m^3 and a solar heater made from black-painted corrugated metal sheet covered with a sheet of glass. The incorporation of the solar heater has resulted in a fuelwood saving of about 15%. Cost analyses show that, assuming the system has a useful life of seven years, that it can be used 123 days a year and that the interest rate remains at 15%, the solar heater's energy cost is US$ 0.29/kWh compared with US$ 1.07/kWh for wood.

In Brazil a hybrid system using solar energy to complement the burning of butane gas has been developed to dry cocoa beans. The construction, resembling a green-house, has a timber frame covered with 3 mm thick glass. Trays with a capacity of 0.7 m^3 at a 3-4 cm depth of beans are placed on wire mesh platforms. In this system 500 kg of cocoa beans can be dried in six to ten days, with the drier reaching a maximum temperature of 25°C above ambient. Drier cost is US$ 1000.

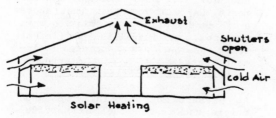

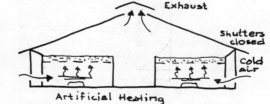

Hybrid glass roof cocoa drier.

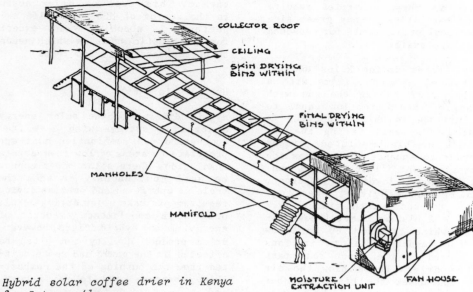

Hybrid solar coffee drier in Kenya for 3 tonnes/day.

In Kenya, the Tropical Development & Research Institute, UK, has developed a hybrid solar drying system capable of maintaining a daily throughput of 3 tons of dried parchment coffee. The system consists of 3 separate stages: initial skin drying, intermediate sun drying on tables (a prerequisite of Kenyan coffee) and final drying. The skin drying and final drying stages take place in a purpose-designed building which incorporates a solar collector within the roof. Since 90% of the co-operative factories are without grid electricity at present, motive power is provided by a diesel-driven engine directly coupled to an axial-flow fan capable of maintaining an air flow of 8 m^3/s. The waste heat from the engine is utilized effectively to heat the air, thereby enabling drying to proceed in all but the worst weather and at night.

The Production and Economics of Solar Driers

Solar driers can be divided into two categories:

1. Cheap, simple, natural draught driers for small farmers

2. Advanced driers with fans and backup systems for cashcrops. The value of cashcrops is so high that the risk of failure because of weather conditions cannot be accepted.

Studies show that a number of simple, direct-mode solar-driers can be constructed economically using local materials.

In calculating the economics of cash crop solar driers in comparison with conventionally-fuelled systems, it is necessary to take into account the drying cost of various products in relation to their market cost. To be economic, it is essential that the cost of drying the product is substantially less than the cost of the product itself.

An assessment of the economics of systems is derived in terms of the cost per unit mass of the product or per unit of derived useful energy. These can be arrived at by using the annual cost accounting method in which the initial investment, lifetime of the system and the annual rate of interest is taken into account. By dividing the annual cost by the annual product yield or by the annual useful energy, one can derive the cost per unit mass or the cost per unit of useful energy. By taking the salvage value as 35%, and the maintenance cost as 15% of the initial investment cost respectively, the calculated cost of drying typical products in solar driers compared with the products's cost can be found. (See Table).

When calculating the cost benefits of solar drying systems it must always be borne in mind that solar energy systems are usually only effective up to a certain stage. In many hybrid systems, for example, it has been found that solar energy can be economically included in the design of a system only for base heat loads and not for peak heating requirements. Calculations show that above a 20% contribution from solar energy, the cost of installing sufficient solar absorbers and associated equipment increases at a cost disproportionate to its benefits.

At present most specially-designed driers use fossil fuel or woodfuel to provide constant temperatures in drying barns or sheds. However,

Approximate cost of Solar Drying

	Product	Drier drying wt. US$/Kg.	Cost of drying per unit dry wt.US$/Kg.
Raw paddy	Indirect mode Bin drier	0.428	0.031
Maize	Indirect mode Chamber type	0.152	.0059
Potatoes	Mixed mode Chamber drier	0.136	.0055
Cassava	Mixed mode Chamber drier	0.374	.056
Cocoa	Direct mode Cabinet drier	2.95	0.011
Coffee	Direct mode Tray drier	3.32	0.0021
Tobacco	Indirect mode drier	3.31	0.167
Timber	Solar kiln	244 per m3	7.3 per m3

For a detailed analysis one has to, however, take into account the cost of insurance, taxes, or government subsidy etc. depending on the local economy.

Source: M.S. Sodha and N.K. Bansal 1984.

with the increase in the cost of fossil fuels and the reduced availability of woodfuels in many areas, such drying techniques are becoming less economically attractive to farmers in rural areas and the use of supplementary solar driers is increasing.

Dissemination Factors

There can be little doubt that solar drying systems could make a sizeable impact on the economies of developing countries but, so far, available systems are not being used to their full potential. There are a number of reasons for this mismatch of potential to use.

In general, farmers of developing countries feel it is more important to produce more food than to reduce wastage. As a result, the use of improved production technologies such as combine harvesters, fertilizers and pumps are emphasized at the expense of solar driers. Compounding the problem is the low price of most basic food products in developing countries which reduces the likely benefits of solar driers in relation to their costs. This is not the case with many cash crops, however.

In addition to these obstacles, there are the problems of the severely restricted financial resources of most farmers in the developing world, a low level of awareness of the exist-

ence and value of solar driers, and the fact that solar driers which are useful to the farmer the year-round have yet to be developed.

In order to disseminate solar drying systems more widely, government-sponsored demonstrations of available systems could motivate the private sector to adopt the technologies as they are proven in the field. Such demonstrations, provided they are successful would enhance public confidence in solar driers.

One of the greatest drawbacks to the widespread dissemination of solar driers in the developing world at present is the lack of experience with such systems under field conditions. Surveys have shown that most solar drier research, development and demonstration programmes now under way are still made under laboratory conditions and are not representative of the scale needed for widespread adoption of the technology. As a result of this shortage of experience, solar driers still have a relatively immature status - there is a lack of proven performance data and only a limited range of economically attractive systems exists. Hence, few drying operations in the developing countries use available solar technologies. In order to overcome these barriers, large-scale demonstrations are needed to gain operating experience and to engender confidence in those ultimately expected to use the systems.

Bibliography

Bamrungwong, S. et al. (1981) A Pre-
liminary Study on Solar-Assisted
Tobacco Curing in Thailand.
Department of Mechanical Engineering,
Chiang Mai University, Thailand.

Boonthumjinda, S., et al. (1982) Solar
Rice Dryer: do it yourself handbook.
Asian Institute of Technology,
P.O. Box 2754, Bangkok, Thailand.

Grainger, W.H. et al., (1981)
Small-Scale Solar Crop Dryers for
Tropical Village Use - Theory and
Practical Experience.
Appropriate Technology Group, Applied
Physics Department, University of
Strathclyde, Glasgow G4 ONG, U.K.

Gustavsson, G. (1982) Solar Assisted
Grain Drying in Hot and Humid Areas.
Swedish University of Agricultural
Sciences, Department of Farm Build-
ings, Box 624, 220 06 Lund, Sweden.

Lawand, T.A. et al. (1975) A Survey
of Solar Agricultural Dryers.
Technical Report T99. Brace Research
Institute, Macdonald Collage of McGill
University, Str. Anne de Bellevue,
Quebec, Canada, H9X 1CD.

Puri, W.M. & Costello D.R. (1982)
Solar Crop Drying.
A study prepared for the World Bank.
Energy Department, The World Bank,
1818 H Street, N.W., Washington,
D.C. 20433, U.S.A.

Sarr, M. (1982) C.E.A.E.R. Solar
Fish Dryer Project.
Solar Energy Research Institute of
Food Technology. B.P. 2765 Hann Dakar,
Senegal.

Sodha, M.S. & Bansal, N.K. (1984)
A Project on Solar Crop Drying.
Indian Institute of Technology, Hans
Khas, New Dehli-110016, India.

Study Group for Development Planning
(1979) Devices for Food Drying -
State of Technology Report on Intermediate Solutions for Rural Applications.
German Appropriate Technology Exchange
(GATE) GTZ., Dag Hammarskjöld-Weg
1, D-6236 Eschborn 1, Federal Republic
of Germany.

Trim, D.S. et al. (1984) A new Parch-
ment Drying System for Co-operative
Factories.
Tropical Development and Research
Institute (TDRI), Culham, Abingdon,
Oxfordshire, OX14 3DA, U.K.

Yacink, G. (1981) Food Drying.
Proceedings of a workshop held at
Edmonton Alberta, 6-9 July 1981.
International Development Research
Centre (IDRC), Box 8500, Ottawa,
Canada, K1G 3H9.

20 SOLAR COOKERS

Introduction

The principle of harnessing solar energy to cook food has been known for over one hundred years yet, since solar cookers were first developed, the technologies used have changed only little.

Solar cooking devices fall into two main categories - solar ovens and direct focussing solar concentrators. Solar ovens are essentially boxes consisting of wooden bases and sides with glazed opening lids. A solar oven's interior is well insulated to prevent heat loss and often a secondary lid is incorporated with a mirrored inner surface which can be angled to reflect solar radiation through the glazed lid and into the box.

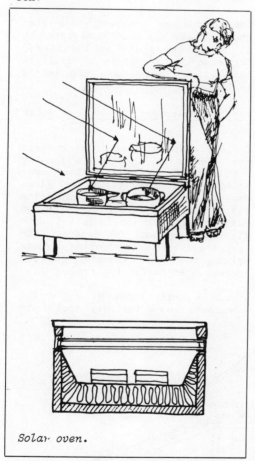

Solar oven.

There are many different designs of concentrating cookers although almost all designs developed are based on the use of a parabolic dish reflector measuring 0.5-1.0 m in diameter which, when directed at the sun, reflects and concentrates solar radiation on to a central platform (raised above the reflector's surface) on which the cooking vessel is placed. Vessels designed for use in solar cookers are generally of light-weight construction with a matt-black finish and a well-fitting lid in order to maximize solar radiation absorption.

Problems

Looked at simplistically, solar cooking devices appear to offer almost unbounded potential in areas of high insolation, fuelwood scacity, commercial fuel scarcity and high cost. This combination of factors is experienced regularly in many of the more remote parts of the developing world which would thus be expected to benefit greatly from the widespread introduction of solar cookers.

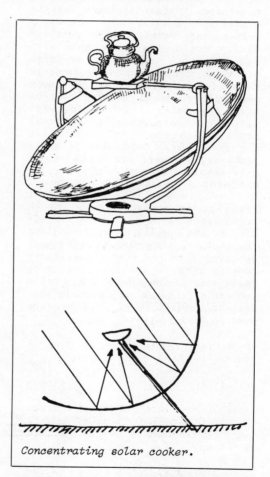

Concentrating solar cooker.

However, solar cookers developed to date face many problems. Experience shows that few people or communities in the developing world use solar cookers regularly and longterm results from field trials are, almost without exception, disappointing.

Not the least of these problems is the change in traditional and social patterns that the use of solar cookers demands. Allied with the high cost of available systems these problems have led some researchers to conclude that, even if solar cookers were distributed free to rural people, these devices would still not be used.

In the main, it is the high cost and unreliability of solar cookers that presents the biggest block to their widespread use in developing countries.

People in remote rural areas require reliable cooking devices in order to prepare sufficient amounts of foods acceptable to their diet at times fitting into their day-to-day working and social patterns. Solar cookers, for the most part, cannot be regarded as reliable in this respect, since they can only be used in periods of moderate-to-high insolation (they are useless in cloudy or rainy periods and at night) which eliminates their usefulness for the traditional cooking times of early morning and evening. Their unreliability and inability to cope with all forms of cooking at any time of day or night has resulted in users reverting to traditional and conventional forms of cooking systems with the solar cooker being used solely to supplement traditional stoves when conditions for their use are right. This also presents problems, however, since few people will consider making a sizeable investment in a system that cannot be used to wholly, or at least largely, to replace dependence on conventional cooking means. In addition, the special vessels required (matt-black, light-weight containers as opposed to the traditional heavy cooking pots) - the need for which again adds to the cost of using solar cookers - cannot always contain sufficient quantities of food for the average family's needs.

On a technical level solar cookers are unable to match the cooking speeds of conventional cookers and require the cook to be in almost permanent attendance to angle it periodically to face the sun. This reduces time available for other chores even though the use of cookers does reduce the time previously allocated for collecting firewood.

Solar cookers are bulky, sensitive to strong winds, difficult to store and thus prone to theft. They require an entirely different planning process; the cooker can only take certain sizes of vessels; stirring is difficult and the glare hurts the eyes. In addition, since solar cookers work at maximum efficiency in bright sunlight conditions, unless a shelter is built for the cook alongside the solar cooker, lengthy periods spent attending to the cooking with one of these devices becomes extremely unpleasant.

One application, however, where solar cookers have met with some interest is in larger-scale applications, for example in schools or military camps. Such units are now tested in India.

The Production and Economics of Solar Cookers

Certain components of solar cookers and ovens are obtainable locally but most of the more refined components such as reflectors and glass have to be imported. The resulting cost increase (through importation duties and transport costs) acts directly against the economic attractiveness of solar cooking devices making them too expensive for the average user in rural areas.

Even the price range of US$ 7 for a simple, low-efficiency cooker to

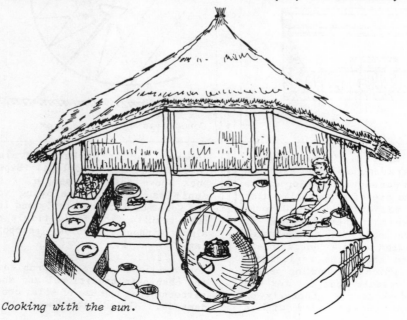

Cooking with the sun.

US$ 18 and possibly US$ 50 for more advanced systems is too high for the average farmer of the developing world - especially when one takes into account that the solar cooker cannot perform all the functions expected of traditional cooking stoves.

Dissemination

The only situation where significant dissemination has taken place is in urban living middle-class families in India. These families have servants who are ordered to use solar cookers when there is an electricity cut (which is quite common in India), or when kerosene is difficult or expensive to get. For example, it is estimated that 5000 solar cookers have been sold in Gujarat in Western India alone, but most of these are in use in urban areas, primarily acting as supplementary cooking devices to conventional and traditional systems.

The solar oven, which requires much less attention that a concentrating solar cooker, has thus been more accepted and somewhat more widely used.

Future Developments

To overcome the inability of the cooker to work effectively at all times of the day and night, as with traditional cookers, heat storage systems have been included in some prototype designs but since this inclusion increases prices still further, such so-called improvements are not seen as the answer to the problem of developing cookers with sufficient economic appeal.

One research team, the Philips company in the Netherlands, has developed a solar-heated "hot plate" consisting of a closed unit containing a salt which melts at 150°C and gradually releases stored heat when solar radiation levels are insufficient for direct solar heating

Conclusions

From experience gained in the field, it has been found that despite the theoretical usefulness and major potential of solar cookers in reducing the dependence of rural populations on wood and commercial fuels, practical, technical and social problems with the devices are encountered on such a scale that solar cookers are now largely regarded as being generally inappropriate, in their present form and state of development, for the purpose for which they were intended.

Bibliography

Gupta, R. (1984) A Sunny Lunch? Report No. 68. Centre for Science and Environment, 807 Vishal Bavan, 95, Nehru Place, New Delhi 110019, India.

Langerhorst, J. et al. (1977) Solar Energy - Study of the Difficulties Involved in Applying Solar Energy in Developing Countries. Prepared for the Minister for Development Cooperation, Ministry of Foreign Affairs, Netherlands.

Lippert, J. (1982) Solar Cookers Face Many Problems. Article in VITA news October 1982. Volunteers in Technical Assistance (VITA), 1815 N. Lynn Street, Suite 200, Arlington, VA 22209, U.S.A.

Seufert, C. (1978) Survey of Solar Cookers. German Appropriate Technology Exchange (GATE) GTZ, Dag Hammarskjöld-Weg 1, D-6236, Eschborn 1, Federal Republic of Germany.

Solar cooking system with heat storage in hot-plate.

1. *Collecting heat.*

2. *During transport.*

3. *Cooking.*

21 PASSIVE HEATING & COOLING OF BUILDINGS

Introduction

Substantial quantities of energy are used to cool and heat buildings, in developing countries. New buildings use more energy per square metre than those of the past since they have energy-intensive air-conditioning and heating systems. Historically, architecture was required to protect inhabitants from climate by natural means alone and it is possible to build, bearing the climate in mind, houses designed to keep people cool in a hot climate and warm in a cool climate.

Turning away from climate-sensitive design and construction techniques has proved costly. Consider the typical modern office building - with glass facades and mechanical "climate-control" systems in use every day of the year - its energy appetite is enormous. The rapid spread of air-conditioning has greatly increased the energy use of these buildings in recent years and, since many developing countries lack both the engineers and spare parts needed to keep systems running, air-conditioners are often out of operation leading to uncomfortably high internal temperatures. It is estimated that the heating, cooling and lighting of buildings accounts for up to 10% of the commercial energy used in many developing nations and a major future challenge facing designers is how to improve housing conditions without compounding an already severe energy problem. Most building designers are becoming familiar with passive heating designs including systems for collecting, regulating, storing and the distributing solar energy. Natural cooling, the reverse of passive heating, involves heat gain control and the dissipation of heat to natural heat sinks, the sky, the atmosphere and the earth.

All over the world traditional architecture incorporates simple passive cooling techniques effectively. In New Mexico, for example, houses are built of adobe - a heat-absorbing material that heats the house during the night and cools it during the day. In the Philippines, raised open sided pole and thatch buildings allow ample ventilation and protection from the heat, and in Iran for thousands of years cooling towers have been used to draw air into buildings, providing ventilation and relief from the hot summer climate.

One of the beauties of passive solar design is diversity. Although the basic principles are simple, they can be applied in a great number of ways to suit the local climate and culture. Building designers have used these principles for centuries and they can still be used today.

An important aspect of this flexible approach to design is that passive systems need not be 100% passive. In may cases, adding "active" features such as a fan that moves heated or cooled air to other parts of the building can make some climate-sensitive buildings more effective at only a small additional cost.

Most recent passive designs have been aimed at the wealthier sections of the world's community and the cooling needs of the poor majority in the Third World have received almost no attention. In many developing countries, past efforts to upgrade traditional housing actually made the structures less habitable. The tin roof that has spread through much of Africa, for instance, is inexpensive and long-lasting, but it is also less effective than a thatched roof in combating heat

Traditional Mexican house in hot, dry climate.

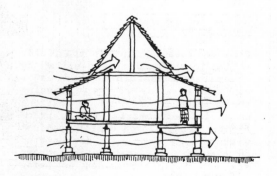

Traditional Malay house in warm, humid climate.

ventilation, and the use of locally-available insulating materials could greatly improve comfort and such changes could be implemented by the buildings' owners who, in developing countries, tend to carry out much of their own construction work.

The problem was summed up by the United Nations at its conference on New and Renewable Sources of Energy in 1981 where it was stated: "It is vitally important to select the right (building) technique to match the climate of a specific area. The Technical Panel would like to stress the advantages of using systematic bioclimatic analysis to identify with certainty the passive cooling techniques appropriate to each particular geographic locality".

Technology Factors In Passive Designs

Air temperature, humidity and movement, and the temperature of surrounding surfaces largely determine the degree of human comfort in different climates. The boundaries of human comfort are surprisingly narrow. Temperature boundaries are 20°C to 27°C and humidity boundaries are between 18% and 78% relative humidity – although both comfort zones can be extended by either exposing or protecting the body from the elements (e.g. by clothing).

In addition to the heat and moisture to which the body is exposed, the body itself generates heat. Of the heat generated, about 80% must be dissipated (if the body is to function effectively) by conduction, radiation and by evaporation in the form of perspiration.

The most effective passive solar designs take account of these human needs and ensure that the design and location of the building make full use of available environmental factors to maximize human comfort.

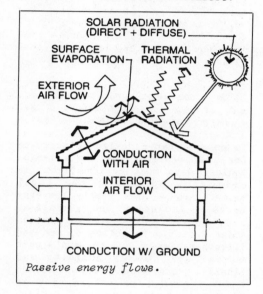

Passive energy flows.

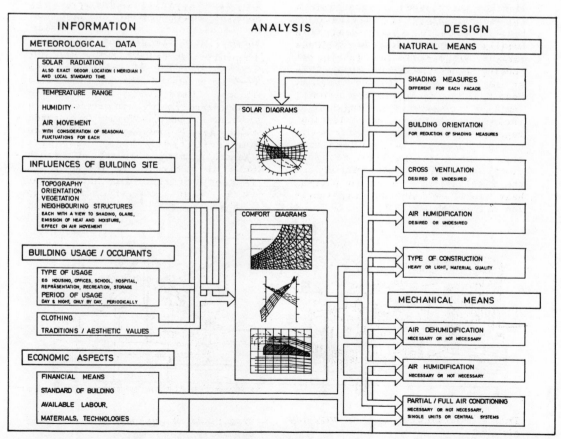

Design strategy for passive buildings. (Reprinted from Lippsmeier 1980).

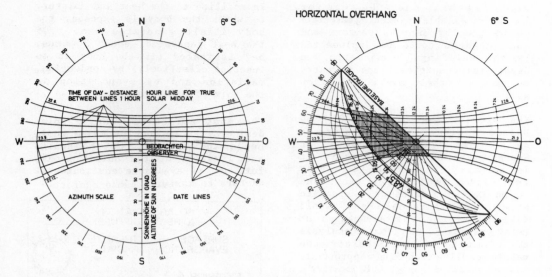

Two examples of architectural design diagrams ("solar diagrams") of the types used for calculations of solar exposure and shadows. (Reprinted from Lippsmeier 1980).

To ensure that a design is adequate for the location, certain meteorological data must be considered in order to understand the local climate. These are: yearly temperatures - including the variations between day and night; year round solar radiation - noting direct and diffuse radiation and the sun´s path in the sky; yearly humidity and finally, yearly wind speeds and directions. Using this information it is possible to construct comfort diagrams which, when integrated with all supplementary information collected, provide a firm basis for locally-appropriate designs. However, designs for a certain locality and climate often involve a great deal

of trial and error before sufficient knowledge is gained.

In many cases local traditional building practices can provide valuable guidance in passive design development. Guidelines towards producing an effective design can be obtained from observations of local building practices and from using a comfort diagram. (See Figure). In order to use this diagram it is necessary to know the local temperatures and relative humidities. There is one zone of total comfort and as soon as local conditions fall outside these limits the diagram prescribes the ideal techniques for effecting cooling or shows if heating is required.

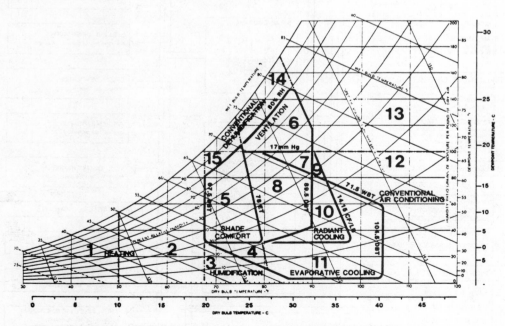

Example of bio-climatic diagram ("comfort-diagram") used for identifying strategies for indoor comfort in buildings in tropical regions. a "Givoni Building Bioclimatic Chart" (Reprinted from Bowen 1981).

Passive Design And Climate

Since most developing countries are located in the area between the 30° north and 30° south latitudes, passive designs of greatest relevance in such regions should concentrate on cooling techniques. In general, these regions can be divided into those with warm, humid climates and those with hot, arid climates and, although certain general principles of design can be applied to each category, individual locality idiosyncracies will necessitate the inclusion of variations in design to suit each site.

In warm, humid climates, where the relative humidity often exceeds 90%, where temperatures are frequently as high as 38°C, and where high rainfall, low-to-moderate wind speeds and medium-to-strong solar radiation levels are experienced, light-weight, open building construction methods are usually most appropriate. The design of buildings in the warm, humid zones puts emphasis on allowing ventilation around, through and under buildings which are orientated north-south to avoid exposure of the larger walls to the sun. Buildings in such climates are generally constructed from permeable, non-heat absorbing materials and use the effect of overhangs and large areas of vegetation to contribute shading to the structure.

By contrast, in the hot, arid regions which experience low relative humidities, high temperature, low rainfall, varying wind speeds and intense solar radiation levels, construction tends to be of the dense, closed variety. In these regions builders have to take account of extreme seasonal temperature variations (often fluctuating between 50°C during summer months and 0°C in winter) as well as diurnal variations of up to 20°C between night and day. To cater for such extreme conditions, buildings are generally placed close together in order to maximize shading; compact structures are built to minimize surface exposure to the sun and maximize radiation reflection; heat-absorbing construction materials are used for heat retention during the day and heat release at night (the time lag effect); protected open areas often with water pools and vegetation for cooling, are incorporated in the layout; techniques are incorporated to enhance ventilation when air temperatures are low or to reduce ventilation when the air is hot or dusty.

There are, of course, many variations within and between these general climatic categories. Upland areas in both categories mostly have more temperate climates but are often exposed to stronger solar radiation levels and experience greater diurnal temperature variations. The low night-time temperatures often experienced in such areas usually result in the need for some form of night-time heating and designs have also to cater for the effects of strong solar radiation levels, cold winds and the frequent formation of dew. Passive solar designs developed to protect against these extremes include aligning fairly compact buildings in an east-west direction with their main openings facing south in order to offer protection from solar radiation in summer and to enhance collection of available radiation in winter.

Composite climatic zones, such as savanna areas, have clearly divided seasons with longer, hot, dry seasons and shorter warm, humid periods. Building designs in such areas are particularly site-specific and few general design guidelines can be formulated.

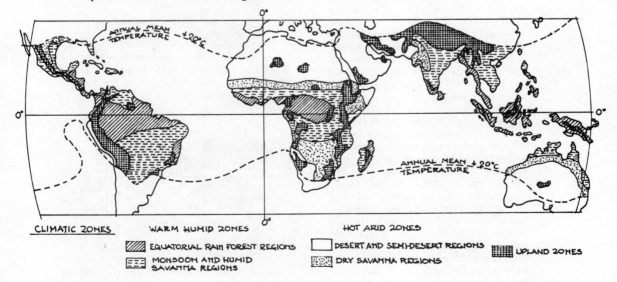

CLIMATIC ZONES   WARM HUMID ZONES   HOT ARID ZONES

▨ EQUATORIAL RAIN FOREST REGIONS   ▢ DESERT AND SEMI-DESERT REGIONS   ▦ UPLAND ZONES
▤ MONSOON AND HUMID SAVANNA REGIONS   ▨ DRY SAVANNA REGIONS

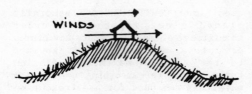

Site selection: windy hill.

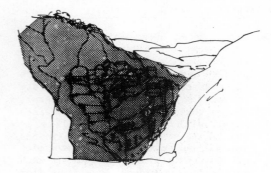

Site selection: shadow in a valley.

Passive Cooling Techniques

Passive cooling designs require features which minimize heat gain from both external and internal sources and induce maximum cooling by natural means. If such features are inadequate for the specific climate experienced, the design should then include auxiliary active cooling systems.

Methods of Minimizing Heat Gain

In order to maximize design efficiencies, the following criteria should be taken into account:

1   Site: Whenever possible buildings should be located on a naturally cool site such as in the shade of a hill or mountain slope. Maximum cooling is often experienced on east or south-east orientated hill sites (in the Northern Hemisphere) which offer afternoon shading from the sun. Sites on or near open water often have low and even temperatures, and cooling is experienced by buildings sited on the leeward side of open water since winds are cooled during their passage over the water surface. In warm, humid climates, buildings sited on hill tops or in open areas are subject to the cooling effect of winds. Valley bottoms and other low sites are cooled by the effect of cold air sinking to the valley's floor. In addition, rivers often occupy valley bottoms and the combined effect of the cooling river water and of vegetation along river banks offering shade, enhance passive cooling.

2   Landscaping: Vegetation is a valuable cooling resource since trees, bushes and grass provide shade, minimize ground reflection and enhance humidification (in dry areas especially). In addition, vegetation can be used to direct and concentrate air on, through and around buildings and, in association with water pools and fountains, vegetation combines cooling with aesthetic appeal. A park actually functions as a cooling device.

Site selection: "cool in the pool".

Site selection: shadow from trees.

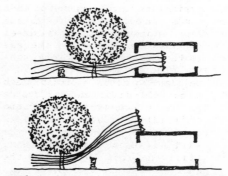

Landscaping: vegetation used to direct winds.

Landscaping: the park is a cooling device.

Creating shadow in the streets.

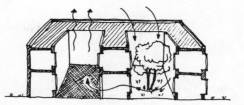

Courtyards with different characters can enhance ventilation.

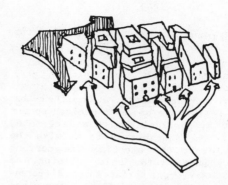

Town planning according to prevailing winds.

3   Town Planning: In hot, arid climates buildings should be grouped to maximize the effects of mutual shading. This results in a layout with narrow, shaded streets and courtyards which can act as ventilation channels. If courtyards are shaded and contain vegetation and water pools, the cooled air sinks to lower levels where it can be used to ventilate buildings in the vicinity. In addition, streets and buildings of towns or villages should be oriented in such a way as to minimize restrictions to the passage of cool winds through the streets and also to minimize wall areas exposed to the direction where the sun is hottest.

Compass termites.

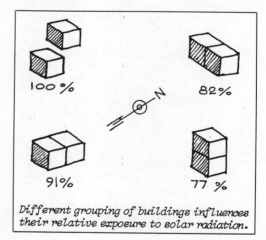

Different grouping of buildings influences their relative exposure to solar radiation.

4   Building Form: An important aspect of building design in hot climates is the volume-to-surface ratio of buildings. Compact buildings expose less surface area to the sun than expansive buildings and, since the roof is usually the surface most exposed to the sun, the correct roof design for a certain climate is essential. The domed roof is often used in hot, arid climates as it exposes a relatively small surface area to direct solar radiation and tends to reflect radiation and enhance ventilation. This roof form also creates a high room allowing good internal temperature stratification. Vents at the top of the dome release heat accumulating in the dome. Domes also elevate buildings to increase their exposure to the cooling effect of winds.

Domes are often used in hot arid climates.

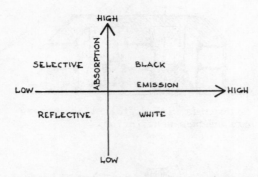

Different colours have different absorption and emission properties of solar radiation.

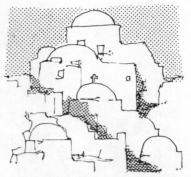

White coloured mediterranean village.

5 <u>Surface materials</u>: Once the building form has been decided on for a particular site and climate, the designer is faced with the task of selecting the correct surface for that building, as different surfaces have different abilities to reflect, absorb and emit solar radiation. Whitewalled buildings with shiny roofs, for example offer good reflective properties whereas darker colours absorb solar radiation. Smooth surfaces offer little resistance to air movement around the building unlike rough finishes which slow air circulation.

Extra roof for shadowing.

6 <u>Shading</u> is essential to minimize heat gain and can be provided in many ways by roof overhangs, balconies, loggias, pergolas or some other kind of horizontal or vertical screening.

7 <u>The shell</u> of the building (the walls, roof and floor): Heavy constructions offer the time-lag effect. Proper use of high mass absorbing materials in hot, arid climates can cool the house during the day by retarding the flow of heat. Lightweight metal roofs used in hot climates should use an insulating material or an airspace to slow heat-flow into

Different ways to shadow windows.

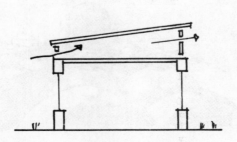

Double roofing for heat removal.

Balconies for shadowing.

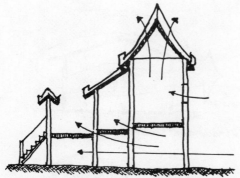

Malay house letting the wind through.

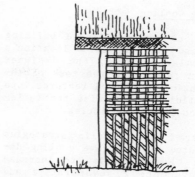

Permeable wall.

rooms. The use of permeable construction materials (eg. thatched
roofs, bamboo walls, wooden floors)
enables sunlight to be screened,
while at the same time allowing
the circulation of ventilating
air. Glazing can also be incorporated to let light in but keep
heat out although it is important
to always shade glass surfaces.
Double roofs and walls insulate
buildings from solar radiation,
and ventilation between the two
layers cools the outer surface.

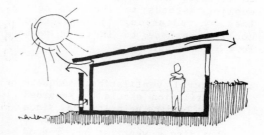

House with double wall and roof.

8 Internal Heat Production: Besides
absorbing external heat, heat
is generated through activities
within the building. For effective
cooling to take place, heat generating activities should therefore
be carried out away from the main
living area. As a result, the
building designer should try,
if possible, to isolate heat sources like kitchens. It is also
important to find ways to allow
daylight to enter without allowing
the associated heat into the buildings (for example, by shading
all openings).

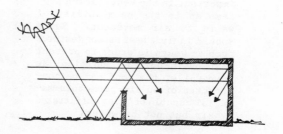

Indirect lighting from the sun.

9 "Zoning" is another route to
achieve a better climate
indoors. The concept is to divide
the house into different parts
(zones) that are used at different
times of the day. Heavy parts
of the building can be used for
daytime comfort while a light
part can be used for comfortable
night-time use.

Blinders for sun-protection.

"Zoning".

Outdoor kitchen.

Passive Cooling Methods

Having chosen a suitable building form and lay-out for a location's particular climate, the designer is then faced with the task of incorporating individual features into the building to maximize the building's cooling capabilities.

There are four main cooling strategies associated with passive cooling designs; 1) ventilation to increase air movement; 2) evaporative and desiccant (drying agent) cooling to regulate humidity; 3) radiative cooling methods to increase heat loss by radiation; 4) conductive cooling methods to increase heat loss by conduction.

1a   Natural ventilation: Adequate ventilation should be regarded as a leading priority and, since windows play a major role in a building's ventilation, a good deal of thought should go into the design and siting of these features. Roof design is also important in the ventilation process as is the permeability of walls to air movement. Roofs should include heat escape features such as vents in order to prevent heat build-up. In addition, as with walls, the roof can be constructed of an air-permeable material capable of being infiltrated by air but not by water.

1b   Induced Ventilation: In some cases, natural ventilation is insufficient for adequate cooling, and induced ventilation features should also be designed into the building. A "cupola", for example, uses the properties of air pressure generated by air movement over the roof to enhance ventilation through the building. In the same way, a thermal chimney uses naturally- occurring air pressure differences between cooler lower layers and warmer upper layers of an air mass to induce a draught effect. These effects can be combined by putting a ventilator cap on a thermal chimney. If, in addition, a solar collector is incorporated into the chimney, an ordinary chimney's updraught can be enhanced. The Egyptian "malkaf" and the Persian wind tower are examples of ventilation-inducing features of traditionally designed buildings. Some wind towers act as wind catchers in the morning and as thermal chimneys in the afternoon.

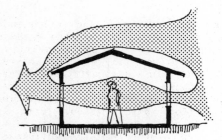

Window ventilation.

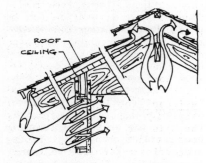

Wall and roof ventilation.

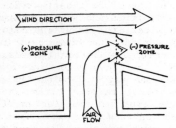

Increasing ventilation with "coupola".

Increasing ventilation with ventilator cap.

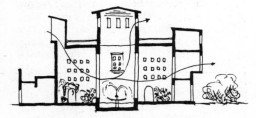

Egyptian house with "malkaf".

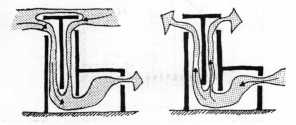

Persian wind towers.

2a   Evaporative Cooling: In hot, arid
climates, evaporative cooling
techniques are extremely effective. The simplest system
of this kind consists of water-
soaked cloths or water jars placed
in areas of maximum ventilation
(for example a window). Warm air
passing over the water, picks
up moisture and, in the process,
loses some of its heat. In more
sophisticated systems, water pools,
fountains or vegetation can be
placed in areas of maximum ventilation, or water can be sprayed
on a building or allowed to trickle
down windows to induce the evaporative cooling effect.

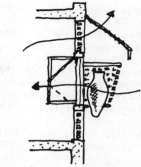

Evaporative cooling with water jar.

*A water-soaked cloth in the window
cools the incoming air.*

2b   Desiccant Cooling: In areas of
very high humidity, increasing
the moisture content of the air
does little to increase its cooling
properties since air of high moisture content prevents the body
cooling itself by perspiring. In
such areas, desiccant cooling,
the emplacement of moisture absorbing materials such as salt, coconut
husk or charcoal in the building,
is a useful traditional strategy.
The technique has the disadvantage
that the material has to be replaced at regular intervals to
maintain the cooling effect.
Salt barrels have been used in
the past but such a method has
the drawback that the salt must
be disposed of once saturated
- however, regenerative passive
systems also exist. Passive cooling
in regions of high humidity remains
a problem and most research is
now focussed on systems with some
active component.

*Evarporative cooling with indoor fountain
and vegetation.*

3   Radiative Cooling: Radiative cooling occurs when interior areas
are exposed to the heat sink of
the clear night sky. To induce
this form of cooling, the roof
can be constructed in such a way
as to allow opening at night,
but more often, in traditional

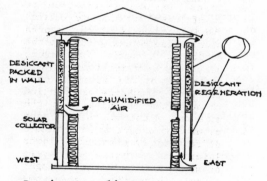

Dessicant cooling.

Radiative cooling.

buildings especially, indirect cooling is carried out by incorporating into interior design large bodies of masonry or water which can be exposed to the night sky. Masonry mass is the key to such examples as the adobe buildings in New Mexico which combine the time-lag effect with night-time radiative cooling. The roof pond system with movable insulation is a very efficient modern system utilizing the same principles. Much work is being put into the development of selective surfaces for heating and cooling in order to find a surface that emits more than it absorbs. "No heat", an aluminium sheet with cooling ability, is such a product from Granges Aluminium (Sweden).

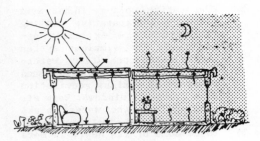

Roof pond for cooling.

4   Conductive Cooling: The earth is where mankind first sought shelter and for a good reason. Earth acts as a time-lag device. Underground or earth-integrated construction is the most common way of using this cooling effect and it has been used in many cultures. In more sophisticated designs, air is cooled before entering a building by channelling it through subterranean piping. The cooling properties of the earth can be enhanced by planting it with long grass, by covering it with white asphalt or by spraying it with water.

Aluminium roof with "no heat" surface.

Advanced Passive Cooling Methods

In order to increase the effectiveness of passive cooling techniques they may either be used in combinations or together with an active component (e.g. a fan) to produce an advanced cooling system. These kinds of cooling techniques may be applied to housing but they can also be used in smaller applications such as the cooling of refrigerators or electronic equipment. Examples of such techniques include:

Underground house.

1)   Nocturnal ventilation.

If an energy "storage" in the form of a rock bed is positioned beneath a house and ventilated at night with cool air, during the daytime it can be used to cool ventilating air which is then passed into the house. A variation of this design is the "Australian pebble bed" which incorporates additional evaporative cooling properties of sprinkling cold water, in addition to passing cold air, through the rock bed.

Earth-tube cooling.

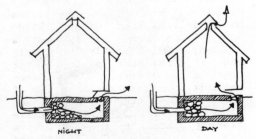

Cooling with rock storage.

2) The evaporative charcoal cooler.

This is a food storage box whose walls are covered with charcoal sandwiched behind chicken wire. Cooling is effected by water from a small tray dripping down cloth wicks and evaporating off the charcoal surface.

3) The desert cooler.

This simple device consists of a wick which draws water from a vessel for evaporation. Its effectiveness can be greatly increased by incorporating a fan to increase the air flow.

4) The two-stage evaporative cooler

Here the air is initially cooled by passage through tubes whose outsides are cooled by the evaporation of water. The air then passes on to a conventional wick-based cooler.

5) Night time radiator cooler.
The equipment to be cooled is positioned under a circulatory water system. Cooling water nearest the equipment becomes warm and thus rises to be replaced with cooler water. At night the whole resevoir is re-cooled by circulation through the radiator.

6) Dehumification cooler.

This consists of two desiccant beds. While one of these is used to dehumify the room´s air, the other is regenerated by hot air from a solar collector.

Passive Solar Heating Techniques

Although the greatest proportion of developing countries lies within the "sun-belt" of latitudes 30° north to 30° south, where cooling is generally regarded as being of greater priority than heating, there are areas both inside and outside this belt where heating takes precedence over cooling, especially in mountainous regions at high altitudes.

Passive solar heating relies on capturing the heat of solar radiation as it arrives at and enters buildings. Truly passive heating designs are capable of storing available heat at times when heating is not required and releasing it when temperatures fall, without the use of fans, pumps or other mechanical devices. However, a number of passive heating systems incorporate active components, like those mentioned above, to enhance the heat storage and circulation properties of passive solar buildings.

Evaporative charcoal cooler.

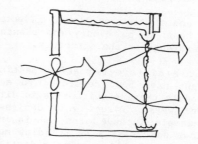

Desert cooler.

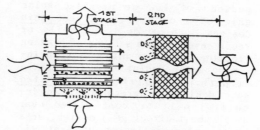

Two stage evaporative cooler.

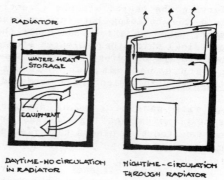

Radiative cooler.

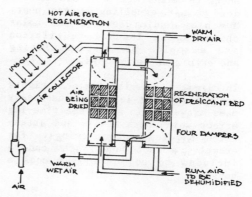

Desiccant cooler.

For optimum performance, a passive solar design should:

- have large areas of glazing facing the sun to maximize the capture of solar radiation

- include features to regulate heat intake in order to prevent the building overheating

- consist of sufficient mass to allow heat storage and release to take place

- contain features allowing collected heat to be evenly distributed throughout the structure.

Several different passive heating designs have been developed to accommodate these requirements in different ways. Direct gain designs, for example, include large, sun-facing windows and heavy mass building materials which allow solar radiation to penetrate the building where it is stored and then released during periods of low or negative solar gain. By contrast, indirect gain designs incorporate heavy walls or roofs covered with a layer of glazing (forming a trombe wall) to trap incoming solar radiation and heat the structure's interior. In isolated gain designs, a greenhouse, an insulated wall or roof glazed over as in the indirect gain design, acts as a solar collector of heat air which is then transferred to the interior. By using the properties of isolated gain designs and by transferring the heated air not directly into the building but into a heat store (such as a rock bed under the building) which then releases stored heat to the building during low temperature periods, an indirect, isolated gain design can be produced.

As with passive cooling designs, some passive heating designs are more suited than others to different climates.

In hot, sunny climates extra care must be taken to regulate the solar energy intake and to ensure that adequate means are included in the design for expelling excess heat. Movable shading devices and solar chimneys for induced ventilation are examples of heat-regulating and expulsion techniques.

In climates experiencing cold nights and hot days, it is best to use heavy mass materials to achieve a time-lag effect which both delays and attenuates the temperature swings. For example, a roof pond system can be included to both heat and cool a house.

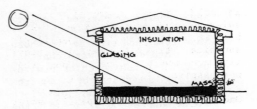

Direct solar heating.

Trombe wall on house in Ladakh.

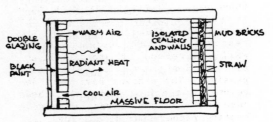

Indirect heating with Trombe wall.

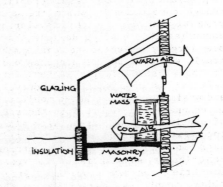

Isolated solar heating with green-house.

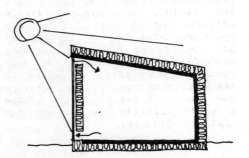

Isolated solar heating with solar wall.

In cold, sunny climates, passive solar heating techniques work well, but in cold climates with little sunshine, it is often better to heavily insulate the house rather than use passive solar features.

One difficult aspect of passive solar heating is to decide the ratio of glass area to the amount of mass in areas of varying insolation. The problem can be solved by using an energy balance calculation. Both manual systems and computer programmes are available for this purpose but it is first necessary to know the local climatic data; the areas and insulating values of windows, roof, walls, and doors; indoor temperature and amount of air-change through ventilation; internal heat load from electrical equipment; hot water use and the number of persons using the house; the orientation of windows; and whether heat-exchangers are used for ventilation and, if so, their efficiencies.

Continuing Technological Development

Substantial progress has been made in passive solar technologies and design over the last 10 years and features for such designs are now well-understood by architects and building designers. But, since most passive solar designs have been developed in the developed world, these have tended to concentrate on passive heating rather than on cooling.

In general, therefore, passive cooling design developments, which are more urgently needed in the developing countries than heating designs, have been largely neglected and development trails well behind the progress made in the passive heating sector. This deficit has been realized, however, and designers are now being motivated to channel more effort into the area of passive cooling, especially since the growth of energy-intensive buildings in the developing countries contributes substantially to these countries' energy problems. Individual countries, especially in the developed world, are now making major efforts to formulate climatically appropriate, economic building designs but, so far, there has been little activity in this field in the developing countries despite the substantial potential for such building methods in these areas. Not only would passive solar buildings in the developing countries reduce energy use but such designs would create a more habitable environment for their populations.

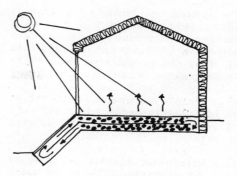

Indirect isolated solar heating.

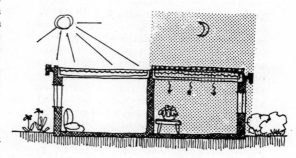

Heating with roof-pond.

The Economics of Passive Solar Heating And Cooling

Most passive solar buildings are cost-effective in relation to buildings with commercially-powered heating and cooling systems for the simple reason that passive solar buildings incorporate relatively simple and inexpensive designs and construction techniques at the expense of costly energy-consuming technologies. By combining design concepts that have been known for centuries with modern building materials and calculating technologies, builders are now able to construct buildings consuming far less fuel and power than conventional buildings at little or no additional cost.

Dissemination Factors

Architects, builders and engineers working together over the last decade have instigated a transition to climate-sensitive, fuel-conserving building practices, and the simple principles developed are now being used in constructing passive solar buildings, using readily available materials at competitive costs.

The transition, however, is gradual and complex, requiring not only further improvement of designs and materials but also the re-education of many designers and builders. These participants in the building industry are only hesitantly moving away from entrenched, traditional building attitudes towards the more rational design concepts employed in passive solar construction.

This change in attitude is slow and still needs to reach the point where solar heating, cooling and energy-conserving designs are regularly integrated into design concepts. Such designs have also yet to be fully developed to the stage at which they are mass-producible at low cost. This stage is expected to be reached once the results of continuing research work are more widely disseminated, at which point architects, builders, governments and government ministries (with responsibility for planning and construction programmes) will have sufficient information necessary for producing long-term programmes geared to energy conservation.

In the main, the obstacles blocking the widespread dissemination of passive solar building in the developing world, centre on a conservatism of attitude and a lack of knowledge of available designs. Too often, new buildings constructed in the developing world are expected to include those features generally seen and experienced in the developed countries. The fact that many architect and construction firms active in developing countries are based in industrial countries is also of importance.

As a result, those connected with the building industry and building programmes in developing countries are often unaware of the passive solar designs now becoming widely available in the developed world. Without readily available information, developing world architects and builders have little option but to design using principles established prior to the relatively recent development of passive cooling concepts.

However, in many cases, even when such modern concepts are made available to designers in the developing world, the influence of established energy-intensive construction has become deep-seated in the developing world and designers are not necessarily concerned with energy-efficiency, but more with moving away from traditional building practices towards the so-called "modern technologies". Since most passively-cooled building concepts centre on improved versions of traditional buildings which have developed along passive lines for centuries, developing world designers tend automatically to shy away from passive designs in favour of energy-intensive designs. Training and re-training are therefore important factors in dissemination of those concepts.

However, it is gradually being more and more accepted throughout the world of building design that passive solar cooling and heating designs are effective and environmentally-appropriate, and that new building work incorporating passive features can be carried out at little or no additional cost. Restructuring older buildings to include passive solar features is, however, more difficult.

Despite the relatively slow growth of passive solar design implementation throughout the world, a limited number of developed countries have embarked on the widespread construction of such buildings. In the USA, for example, there are already 60 000 to 80 000 passive solar houses in existence and it is estimated that about 11% of all new US housing in 1983 will include at least some passive solar features. In Sweden, energy-efficient buildings are now widespread and accepted as part of building practice. Many Swedish buildings use up to 90% less energy than similar, well-constructed conventional (non-passively designed) structures, through a combination of heavy insulation and other energy conserving techniques.

Most of the developed world's passive buildings are designed to harness available solar energy for heating, and only in Australia, Israel and the USA is there any regular frequency of passively-cooled solar building work continuing. In the developing world, examples of modern passive cooling construction are scarce, despite the fact that over the centuries traditional building techniques have produced styles appropriate to specific climates. If the countries of the developing world were to combine the more modern concepts and materials with traditional climate-conscious designs, not only would they benefit from the lower consumption of costly fuels, but their inhabitants would gradually gain access to more acceptable living conditions.

Bibliography

Bowen, A. et al. (1981) Passive Cooling.
Proceedings of the International Passive and Hybrid Cooling Conference.
American Section of the International Solar Energy Society (ISES).
250 B McDowell Hall, University of Delaware, Newark, DE 19711, U.S.A.

Dunham, D. (1984) Building for the Maritime Desert.
Volunteers in Technical Assistance (VITA)
1815 N. Lynn St. Suite 200, Arlington, Virginia 22209-2079 U.S.A.

Evans, M. (1980) Housing, Climate and Comfort.
The Architectural Press Ltd., London.

Flavin, C. (1980) Energy and Architecture: The Solar and Conservation Potential.
Worldwatch Paper 40.
Worldwatch Institute, 1776 Massachusetts Avenue, N.W., Washington, DC. 20036, U.S.A.

Lippsmeier, G. (1980) Building in the Tropics.
Callwey Verlag, München, West Germany.

Reseach & Design (1979) Passive Cooling, Designing Natural Solutions to Summer Cooling Loads.
Research & Design. Vol. II. No. 3, Fall 1979.
The AIA Research Corporation, 1735 New York Avenue, N.W., Washington, D.C. 20006.

Taylor, J.S. (1983) Commonsense Architecture.
A Cross-Cultural Survey of Practical Design Principles.
W.W. Norton & Company, Inc. 500 Fifth Avenue, New York, N.Y, 10110, U.S.A.

Temple, P. & Norris D. (1982) Design of Demonstration Passive Solar Buildings.
Associates in Rural Development, Inc., 362 Main Street, Burlington, VT 05401.

Section IV

HYDRO, WIND AND
WATER POWER

22 SMALL-SCALE HYDROPOWER

Introduction

The principles of using water to generate mechanical and electrical power are well-known and are in widespread use throughout the world. Currently, hydro power represents one of the most practical and effective ways of using renewable energy resources on a large scale and the economics and environmental problems of such systems are well-known.

Recently the utilization of hydro power on a smaller scale has again been attracting interest. <u>Small</u> scale hydro can be divided into three size classes. Small usually refers to those schemes with generating capacities of less than 10 000 kW. <u>Mini</u>-hydro plant has rated capacities in the 100–1500 kW range and <u>Micro</u>-hydro refers to those schemes with outputs of less than 100 kW. Small-hydro installations are usually constructed and operated in the same way as large-scale installations, however, micro-hydro is approached in a completely different way and this will be concentrated on in the following discussion. The art of micro-hydro design is to be able to select the appropriate features from different hydro-power technologies that minimize cost and still deliver reliable, high quality electrical power.

The improved economy of larger plants has gradually favoured large-scale development, but there may often be mitigating circumstances which enable small-scale hydro to achieve economic parity with large-scale systems. For example, if electric demand centres are widely dispersed, the cost of constructing extensive transmission facilities may exceed the cost difference between small-scale and large-scale construction.

In addition, since the use of standardized power plants and local materials and labour are more applicable to small-scale sites, the net cost savings may be sufficient to justify a preference for small scale development on both a national economic and regional economic basis.

Small-scale hydro is economically competitive with small-scale fossil-fuel/steam-electrical plants, particularly if the hydro sites are located near electricity demand centres (a likely prospect in villages or rural agricultural areas). Also, small-scale hydro-development is often considered to be more environmentally favourable than both large-hydro and fossil fuel-powered plants.

Much of the impetus for the development of indigenous resources stems from a societal desire to achieve self-sufficiency in energy production in order to avoid rising prices of imported energy. The net social impact at the local level of such power plants may be significant, especially if small-scale hydro development serves as an economic and social catalyst for more extensive economic development.

As a result, an increasing number of development programmes are now looking at small-scale hydro power technologies, especially with a view to fabricating the necessary system components locally in order to keep system costs as low as possible.

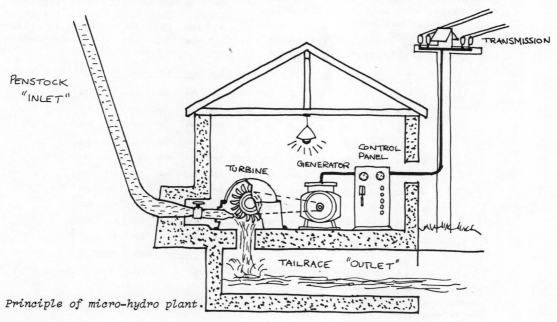

Principle of micro-hydro plant.

Environment

Most hydro power systems in use today are large-scale and are unfortunately hampered by many environmental problems.

Large-scale hydro power inevitably necessitates the building of dams in order to create a reservoir. There can be few technologies that so visibly and dramatically alter the face of a landscape as large dams and artificial reservoir lakes, so the decision to build such a power system carries with it a multitude of associated and frequently neglected hazards.

To begin with, flooding a valley usually necessitates mass population migration and local authorities are forced to cope with all the resettlement problems that such an exodus brings. In Egypt, the creation of Lake Nasser displaced some 80 000 people; in Ghana, 75 000 people had to be evacuated when Lake Volta was created, and the establishment of Lake Kariba in East Africa led to the resettlement of 57 000 people.

The human problems created by the establishment of large hydro power schemes are compounded by a number of environmental problems. Traditional fishing practices are disrupted by the construction of large dams, even when fish ladders are installed. Such dams can prevent fish migration to traditional fishing areas where, as a result fish stocks are decimated. In addition the reservoir and associated irrigation channels provide ideal breeding sites for snails that transmit schistosomiasis – a dehabilitating and often fatal disease that currently afflicts about 200 million people in tropical countries. Large dams also often reduce the previously-experienced, regular flooding of surrounding land which brought rich silt to the areas and also washed out salt from the soil before the dam was built.

Reduced silting results in the need for costly artificial fertilizers to be applied to the land, providing that it is still suitable for cultivation. Salt content build-up caused by reduced washing-out frequently leads to previously fertile land becoming infertile.

Silt, which previously reached surrounding land via flood water, now remains in the reservoirs and often causes major silting problems in the hydro power turbines. The Sanman Gorge Dam on the Yellow River in central China, for example, has lost about 75% of its 1000 MW power generating capacity due to sediment build-up.

Part of the reasoning behind the condoning of large-hydro power schemes by local and national authorities centres on the belief that local populations will benefit from the newly-established industry that such power supplies are expected to attract. In a general sense this is naturally true, given the relationship between electricity and industrial development. Unfortunately, the new industry and the supposed associated improved living standards that such industry brings, rarely affect those displaced by large-hydro schemes since energy-intensive industry only provides employment for a skilled workforce. Few local people possess the necessary skills required to qualify for jobs created by the new industries in their area. In Sumatra, for example, the US$ 2 billion Asahan aluminium production plant and hydro-electric scheme employs only 2100 of the Island's 30 million people.

Micro Hydro

Site and Size Selection

In the case of micro hydro, quite often the feasibility study alone (if done by a large international consultancy firm) for one scheme costs more than the total implementation. Therefore, it is important to find methods of site selection that can be done with more modest inputs, preferably local.

Head m.	0.2	0,4	0,6	0,8	1,0	1,5	2,0	2,5	3,0	3,5	4,0	4,5	5,0	Flow m³/s
2	3	6	9	13	16	24	31	39	47	55	63	71	78	
4	6	13	19	25	31	47	63	78	94	110	126	141	157	
6	9	19	28	38	47	71	94	117	141	165	188	212	235	
8	13	25	38	50	63	94	126	157	188	220	251	283	314	
10	16	31	47	63	78	118	157	196	235	275	314	353	392	
12	19	38	57	75	94	141	188	235	283	330	377	424	471	
14	22	44	66	88	110	165	220	275	330	385	439	494	549	
16	25	50	75	100	126	188	251	314	377	439	502	565	628	
18	28	57	85	113	141	212	283	353	424	494	565	636	706	
20	31	63	94	126	157	235	314	392	471	549	628	706	785	Electric effect kW

Table showing the power output (80% efficiency) for a microhydro plant with different head and flow.

Source: Staffan Engström, 1983,
Sma Vattenkraftverk, (In Swedish only).

There are three factors of interest. The head (the vertical height from the turbine up to the point where the water enters the intake pipe) in m, the flow (the quanitity of water flowing past a point in a given time) in litres/s and the need (what the energy needs are now and what they will be in the future) in kW.

The head can be measured in three different ways.

1   With a surveyor's levelling instrument and scale.

2   With a carpenter's level together with scale, wooden boards and plugs.

3   With a pressure gauge. For this you need a long hose and a pressure gauge.

In the case of reaction turbines, the head might also include the suction head of the draught tube below the turbine down to the tail race water level.

The flow rate can also be calculated by several methods.

1   With a bucket (in small streams) a rough estimation of flow rate (litres/s) can be made from the time taken for it to fill up.

2   With a weir. A weir is built in the stream and the flow width multiplies by the height are measured and the resulting flow rate looked up in a table.

3   The float method (used in larger streams) measures the cross-section of the stream and the speed of the water.

4   With the dilution gauging method, a known quantity of salt is poured into the river and at a point further down stream the conductivity of the water is monitored with a probe or a meter (for difficult sites).

It is essential to have a good estimate of the seasonal variations in river flow that can be expected. One way is to work out the catchment area (from a map) and estimate the flow duration by comparing with known records in the area. Another is by questioning local inhabitants.

The energy need. It is always economically sound to have a "productive" need, e.g. a workshop, to carry a large portion of the installation cost (so that loans can be repaid). In addition, energy needs can be divided into current needs and future needs. It is also important to understand the diurnal and seasonal varia-

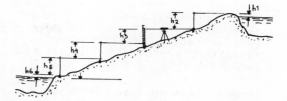

Measurement of head.

Weir for flow measurements.

Depth (mm)	Flow Rate litres/sec
(mm)	
20	5.1
30	9.5
40	14.6
50	20.5
75	38.0
100	58.0
125	81
150	107
175	135
200	165
225	200
250	225
300	300
400	460
500	640

Flow rate for each metre of opening of weir.

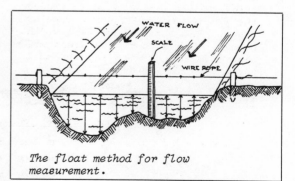

The float method for flow measurement.

MONTHLY AVERAGE FLOWS AT SITE

An example of monthly average flow at a particular site.

tions in energy needs, to be able to compare them with seasonal flow rate variations, and to calculate installation size as a result of this information.

Losses occur in every stage of a water-power scheme and must be accounted for in the process of sizing. Head losses are the open- channel losses and penstock-losses, caused by friction when water flows in different structures. Efficiency losses are turbine losses (efficiency usually 60-90%), and generator losses (80-90% efficient) plus eventual gearing or beltdrive (about 95% efficient) losses. These combined will result in efficiency of about 50-60%. Finally the load factor, i.e. the proportion of the available energy that can actually be used (a load factor of 50% is considered good) must be considered.

Site Preparation

The civil works required in the construction of a small-scale hydro installation are a considerable part of the total cost and it is important to keep them as simple and cheap as possible. This is also an area where the use of local labour and materials can be maximized so that the costs are kept to a minimum.

The intake area is where water is diverted from the river into the head-race canal. It usually consists of a small diversion dam diverting water from the river into the canal. The head-race canal takes the water from the intake area to the forebay. Except in case where the water is free from silt, a settling pond should be included. This is simply a submerged section of the intake canal. The forebay is situated at the end of the head-race canal and comprises a small basin with a trashrack in front of the penstock inlet and a spillway. The trashrack is required to prevent debris from entering the turbine and destroying it. The penstock is a tube that connects the forebay with the turbine. The spillway serves to discharge excess water. The tail-race takes care of the water after the turbine. There are gates at the penstock and the intake to facilitate maintenance and cleaning.

Intake area; The cheapest solution is when a dam can be excluded, and only a simple intake is made. If the water level has to be raised, the simplest solution is a temporary diversion structure, a low wall of stones across the stream, that directs a portion of the flow into the intake. A temporary division easily washes away during periods of heavy rains and flooding. But the villagers can easily reconstruct the diversion after the flood waters have subsided.

Dams often require high-cost materials such as concrete and steel, but there are also cheaper alternatives such as earth dams. These often need a concrete core wall and must be provided with separate concrete spillway lines, as water can never be allowed to flow over the crest. Crib dams are very economical in timber country as they only require rough tree trunks, cut planking and stones. Concrete and masonry dams more than 3 m high should not be built without the advice of a competent engineer who has experience in this field.

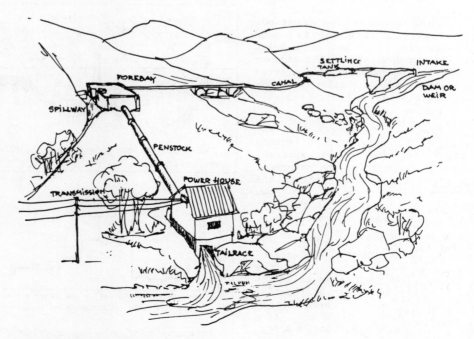

A micro-hydro installation.

Headrace. At many sites a headrace
(channel) that runs parallel to the
river in order to build up a sufficient head is required. The cheapest
way of achieving this is usually
an unlined earth canal, similar to
the traditional irrigation canals.
Occasionally, if it passes through
soils that are porous or is built
above the terrain, it may be lined
with concrete or be constructed from
masonry. Settling ponds add extra
cost, and are sometimes excluded,
in that case the settling that occurs
tends to do so in the headrace itself
which then has to be cleaned regularly.

Forebay. The headrace terminates
in a forebay, the foundation of which
can be made of rubble masonry - concrete only having to be used for
the upper part of the forebay structure. A spillway, a wooden sluice
gate to shut off the water and a
trashrack in front of the penstock
must also be included. To eliminate
bends in the penstock and to minimize
its length, the forebay is often
placed as close as possible above
the powerhouse.

Penstock. At sites with heads exceeding 6 m usually conventional,
relatively costly steel pipes are
used, but for sites with lower needs,
200 litre oil drums or wood can be
used. Wooden penstocks can have a
circular or rectangular cross-section.
Plastic can also be considered as
a penstock material. Water flow
is controlled by a gate valve in
steel pipes or by a board in drum
or wooden constructions. It is important that the penstock is well
anchored at each end, and that it
is firmly supported.

Powerhouse. The powerhouse must
be situated so that it is not flooded
and the foundations must be constructed and placed so that they
are not washed out by the tail-
water. The two bearings that support
the turbine axis should be secured
to a concrete foundation. The powerhouse can be built very cheaply and
simply, but it can also be made to
serve as a mill or a workshop containing a variety of tools.

Turbines

Turbines are divided broadly into
three groups corresponding to high,
medium and low heads. Another important factor is efficiency at partial loads, i.e. does the turbine
perform well over a broad range or
does it have a narrow peak efficiency
at a certain water flow rate. A
third factor of special interest
to developing countries, is whether
the turbine is cheap, robust and
simple to produce and maintain.

With high head hydro systems, the
Pelton Wheel design is usually regarded as the best choice of turbine
(see Figure). The Pelton Wheel consists of a number of buckets mounted
on a wheel into which fine jets of
water are sprayed in order to make
the wheel rotate. They are best
suited to high heads (40 to 1000
m) and have a high power-to-weight
ratio (thus reducing costs) and a
high efficiency over a broad range
of flow rates. Such turbines are
relatively easy to produce although
a fairly high level of engineering
skill is required in the casting
of the wheel.

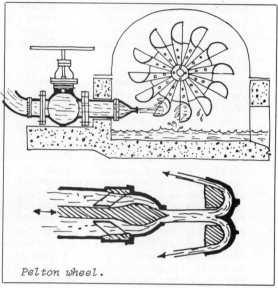

Pelton wheel.

A variation of the Pelton Wheel is
the Turgo Wheel in which the Pelton
Wheel´s double cups are replaced
by a series of curved vanes (see
Figure). The Turgo Wheel´s characteristics are similar to those of
the Pelton but the Turgo can also
be used for lower heads, or in the
15-300 m range.

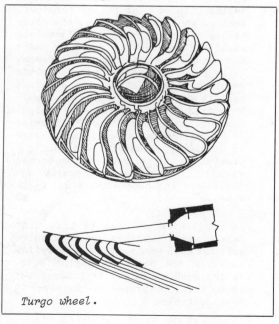

Turgo wheel.

The oldest type of turbine suited to medium heads of the 2–200 m range is the <u>Francis turbine</u> (see Figure). With this type, water enters from the periphery of the turbine via a spiral casing, passes through adjustable guide vanes, strikes the twisted blades of the turbine wheel and is forced out from the centre of the wheel. The design has worked well with large hydro systems and is also suitable for smaller systems. Local fabrication is difficult since the design's complex shape requires advanced and expensive casting processes. For heads of less than 10 m the outer spiral casing of the design is changed to produce an Open Flume Francis. However, open flume types have the disadvantage of losing efficiency when conditions are not ideal.

More suitable small-scale hydro system turbines for use with medium heads of the 2–60 m range are the cross flow turbines known as <u>Ossberger, Mitchell or Banki turbines</u> (see Figure). In these systems, a water jet is directed at blades above the rotor axis, the water passes through the rotor and strikes a similar set of blades below the axis before leaving the turbine. With efficiencies of 70–80%, these turbines are 10–20% less efficient than the other turbine types described here but, since they are cheap, robust and can be manufactured locally, this loss of efficiency is more than compensated for by their suitability to environments and conditions in developing countries.

In low head hydro systems of the 1–30 m range, <u>propeller-type turbines</u> are considered most suitable (see Figure). These consist of a rotor with guide vanes or a spiral casing through which the water is channelled. In some propeller turbines, such as the Kaplan turbine, the rotor blades and guide vanes are adjustable enabling the turbine to work under a broad range of flow conditions. In types with fixed rotors and guide vanes, operating efficiency declines rapidly when conditions are not ideally-suited to the turbine's design. However, the fixed type of turbine is cheaper and more easily fabricated in developing countries due to their simpler specifications. A compromised version incorporates manually adjustable rotor blades and fixed guide vanes.

In conclusion, it can be said that Pelton Wheel turbines are most suitable for high heads whereas cross-flow turbines are best for medium heads in the 6–60 m range. In heads lower than this, a propeller turbine is most suitable.

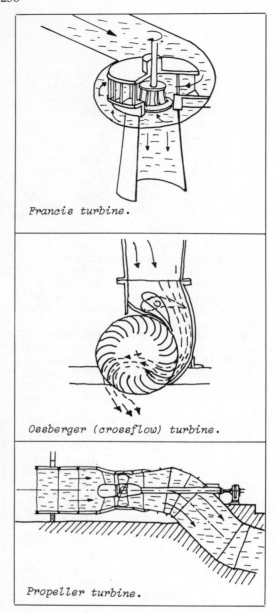

Francis turbine.

Ossberger (crossflow) turbine.

Propeller turbine.

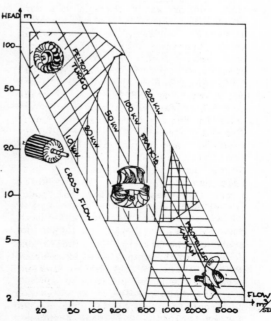

Diagram showing the application ranges at different turbines.

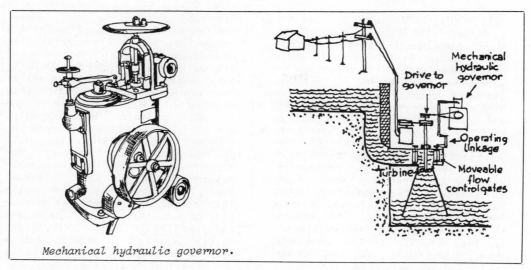

Mechanical hydraulic governor.

Electrical Equipment

For very small systems (under 1000 W) DC is usually preferred. Such a system can be very simple, batteries are incorporated and regulation is not a problem. In larger systems which require a governing system it is cheaper and more practical to use AC. Single phase, 120/220 V is often used up to between 10 and 20 kW and three phase 380/660 V up to 200 kW. All these systems are cheap and simple to install. For bigger units it is necessary to switch to high voltage systems for distribution which is more difficult and expensive.

A water turbine requires a governing system so that a change in load (varying electrical demand) does not result in a change in turbine speed and thus a change in the supply frequency. The usual governing method is a mechanical governor which regulates the flow of water through the turbine. For small hydro plants such a system is relatively expensive. An electronic load controller governs the turbine speed by adjusting the electrical load on the alternator. As lights and electric appliances are turned on and off the electric controller varies the amount of power that is fed into a "ballast" load. The adjustments are made instantaneously and there will be no perceptible change in frequency. In China many turbine/generating sets have manual governing systems which limit the uses to which the power can be put.

The recent advances in the state-of-the-art in electronics can have a positive impact on micro-hydro design. There are four major component areas in the power-plant, the generator (which changes the rotating motion to electric energy AC), the exiter system including a voltage regulator (it supplies DC power to the rotating field of the synchronous generator),

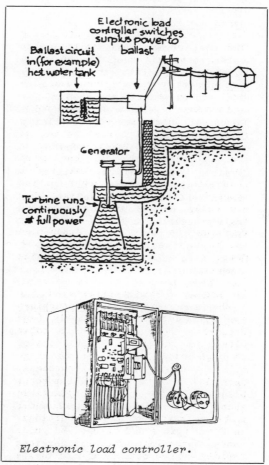

Electronic load controller.

the power control and protection equipment, and the low energy electrical appliances recently developed for photovoltaics.

New generators use lightweight materials, better insulating materials, and more efficient cooling. They are more reliable and have a longer life. Voltage regulators regulate by sensing AC generator output voltage and then controlling the amount of DC current supplied to the exiter field. They have no moving parts and maintain the output voltage of a generator as constantly as possible. Exiters are used because the generator field usually requires more power than a voltage regulator can provide. The use of solid state relays can provide improvements in protection and for control, in which the most important step is to eliminate the hydraulic speed governor, (which sometimes accounts for as much as 25% of the total cost). An electronic load governor is most appropriate in the small sizes up to 50 kW. However, the concept of constant load integrated with a crude mechanical water control is being introduced for units greater than 50 kW. The idea is that the electronic load controller will respond to all sudden changes and the mechanical control system is used for slower daily and seasonal changes.

Hydro for Mechanical End-uses

The most common traditional use of water power has been the operation of mechanical mills. In the developing countries there are many mechanical end-uses which are both cheaper and easier to perform without needing to convert the power to electricity. The majority of these uses are in the processing of local crops.

Typical processing capabilities of a standard Nepalese mill include grain milling, rice hulling, and oil seed pressing. Other common agro-processing end-uses are coffee hulling, sugar cane crushing and wool processing. Crop drying is also a possibility and in Nepal a heat generator is being developed for this purpose. Water power is also used for saw-mills, and water driven water-pumps is another interesting field.

Nepal is one of the best success stories to date as regards obtaining mechanical power from watermills and thus provides a good case study. Traditionally thousands of small water mills, having a vertical-axis wooden water wheel directly driving millstones have been used for milling grain. More recently, diesel- powered mills have been introduced which can hull rice, mill grain and expel oil from seed. Despite the high capital cost of diesel-powered mills and that of diesel fuel in remote areas, these mills have proved popular. It was to provide an alternative source of motive power to these mills that Balayu Yantra Shala Ltd., a private machine shop in Kathmandu established under a Swiss aid programme, developed a reliable, low-cost cross-flow turbine, to drive agro-processing machinery directly. Later, Butwal Engineering Works Ltd. also started to produce and install this kind of equipment. The technology has been successful and over 100 sets of this equipment have been installed, most owned by private mill-owners.

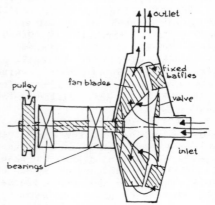

Heat generator.

Modern water mill in Nepal.

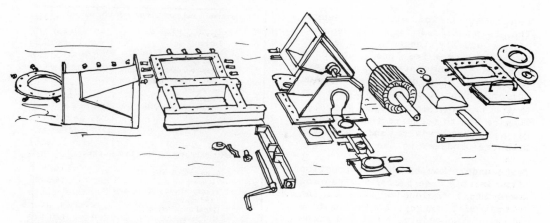

Locally produced parts for the Nepali crossflow turbine.

The cross-flow turbine was chosen because it was easy to fabricate and could accommodate a wide range of flows and heads with only minor design modifications. The flow is controlled manually with a single guide vane. The turbine, housing, transition piece between the penstock and turbine, and the draught tube are fabricated from steel plate. A hydraulic press is used to form the runner blades, guide vanes, flanges of the housing and any other components formed by bending. Sealed, self-aligning bearings are imported from Japan. The penstock pipe is made from 2.5 mm thick mild steel plate in two m lengths. A single frame has been developed on which all the machinery can be mounted. The agro-processing machinery is imported from India. V-belts are used for transmission and the lay-out is compact so that only one shaft on three bearings is needed; the shaft and belts are also kept as short as possible.

Nepali cross-flow turbine.

The cost of implementing a water-powered mill is clearly site-specific. Although the total cost of a mill might seem high (US$ 8750), the owner of a water mill charges less than the diesel mill owner and can still repay his loan in four-seven years. Most mills installed by Butwal range from 8-12 kW so the cost is in the range of about US$ 1000/kW.

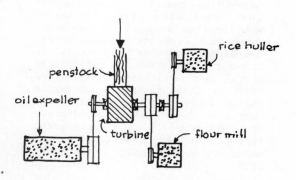

Set-up of machinery in a water mill.

Implementation of Micro-hydro Plants

There are many stages in the implementation of a micro-hydro plant. These will be examined by means of a Nepali case study.

Typical requirements of a prospective hydro-plant owner are as follows: First he has to see if there is a suitable local site by estimating head, flow rates and energy needs. Then he has to obtain the necessary licence to build and operate a micro-hydro unit (from the Department of

Cottage Industries). He then has to purchase the land and the right of way for the head race, power house and tail race. Also he has to arrange for water rights (it is more practical not to work under high heads as this minimizes conflict over irrigation and water rights). He has to locate sufficient cash to cover at least a part of the total cost of the installation and apply for a loan for the balance. (For financing, most customers turn to the Agricultural Development Bank of Nepal which offers an interest rate of 11% and a loan repayment period of seven years. The loan officer from the nearest office usually inspects the site). He has to organize local labour to undertake necessary work i.e. excavation of the head race and the collection of locally available materials, e.g. stone and gravel). Then he has to organize the transport of all the hardware and materials from the producer to the site.

Tasks undertaken by the construction firm (in this case Butwal Engineering Works Ltd.) include working out an initial site survey, (this requires a team of two people, a skilled mechanic and a mason), final site selection and price quotation. The team restricts its surveying equipment to the minimum required for the task, a hand-held sighting level, a vinyl ribbon staff, a 30 metre measuring tape, and several square metres of plastic sheet for flow measurements. The exact location of installation is determined and the design of the hardware specified. Generally a year elapses between the time of the survey and the placement of the order - the time is needed for land purchase and legal arrangements. The team returns to the workshop to design, fabricate or purchase the turbine, penstock and other hardware, and then packs all the machinery for transportation. The team provide technical guidance and install the plant which usually takes three-four weeks. Most of these mills are in the size range of 8-12 kW and operate under heads of from 5-15 m. After completing the installation, the team makes all final adjustments and operates each of the machines under full load for several hours. The operator or owner is also required to run the mill under the supervision of the team until they are satisfied that he understands the operating procedures and safety requirements. They also teach him how to maintain the machine and obtain spare parts from the workshop (the turbines and machinery are produced in the country).

Machinery and services provided by Butwal	
Survey and design	200
Turbine	1,100
Mounting frame, pulleys, belts, and misc. hardware	800
Agro-processing machinery (flour mill, rice huller, and oil expeller)	1,700
Penstock	1,000
Installation (est. 30 days)	500
Sub-total	$5,300
Additional costs incurred by mill owner	
Land and canal right-of-way	400
License	40
Transportation of machinery and materials	600
Canal excavation	700
Mill house	700
Workers (fitters, masons, etc.)	500
Cement ($14/bag)	400
Sand, stones, and other materials	100
Sub-total	$3,440

Cost breakdown for a typical water-powered mill, about 8-12 kW, (presented in 1981 U.S. dollars).

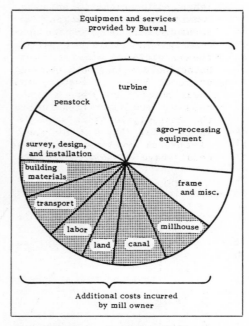

A graphic representation of the breakdown of the total cost.

Source: Inversin, 1982.

Economics

Until recently the conventional wisdom shared amongst aid agencies and multilateral funding bodies (e.g. the World Bank) was that mini, and micro hydro plants could not be a viable economic investment in developing countries because of the high capital costs and the low load factor. This is because they have based their ideas on a scaled down large hydro approach. It has now become more and more apparent that this approach is inappropriate, in particular to very small schemes.

It has previously been thought that although small-scale hydro installations (down to some 500 kW), have reasonable installation costs under US$ 2500/kW, as the units get smaller the costs rise inversely. On the other hand case studies from actual micro-hydro installations have shown prices to be much lower (US$ 400-800/kW in Pakistan and Nepal, for 10 kW micro-hydro) where "low technology" features such as locally made turbines, unlined power canals, rock pile intake structures etc. have been used.

Three factors must be considered in order to reduce micro-hydro costs:

1 The establishment of a local management and engineering capability – covering planning, implementation and service. It is not economic to use international consultancy firms to carry out feasibility studies, design and contract supervision. Planning should be done first by the potential owner and then by a local survey team together with the customer. The team is usually employed by the local micro-hydro producer, which also takes care of the plant´s implementation, the training of operators and future service arrangements.

2 Capital costs can be kept down by building a design that uses local material and labour as much as possible, especially in the civil works, minimizing penstock length and concrete use and using locally produced equipment (e.g. turbines, penstocks and agroprocessing equipment). In Nepal all the excavations (headrace, powerhouse foundations and tail race) are done by local labour, and local porters complete all necessary transportation before the installation team arrives. In this way only the bearings, electrical components and the metals have to be imported.

3 Achieving a high load factor. This means that the power should not be used for lighting only. Few people can afford a domestic electrical connection and projects for this purpose only achieve very low load factors. If the supply of power is to achieve real development for rural communities it must increase local productivity. There are many possibilities for powering local village industries, and the majority of these uses are for the processing of local crops. In summary, ways must be found to integrate different needs (industrial, domestic and others) so as to achieve a high load factor and therefore good project economy.

It has been demonstrated in many places that small-scale hydro power can make a very important contribution towards rural development and that such hydro plants can produce power more cheaply than other alternatives such as diesel generators or high voltage grid extensions. It must be understood, however, that micro-hydro economy is very site specific.

Micro-Hydro in the Developing Countries

China. The one country where micro-hydro plants have been widely disseminated is China where some 85 000

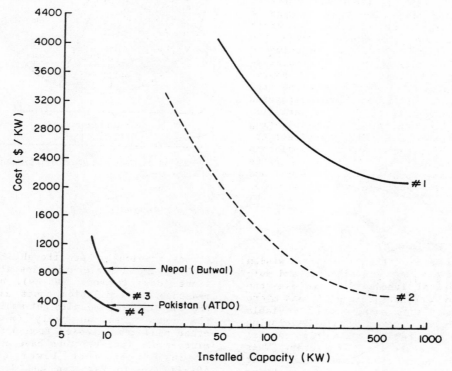

Specific capital costs of mini- and micro hydro power plants, using different construction and design principles. Curve 1 shows the situation for plants built according to scaled-down conventional hydro-power technology. Curves 3 and 4 show actual costs of locally constructed microhydro plants in two countries. Curve 2 indicates approximate costs for installations using features from both approaches. (Source: Jackson 1982).

plants have been installed over the last 20 years. They have played an increasingly important role in supplying energy for the rural areas of China, particularly in the south and the south-west. There is evidence that the installation of such hydro electric plants has resulted in a remarkable increase in the productivity of rural communities. A very important reason for the success of small-scale hydro power in China has been the backing, at all levels of activity, by the Government. 95% of the installations are smaller than 500 kW and the average size is less than 100 kW. Hydro power is often incorporated as part of an integrated water management scheme including: local water storage, irrigation, flood control, navigation, fisheries and electricity. Most of the turbines are of the Francis-type and the governing is done manually. There are some 100 turbine and generator factories.

Nepal. The introduction of small hydropower units in Nepal is generally regarded as a very successful example of renewable energy implementation in a developing country. Some 100 plants have been installed and many more are under construction. Most of the units are used for mechanical end-uses, rather than power production. Two companies (Balayu Yantra Shala PVT Ltd. and Butwal Engineering Works PVT Ltd.), set up by Swiss aid and by a group of Protestant missionaries, but run by Nepalies, are making and installing cross-flow turbines to drive agro-processing machines (corn mills, rice hullers and oil seed crushers). As these end-uses are directly income generating for the rural population they are on a sound financial footing and this approach could certainly be used in many remote areas in the developing world. For the microhydro plants for electricity production that have been built, all the parts are made in Nepal except bearings and generators. Generators are imported from China.

Pakistan produces the cheapest microhydro power in the world. The cost per installed kW ranges from about US$ 350-500/kWh (in 1981). This low cost is due to the utilization of local materials, designs suited to local situations and community involvement at all stages of implementation and utilization. The installations range from 5-15 kW in size,

Micro-hydro plant in Nepal for power production.

and approximately two dozen plants have been installed to date with a similar number of plants under installation in the up to 50 kW size range. The Appropriate Technology Development Organization (ATDO) is the organization behind these projects.

Sri Lanka has about 1000 tea estates, many of which used micro-hydro fifty years ago. Many of these plants are now being reactivated. ITDG from the UK is helping local engineering companies to install new turbine-generator sets and to manufacture turbines (Pelton-wheels) locally.

Colombia has a modest government-run mini-hydro programme but there is plenty of scope for independent initiatives by rural development organizations. ITDG has installed a small Pelton driven saw mill in Colombia, and the development organisation "Las Gaviotas" has started production of small scale turbines.

Countries such as Burundi, Ecuador, India, Indonesia, Malaysia, Peru, and the Philippines have recently launched mini hydro programmes, which, if successful, may lead to significant development in the rural areas. In addition, several aid agencies now have put small-scale hydro power projects high on their priority list. The number of studies, training courses, pilot projects and commercial installation is therefore rapidly rising.

Special Application of Small-Scale Hydro

Electrical Heat Storage Cooker: Although using electricity as an alternative source of energy for household cooking would be an ideal way to save fuelwood, there is a problem in that high amounts of energy (70-1500 W), are required for short periods of time and the cooking of the evening meals often coincides with the peak lighting load. This puts a heavy strain on a small independent electricity grid, resulting in a low load factor and high peak loads.

For micro-hydro (or other renewable energy sources) to be a technically and economically viable source of energy for cooking, requires a storage device that will allow energy that is generated between the normal cooking times to be used during these times. The full capacity of the generating equipment could then be utilized by delivering energy to the storage system continuously. This would result in a higher load factor as well as a reduced peak load.

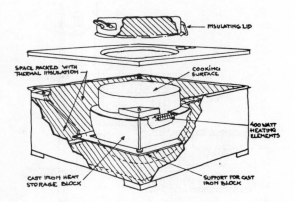

Electric heat storage cooker.

As hydro-electric power use expanded in the US and Europe during the first half of this century, commercial heat storage cookers, which met the above requirement, were developed and marketed. These were essentially electrical elements embedded in well-insulated cast iron blocks which stored heat until it was needed. These became popular because of a tariff structure which encouraged the customer to use, at a fixed tariff, all the power to a maximum limit. The last company known to have manufactured storage cookers discontinued marketing in Norway in the early 1950s.

ITDG has developed a simplified version of the type used in Norway. The heat is stored in an insulated 40 kg cast-iron block - it consumes 200 W continuously and when required can give a heat output of up to 4 kW. These cookers are being used at El Dorion, in Colombia, in conjunction with a micro-hydro plant. The plant drives a saw-mill (a 25 horse power Pelton-wheel) and a 18.5 kilowatt generator. The generator is permanently connected to the load controller giving precise speed control of the saw. As the saw starts to cut, power is automatically shifted from the ballast load (the storage cookers), so that full power is available for cutting. The system enables a large number of cookers to be used on a very limited power supply. The complete system, comprising turbine, saw, lighting and cookers is now running well. Performance figures indicate an overall efficiency of well over 80%, but there is also the additional costs incurred of purchasing flat-bottomed cooking pots and kettles which are required for cooking on hot plates. Robert Yoder at Cornell University has tried to solve this by designing a storage cooker which incorporates stones as a heat storage medium instead of a heavy, expensive, iron mass.

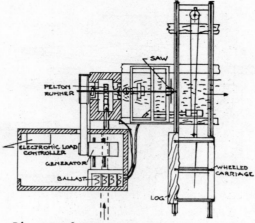

Diagram of water-powered saw-mill and generator.

Hydro Powered Battery Chargers:

The smallest hydro power plants commercially available are used as battery chargers. The systems are often complete battery charging systems which include the control box, wiring harness, generator, meters and a turbine (usually a Pelton or a Turgo wheel). The only other thing required is to "add water", this usually being supplied through a plastic hose of 3-5 cm^2 in diameter. Sites can be utilized with heads as low as a 0,75 m, but around 8 m is more appropriate and higher heads are even better. The price for a complete system is about US\$ 350-550. The batteries can be charged directly on the spot or by using transformers and cables for long distance charging. At a head of 8 m a hydro plant of this size produces about as much power as five standard 35 W photovoltaic panels for a much smaller investment cost.

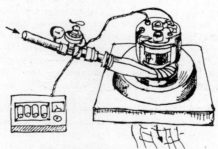

Hydro powered battery charger.

Water-powered Pump Technologies

In places where there is a river or a stream, water-powered pumps can be used to pump the water up from the stream, to houses or villages situated above the river. There are several concepts available:

1. The horizontal axis water wheel. This has a large diameter wheel with buckets or vanes mounted on the periphery to collect the water as the wheel passes through the water flow. The wheel is driven

Noria water wheel.

by the action of flowing water on the buckets which fill and release their contents to a trough or channel at the top of the wheel. Such systems are known as Norias in China and a 4 m diameter Noria, recently tried in Sudan, produced a water output of 38 000 litres of water per 24 hour operating period. A version of the Noria, developed at Loughborough University in the UK, incorporates a coiled pipe around the water wheel, which is partially immersed in water and pumps water to at least 7 m as it is rotated. No data are yet available on the efficiency or output of this pump.

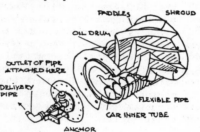

Stream-powered rotating coil pump.

2. The vertical axis water wheel consists of three or four blades rotating around a vertical axis which supplies power to a centrifugal pump to raise water through a flexible pipe. When suspended from a raft in a river, the wheel's rotation can be used to either power a pump or to generate electricity. The system is under development at Reading University, UK, where trials with models have produced efficiencies of 30% from systems 2 m in diameter and 2 m high in a 1 m/s stream flow. Such a system is estimated to be capable of generating about 2 kW of power.

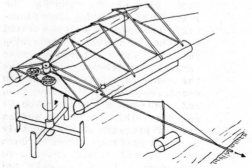

Vertical axis water wheel.

3. The turbine pump is another form of water-powered water pump. Essentially, this device consists of a water turbine directly coupled to a water pump (either axial-flow or centrifugal) which results in a pumping rate directly proportional to the speed of flow. Such pumps have been used successfully in China, especially in Fujian Province, where more than 100 000 units have been installed over the past 20 years. The device's main advantage is its high efficiency. It can be built to supply from just a few kW to over 100 kW with drive heads of 0.5-6 m (the most common is 3-4 m) and delivery head up to 100 m. They are mainly used for irrigation. There can be problems with cavitation and silt, both of which can seriously damage the bearings and other turbine pump components.

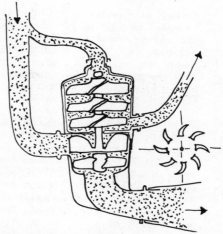

Chinese turbine pump.

4. The hydraulic ram was developed by Montgolfier in 1796 and was used extensively in Europe and the USA in the first quarter of this century. Despite the fact that such pumps require no external power source and run automatically 24 hours a day with minimum maintenance, they have been neglected until recently. The ram utilizes the pressure surge (water hammer effect), which develops when a moving water mass meets an obstruction, to produce power for water lift.

Referring to the figure the ram consists of a drive pipe (F), leading from the water source to the ram body (B). The ram body incorporates three valves, the impulse valve (C), that is equipped with a return spring that lets water escape from the ram body, the air feeder valve (E) that allows air to enter the ram body, and the delivery valve (D) that allows water into the air chamber (A). Water flows through the drive pipe (F) and through the impulse valve (C). As the water velocity increases, the dynamic pressure on the underside of the impulse valve C increases until it overcomes the force of the return spring. Then C moves radidly upward, closing the opening. The pressure of the moving water causes valve D to open and water is driven into the air-chamber A, compressing the air, and discharging water through the delivery pipe G. As the momentum of water in the ram decreases, valve D drops down and closes and the water rebounds somewhat. This creates a sudden drop of pressure in the ram body B, which makes air enter through valve E, and causes valve C to drop down and open quickly and the cycle begins again. The fre-

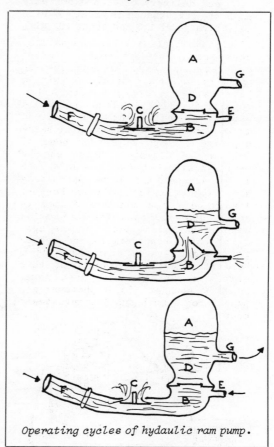

Operating cycles of hydaulic ram pump.

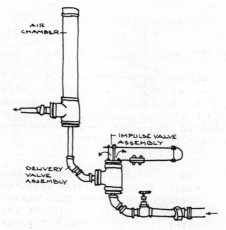

Hydraulic ram.

quency with which the cycle is repeated is regulated by adjusting the spring. Once the adjustment has been set, the hydraulic ram needs no attention, providing that the water flow is continuous and no foreign material gets into the pump blocking the valves.

A hydraulic ram can be installed when the drive head exceeds 1 m and a driving water flow is exceeding 0.1 litres/s. Rams are available for driving flows of up to around 20 litres/s and can deliver water to a head at about twenty times the drive head. The performance of a ram drops with decreasing ratio between driving and delivery head, and normally the performance is about 60% with a ratio of 1:3 and about 20% with a ratio of 1:20 (See Table for the capacities). The hydraulic ram although customarily a device for pumping water, can also be used to compress air.

Manufacturers of hydraulic rams exist in many countries and they are not expensive (in the order of US$ 300 to 2600 according to capacity). In areas where iron pipe and pipe fittings are available, hydraulic rams could be constructed with local materials and by utilizing local skills. The only parts that wear out and have to be replaced are the valve rubbers. Good manuals with complete details are available from ITDG, VITA or the German Appropriate Technology Exchange (GATE).

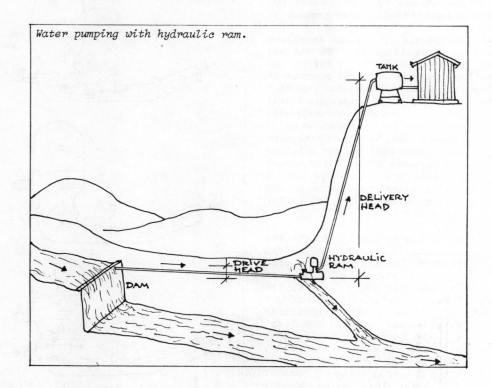

Water pumping with hydraulic ram.

Supply head (m)	Delivery head (m)											
	5	7.5	10	15	20	30	40	50	60	80	100	125
1.0	144	77	65	35	59	19.5	12.5					
1.5		135	96	70	54	36	19	15				
2.0		220	156	105	79	53	33	25	19.5	12.5		
2.5		280	200	125	100	66	40.5	32.5	24	15.5	12	
3.0			260	180	130	87	65	51	40	27	17.5	12
3.5				215	150	100	75	60	46	31.5	20	14
4.0				255	173	115	86	69	53	36	23	16
5.0				310	236	155	118	94	71.5	50	36	23
6.0					282	185	140	112	93.5	64.5	47.5	34
7.0						216	163	130	109	82	60	48
8.0							187	149	125	94	69	55
9.0							212	168	140	105	84	62
10.0							245	187	156	117	93	69
12.0							295	225	187	140	113	83
14.0								265	218	167	132	97
16.0									250	187	150	110
18.0									280	210	169	124

Table showing the capacity of hydraulic rams. Litres pumped in 24 hours per l/min. of drive water for different drive and delivery heads.

Source: Inversin (1979).

Conclusions

Prior to the 1930s small-scale hydro plants were installed in large numbers in many countries and substantially contributed to national development and the supply of energy needs. Later, larger plants were introduced and incorporated into the electric grid system - this often starting with the coupling of many small independent grids together. This history of development has been experienced in most developed countries and may also be repeated in developing countries which have exploitable water power resources. With the advent of cheap oil-fired power and diesel plants, the smaller plants fell into disuse. However, with high oil prices and newer, improved small-hydro systems being developed, a resurgence of interest in smaller systems is being experienced. Providing climatic, geographical, technical and demographic factors fall within certain acceptable boundaries, it can be concluded that the use of small hydro power schemes is already technically and economically acceptable.

Besides their technical and economic advantages, small-hydro plants are environmentally more acceptable than fossil fuel and large-hydro schemes in that they are virtually pollution-free, do not create large-scale environmental damage and can benefit local populations directly both in terms of improved power supply and living conditions, and in terms of increased potential for industrial development and associated employment prospects.

For these reasons, small, mini and micro-hydro power plant systems are considered to have major potential for the developing world and such systems could significantly help in development programmes, particularly those in rural areas far from grid networks.

The dissemination of small-hydro power schemes is continuing throughout the developing world, but still at a relatively slow rate - there are still many countries which have yet to accept the possible potential of these systems.

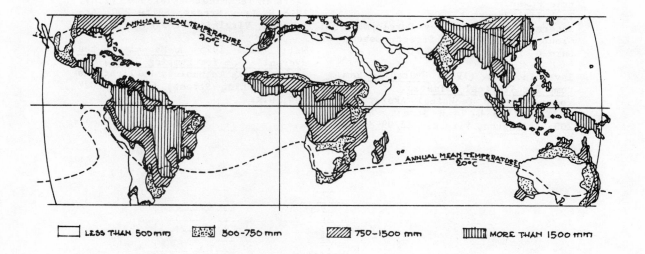

Annual global precipitation.

Source: Lippsmeier 1980 (see chapter 21).

Bibliography

Alternative Sources of Energy (1984) Spectrum: "New Hydro Battery Chargers".
Article in Alternative sources of Energy Jan/Feb 1984. No. 65. 107, S. Central Avenue, Milaca, MN 56353, U.S.A.

Garham, P. (1981) The Development of a Turbine for Tapping River Current Energy.
Article in Appropriate Technology. IT-Publications Ltd., 9 King Street, London WC2E 8HN, U.K.

GATE (1979) Drawings of Hydraulic Ram.
German Appropriate Technology Exchange (GATE), c/o GTZ, Dag Hammarsköld-weg 1, D-6236, Eschborn 1, Federal Republic of Germany.

Hamm, H.W. (1967) Low Cost Development of Small Water Power Sites.
Volunteers in Technical Assistance (VITA), 1815 N. Lynn Street, Suite 200, Arlington, Virginia 22209-8438, U.S.A.

Holland, R.E. (1983) Micro Hydro Power for Rural Development.
Lessons drawn from the experience of the Intermediate Technology Development Group.
Intermediate Technology Industrial Services (ITIS). Myson House, Railway Terrace, Rugby CV21 3HT, U.K.

Holland, R. (1983) Micro Hydro Electric Power.
Technical Papers 1.
Intermediate Technology Development Group (ITDG), 9 King Street, Covent Garden, London WC2E 8HN, U.K.

Inversin, A.R. (1979) Hydraulic Ram Pump for Tropical Climates.
Volunteers in Technical Assistance (VITA). 1815 N. Lynn Street, Suite 200. Arlington, Virginia 22209-8438, U.S.A.

Inversin, A.R. (1982) Nepal. Private-sector Approach to Implementing Micro-Hydropower Schemes.
National Rural Electric Cooperation Association (NRECA). 1800 Massachusetts Avenue NW, Washington, DC 20026, U.S.A.

Inversin, A.R. (1981) Pakistan. Villager-implemented micro-hydropower schemes.
A case study.
National Rural Electric Cooperation Association (NRECA). 1800 Massachusetts Avenue NW, Washington, DC 20026, U.S.A.

Jackson, B. (1982) State-of-the-Art in Mini-Hydro Electrical Design.
Article in Renewable Energy Review Journal Vol. 4 No. 2, Dec. 1982. Asian Institute of Technology (AIT), P.O. Box 2754, Bangkok, Thailand.

Meier, U. (1981) Local Experience With Micro-Hydro Technology.
Swiss Center for Appropriate Technology (SKAT). Varnbüelstrasse 14, CH 9000 St. Gallen, Switzerland.

Tiemersma, J.J. & Heeren, N.A. Small Scale Hydropower Technologies.
Technische Ontwikkeling Ontwikkelingslanden (TOOL), Entrepôtdok 68a-69a, 1018 AD Amsterdam, The Netherlands.

VITA (1982) China´s Turbine-Pump Lifts Water to New Heights.
Article in VITA-news July 1982. Volunteers in Technical Assistance (VITA), 1815 N. Lynn Street, Suite 200, Arlington, Virginia 22209-8438, U.S.A.

Watt, S.B. (1975) A Manual on the Hydraulic Ram for Pumping Water.
Intermediate Technology Publications Ltd., 9 King Street, London, WC2E 8HN, U.K.

23 OCEAN POWER: TIDAL, WAVE & OTEC

Introduction

During the 1970s three types of energy from the oceans, tidal, wave, and ocean thermal energy conversion (OTEC) began to receive renewed interest. Considerable R and D has recently been made into the utilization of OTEC and wave energy whilst tidal energy has already started to be exploited.

Tidal energy is extremely site-specific, requiring mean tidal differences of greater than 4 m and also favourable topographical conditions, such as estuaries or certain types of bays, in order to be economically viable. This is necessary in order to bring down costs of dams etc. Besides two major plants now operating off Murmansk, the USSR (4 MW), and Rance, France (240 MW), a few others are currently being planned, e.g. a several thousand MW plant off the Kila peninsula, USSR. Canada has a 20 MW plant under construction in the Bay of Fundy. China is operating or planning a large number of small tidal power plants with a total capacity of some 8 MW. On a global scale, the most promising sites have already been identified and several have already been subject to preliminary technical and economical evaluation, similar in scope to those being done for large-scale hydropower schemes.

Wave energy is presently subject to research programmes in several industrial countries, particularly Japan, Norway and the UK. Current annual total global R and D expenditure amounts to about US$ 25 million. The energy carried by ocean waves is large; typical mean values are between 20-70 kW/m of wave front. A shoreline of 1 km would thus receive 20-70 MW which converted to electric power, with an efficiency of 50%, would produce 10-35 MW. Wave energy resources are, however, unevenly distributed and the best potential seems to lie in the oceans of high latitudes.

A great variety of technologies for wave power plants have been suggested and in some cases tested. The reciprocal motion of a floating device may be used either to drive a generator directly, or to pump water or compressed air to run specially designed turbines. In addition, a few designs based on the focussing of waves into an on-shore inlet, avoiding the difficulties of floating designs, are now being developed. One of them, the 500 kW Kvaerner MOWC is now tested outside Bergen in Norway.

Several recent optimistic estimates regarding the costs of electricity from wave power fall in the range of US¢ 3-6/kWh. These have proved encouraging enough to merit further R and D programmes; however, other estimates are less optimistic. Several years of R and D may lie ahead before viable, competitive wave power plants become available.

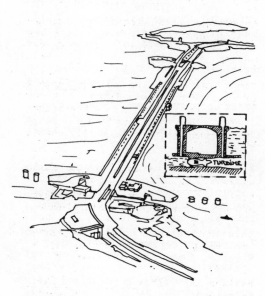

Tidal power plant in Rance, France.

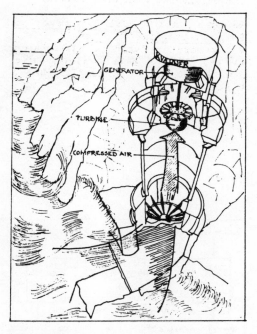

Diagram of Norwegian experimental on-shore wave power plant.

269

Ocean Thermal Energy Conversion (OTEC) utilizes the temperature difference between the warm (26°C) surface waters of low-latitude tropical oceans and the cool deep waters (5-10°C), lying below a depth of a few hundred m, to run a turbine and a generator via a heat exchanger and a suitable medium (e.g. ammonia or low pressure steam). OTEC is thus a method of using solar energy stored as temperature differences in the oceans. The theoretical global energy potential is enormous and not subject to large time variations, as would be the case with wave energy. In practice, however, the potential seems severely limited by the problem of finding suitable sites which are reasonably close to centres of demand. Several technical designs have been proposed and some actually tested since the 1930s. The main problems have been to overcome the high capital costs of the large piping units required to pump cold deep water up to the warmer surface layers, together with the low efficiency of the turbines due to the small temperature differences.

Research programmes are currently underway in France, Japan, the US and a few other European countries. Cost estimates are still very uncertain but range from US¢ 5/kWh (considered highly optimistic) upwards for future large-scale projects in the 100 MW range, with those for smaller units being considerably higher. Several pilot plants are presently being planned or proposed, some in combination with aquaculture or fresh-water production in order to improve on the overall economy; however, it would appear that many years of R and D are necessary before economic viability can be achieved. Current designs cannot compete economically with available alternatives such as land-based diesel or coal-fuelled power plants.

In addition to the three methods of harnessing ocean energy mentioned above, the use of ocean currents and salt gradients, i.e. the difference in salt content between sea and fresh water in river outlets, have also been proposed for energy production. Such methods are, however, still at the level of basic research. Practical application is still far off and meaningful cost estimates are not yet available.

Conclusions

Tidal power is a well-established technology for power production. The economy is very site-specific and few good sites may exist in the developing countries. Exploration is done on commercial terms.

Wave power and OTEC are both in the research and prototype stages. Ongoing development concerns research on cost-effective designs and systems analysis. Current high costs and lack of experience so far prevents any commercial applications, although several research projects in developing countries' waters have been suggested.

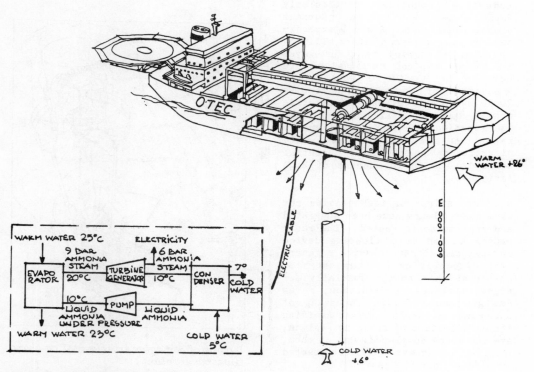

Experimental OTEC power plant.

Bibliography

Davies, P.G. et al (1985) Wave Energy. The Department of Energy's R & D Programme 1974-1983.
ETSU Report R26.
Energy Technology Support Unit (ETSU), AERE Harwell, Oxfordshire, OX11, ORA, U.K.

Flood, M. (1985) The Troughs and Crests of Wave Energy.
Chartered Mechanical Engineer, June 1985, London, U.K.

Lavi, A. (1980) OTEC; A General Introduction.
Energy, No. 5, 1980, Page 469.
Pergamon Press, Headington Hill Hall, Oxford X3 OBW, U.K.

Merriam, M.F. (1978) Wind, Waves and Tides.
Annual Review of Energy, Vol. 3, 29.
Annual Review Inc., Palo Alto, California, U.S.A.

Ocean Thermal Energy for the 80's. (1979) Proceedings of 6th OTEC Conference, Washington, D.C. June 1979.
Applied Physics Laboratory, The John Hopkins University, U.S.A.

Sørensen, B. (1980) Renewable Energy.
Academic Press, London, U.K.

Stephens, H.S. & Stapleton, C.A. (1981) Wave & Tidal Energy.
Proceedings of Second International Symposium, Cambridge, September 1981.
BHRA Fluid Engineering, Cranfield, Bedford, MK 43 OAJ, U.K.

Symposium on Hydrodynamics of Ocean Wave-Energy Utilization. (1985)
Lisbon July 1985.
International Union of Theoretical and Applied Mechanics.
Springer-Verlag, West Germany.

Voss, A. (1980). Waves, Currents, Tides - Problems and Prospects.
Energy, No. 4 1980. Page 823.
Pergamon Press, Headington Hill Hall, Oxford OX3 OBW, U.K.

Waid, R.L. (1979) The Mini-Otec Test.
Lockheed Missiles & Space Company Inc., Sunnyvale, California 94086, U.S.A.

West, M. et al (1984) Alternative Energy Systems, Electrical Integration and Utilization.
Proceedings of the Conference, Conventry. September 1984.
Pergamon Press, Headington Hill Hall, Oxford OX3 OBW, U.K.

24 WIND POWER

Introduction

The first electricity-generating wind machine was developed in Denmark in 1890. Soon afterwards engineers realized that to generate electricity efficiently, fewer and thinner blades were needed compared with the blades used in the traditional windmills and windpumps. These sleek new machines found a wide market in Denmark, the United States and a few other countries during the twenties and thirties. Most were used to supply electricity to farms and other remote operations without access to a central grid. Wind power (and windpumps, see Chapter 26) made a vital contribution to the development of the great plains of the USA and of Australia's outback.

From the twenties onward, rural electrification signalled the demise of wind machines in much of the world. Research went on in Britain, Denmark, France, the United States, the Soviet Union and West Germany and wind turbines with up to 20 m blades and 100 kW of generating capacity were built, but cheap oil and optimism over the development of nuclear reactors made wind machines seem somewhat antiquated.

It took the energy crisis of the seventies to spur a wind power revival. Since 1973, dozens of small wind machine manufacturers have started up and both private companies and national governments have carried out research on larger and more sophisticated turbines, resulting in a large up-swing in the technology of windmills, and the introduction of grid-connected windpower stations in a few countries in Europe and in the U.S.

Jacobs wind turbine, 1930.

There are also several regions where wind electricity may well come to play an important role in developing countries both in supplying electricity in remote areas and supplementing conventional generating plant in the main grid. Development of modern wind energy equipment is however, both difficult and expensive and should be avoided by developing countries. Such countries are advised to consider using the technologies developed by the developed nations.

The current development can be divided generally into:

1   Small wind turbines intended to charge batteries. These have a turbine diameter ranging from 0.5-5 m and a power output of 50 W-2 kW. Some designs have survived since the 1930s; others have been developed during the last ten years and there are several well-proven designs intended for battery charging now available on the market. In combination with new energy-efficient flourescent lamps, wind turbines of this size are well-suited for providing small amounts of energy for lighting and telecommunications in regions not connected to the electricity grid.

2   Medium sized wind turbines to be used in connection with a large electrical grid. These machines have a turbine diameter of 7-17 m and a power output of 10-75 kW. In Denmark wind turbine generators provide a relatively common means of generating electricity and about 1400 wind turbines of this size have been installed by about 10 companies. The Danish 55 kW wind turbines currently produce the most economical wind electricity available at about US¢ 10/kWh. This size of wind turbine is also gaining popularity in the USA, but the US approach is to develop commercial windfarms – there are already about 12 000 wind turbines installed at windfarms, mostly in California – most machines having been manufactured in USA, Denmark and the Netherlands. Some of the machines are as big as 100 kW. Such systems are proving reliable but require a well-organized back-up service facility.

3   Medium sized wind turbines in conjunction with small independent grids. The most common configuration here comprises one or several wind turbines connected to the

Wind Turbines Connected to the Grid

In Denmark and the USA there is a growing practice of coupling wind turbines in the 10-100 kW range directly to the grid. The main aim was originally to save electricity, but now energy credits can be obtained when surplus electricity is produced. According to Danish experience of typical household wind installations, the yearly production of the wind turbine and the owner's electricity consumption work out as virtually identical. 45% of consumption is obtained directly from the turbine whilst the remaining 55% is first sold to and then bought back from the grid, so providing the producer with flexibility of power usage without having to invest in battery storage. If there had been no grid connection, this amount of energy would have been wasted. (One has to be cautious, however, in drawing conclusions from this example since the consumption pattern in a developing country is quite different from that in Denmark).

This situation has led to the development of privately-owned windmills and wind farms which are run as a business selling electricity to local utilities. Hence the development of grid-feeding wind turbines in the 10-100 kW range is proceeding at a rapid pace and probably makes these machines more cost-effective at present than the prestigious megawatt sized giant wind turbines being developed with government funds. The reason for this is that the scale of production and greater experience with medium-sized machines leads to greater economies of scale than those theoretically attainable with MW-sized machines.

Medium-sized wind generators have a rotor diameter of 7-17 m. The system cost (1982 figures) is between US$ 20 000 and 40 000 depending on the size, and the unit cost is US¢ 9-15/kWh. The main companies producing machines of this size include Carter Wind Systems Inc., ESI Inc., Enertech Corp., Fayette Manufacturing Corp., Dynergy Systems Corp., and Century Design Inc. from the USA, Polenko

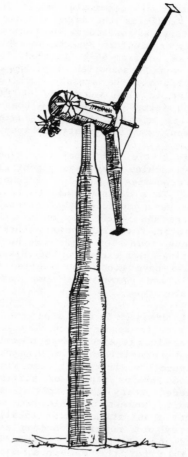

Danish 55 kW wind turbine.

and Bourma from the Netherlands and Vestas, Bonus, Nordtank Micon Wind Turbines Inc., and Windmatic from Denmark.

There are also two companies in USA that produce vertical axis turbines of this size, Flo-Wind and Vawtpower.

Denmark already has about 1400 grid-connected machines. In the USA there were approximately 12 700 wind turbines having a combined capacity of about 1000 MW installed by the end of 1985, most of them in California, where wind energy production is favoured by the tax system. In most cases with such windfarms, wind-generated electricity costs USC 9 to 15 per kWh. As a result, wind energy farms are now increasingly

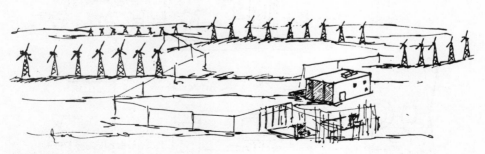

Wind farm.

being built on a commercial basis. In those windfarms the trend towards larger turbines continues. The average capacity of windfarm systems in 1984 was 81 kW, up from 66 kW in 1983. In the Altamont Pass and Tehachapi areas in California the average turbine installed in 1984 was rated at 97 kW. Such farms produced more cost-effective electricity than single large-scale multi-megawatt machines.

There is little doubt that this kind of wind machine does have potential in the economies of developing countries. This is particularly true for the many countries that are oil importers and which rely on diesel generation for grid power. Some of these countries typically have generating costs exceeding US¢ 15/kWh but also have good wind regimes - many island communities come into this category.

The main drawback is that wind generators require regular servicing and frequent minor repairs by experienced, skilled technicians. In Denmark, experience has shown that some kind of wind turbine repair or service is needed every third month. As a result, unless reliable back-up service organizations exist locally to carry out repairs, and unless sufficient spare parts are regularly available for the machines, wind turbines cannot be regarded as reliable power generating systems for the developing world.

Wind Electricity in Small Independent Grids

In such systems electricity consumption fluctuates constantly as does the availability of wind energy. The degree of coincidence of supply and demand can be calculated by statistical means and it has been found that electricity supply with an acceptable degree of reliability cannot be based solely on wind energy. If an extensive grid does not exist, electricity storage (batteries) or a back-up system (diesel) is

required. Loads for remote systems of up to 6 kWh/day equivalent to an average power consumption of 250 W with a duty cycle of 24 hours, can be provided with battery storage.

If a diesel and wind generator are used in conjunction with a grid, the diesel generator should only be used when wind energy is absent. Problems can occur, however, when the diesel generator is called on to change its output frequently as wind energy availability fluctuates. Besides decreasing the oil saving, diesel generation on this basis leads to more frequent overhauls of the generator. Both factors will increase costs. Several methods of overcoming these problems have been tried but there is not yet an established solution. Some development work has still to be done before wind generators can be run in parallel with diesel on a routine basis. Practical experience so far exists only in developed countries. On Block Island, the USA, a 200 kW wind turbine used in conjunction with a diesel generator in a small grid has its power output restricted to allow the diesel system to operate without fluctuation and, at Holyrod in Canada and on Kythnos Island, Greece, four 50 kW wind turbines and five 11 kW turbines (respectively) are shut down when their power output falls below 4 kW in order to allow the diesel plants used in conjunction with the wind generators to operate at outputs in excess of 20 kW at all times. On Arran Island, Ireland, a 63 kW wind turbine has a load dumping device built on it so as to reduce input to the local grid by the wind turbine in order to let a 12 kW peak load diesel generator and a 44 kW base load generator function smoothly.

Work is going on to solve these problems. Chalmers University in Sweden and the Riso Research Centre in Denmark are cooperating on one project and the Dutch at Einhoven University and the Energy Research Centre, (ECN) at Petten, are also working on the problem. Work is also continuing in Ireland, the UK, the USA and West Germany.

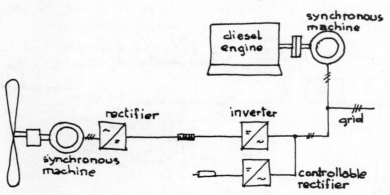

Principle of combined wind/diesel power generation.

Wind Electricity Economics

Wind generator power costs are heavily linked to the characteristics of a wind resource in a specific location. The cost of supplied power declines as wind speeds increase, and the power supplied increases in proportion to the cube of the wind speed.

Matching available energy and load requirements is also important in wind energy economics. The correct size of wind generator must be chosen together with some kind of storage or co-generation with an engine or a grid to obtain the best economy. The ideal application is a task that can utilize a variable power supply, e.g. ice-making or water purification.

Regarding the economics, the choice of interest rate obviously has a major effect on the overall energy cost. With low interest rates, capital intensive power sources such as solar and wind are favoured.

Other factors bearing a strong influence on the economics of wind electricity are the standard of maintenance and service facilities (will breakdowns be promptly repaired?) and the cost of alternative energy supplies in the particular area.

Taking the Danish Windmatic 14.5 m diameter machine as an example, with a mean wind speed of 5 m/s, the machine will produce about 50 000 kWh of electricity/year - the equivalent of using 17 000 litres of fuel in a diesel generator running at 30% efficiency. Assuming an installed cost of US$ 43 750, operation and maintenance costs at 5% of installed cost per year and finance at 10% interest per year over 10 years, the wind machine generates electricity at US¢ 27, 14 and 8 per kWh for mean wind speeds of 4, 5 and 6 m per second respectively. Assuming diesel costs US¢ 40/1, the wind turbine will show a return on investment in 12.9, 6.6 and 3.75 years for each of these wind speeds.

Studies of energy cost to rated power ratios have found that, by and large, the cheapest power is supplied from wind generators with 20-60 kW capacities costing US$ 20 000-40 000 to install and producing power at US¢ 9-14/kWh in 5 m/s winds. Systems in the 3 to 12 kW capacity range cost US$ 3000-10 000 to install and produce power at US¢ 17-25/kWh in 5 m/s winds, while systems in the 0.5-2 kW range costing less than US$ 3000 to install produce power at US$ 0.5-1.0/kWh in 5 m/s.

Wind turbines also appear economically favourable for small loads of about 3 kW/day. Such loads are usually

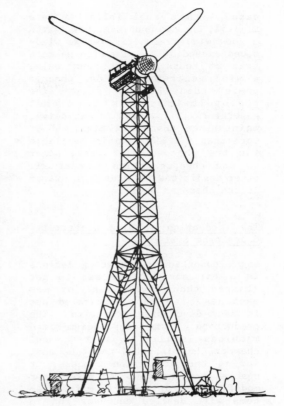

Danish "Vestas" wind generator.

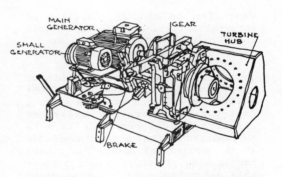

Danish windmills often have two generators.

associated with battery charging and it is worth comparing wind electricity costs against solar and diesel generators. A wind generator for this purpose will cost about US$ 625/m^2 (area swept by the propeller), it will have a maintenance cost of US$ 50/year and the cost of batteries is US$ 275/kWh. A solar photovoltaic system costs US$ 16 per Wp in 1982 and is expected to cost in the region of US$ 8 per Wp in 1990. A diesel powered generator will cost about US$ 2500 and here two cases have to be considered - a low cost case with maintenance at US$ 200/year and diesel at US$ 0.38/1 and a high cost case with maintenance at US$ 400/year and diesel at US$ 0.76/1. If the efficiency of a wind powered generator is taken as 20%, that of a photovoltaic array as 10%, and that of a diesel generator as 10% (low case projection) or 15% (high case) with a fuel inflation rate of 3.5%, one can conclude that, given a good wind regime of average wind-speed exceeding 4.5 m/s with no more than 36 hours of continuous calm, a wind generator provides cheaper energy than the best figures for photovoltaic arrays or diesel generators. If diesel is expensive, maintenance costs are high and if more than 4.2 kWh/m^2.d is available for the photovoltaic array, solar is the cheaper option. Under all other conditions the diesel generator is the cheapest option.

The Production of Wind Electricity Generating Systems

Local poduction of existing designs is a far simpler process to get started than development of new machines, and can be carried out in many developing countries. The production of small and medium-sized machines locally is a great deal cheaper than imported machines and, during the production process, it enables manufacturers to make minor modifications in order to match systems with desired end-uses and to the conditions under which they are expected to operate.

Depending on the availability of materials, rotor-blades can be made locally from laminated wood, steel, plastic or combinations of these materials, whilst some of the machinery components can be made by small engineering workshops. Other parts, including special bearings, gear-boxes, generators and other electrical equipment may have to be imported if they are not available in the country of assembly. Towers can be made of welded steel, preferably galvanized, which can be manufactured in many local engineering works, whilst the foundations can be cast from reinforced concrete on site.

There are, however, a number of pitfalls of which potential local wind turbine manufacturers should be aware. First, it is usually necessary to buy a turbine design from the developer - a process which can prove time consuming and costly. If a design is too expensive to buy outright, a manufacturer may consider taking out a licensing agreement on a particular design but then, before entering into such an agreement, he must first spend time, money and effort evaluating which foreign design he should undertake to manufacture, since agreements arrived at under this procedure are often binding over lengthy periods.

Once a manufacturer has decided on a design and is ready to start production, he is then faced with the problem of promoting and marketing the turbine. In developing countries, the market is usually uneducated in such new technologies and hard to convince of the potential of such systems. Finally, there is the problem of financing such a venture. It is often difficult or impossible in some developing countries to obtain risk capital to introduce even tried and trusted technologies. This problem is compounded when a company requires substantial funds for a new and relatively unknown technology, but it is often an advantage to have a local company acting as agent for a foreign concern. A local company is better placed to promote the turbine locally than a foreign concern; a local company will better understand the requirements of the local market, and the cost of locally-built turbines will be far lower than imported machines, as import duties and tariffs can largely be avoided.

Such advantages naturally presuppose the existence of competent and efficient local manufacturers marketing a good product. Unfortunately experience tends to suggest that there are still relatively few companies in the developing world which adequately fill these requirements. However, some local concerns in developing countries have started manufacturing wind turbines. In India, for example, Joyti Ltd. of Baroda has designed and is producing turbines in the 40 W, 350 W and 1 kW range and Escorts Ltd. of Faridabad is collaborating with MAN of West Germany to mass-produce 10 kW rated wind generators. In China, the Commission of Science and Technology, the Ministry of Farm Machinery and the Ministry of Electric Power have all supported the production of 40-55 kW capacity wind turbines in a number of Chinese engineering works.

India´s first windfarm will be installed in Gujarat, Kandla, India. The windfarm is a joint project between the state-owned Gujarat Industrial Investment Corporation and Western India Erectors Ltd. (WEIL) of Poona. The power will be sold to the Gujarat Electricity Board. Two Danish-designed wind turbines will be used for the project. Danish Wind Technolgy (DWT) of Viborg, Denmark and Micon A/S of Randers, Denmark, have signed licence agreements with WEIL for the manufacture of certain of their turbines in India. DWT has licensed WEIL to manufacture its 18.5 kW Windane 9 model, although the first systems are, being shipped from Denmark until WEIL gets into production. Micon has licensed WEIL to manufacture its 22 kW, 55 kW and 110 kW models in India. Fourteen of the 60 kW and two of the 110 kW units have already been shipped from Denmark.

Conclusions

Very few wind generators are presently manufactured in developing countries. Several countries, however, report that development manufacturing programmes are under way.

The development of small wind electric systems is, therefore, tending to be biased towards systems adapted for grid connection in industrialized countries away from small systems of most immediate use in developing countries. The developed country models may however, also come in useful in developing countries, in order to save diesel fuel in grid systems.

However, 5 kW to about 50 kW wind electric systems used in conjuction with diesel generating sets or for feeding small grid systems appear to be potentially useful in the future for developing countries. These should be studied, developed and tested in order to get good reliable output.

The other application for windpower of considerable potential importance to developing countries is through using small wind generators to charge batteries for such essential purposes as lighting, radio communication, or operating hospital or clinic medical equipment.

For small wind generators to become economically viable with present-day prices, mean wind speeds of at least 3-4 m/s are required, where the wind regime is known, or probably 4-5 m/s where not much detail of the wind regime is known or where the data may be suspect.

Maglarp 3 MW wind turbine in Sweden.

Bibliography

Both, D. & van der Stelt. L.E.R. (1984) Catalogue of Windmachines. Consultancy Services Wind Energy Developing Countries (CWD) P.O. Box 85, 3800 AB Amersfoort, The Netherlands.

de Bonte, J. (1983) Review of Conversion Systems used in Autonomous Wind Energy Systems. Wind Energy Group, Eindhoven University of Technology, P.O. Box 513, 5600 MB Eindhoven, The Netherlands.

Fraenkel, P. et al. (1983) Wind Technology Assessment Study. UNDP Project GLO/80/003 executed by The World Bank. Intermediate Technology Power Ltd., Mortimer Hill, Mortimer, Reading, Berks, RG7 3PG, U.K.

Holmblad, L. & Linders, J. (1984) A 25 kW Autonomous Wind-Diesel System, with Transistor Convertors and Accumulator Storage. Department of Electrical Machines and Power Electronics, Chalmers University of Technology, Göteborg, Sweden.

Lundsager, P. & Madsen, H.A. (1985) Status of Wind/Diesel Systems. Article in Windpower Monthly Vol. 1 No. 6, June 1985. Forlaget Vistoff ApS., Vrinners Hoved, DK-8420 Knebel, Denmark.

Lysen, E.H. (1982) Introduction to Wind Energy. Consultancy Services Wind Energy Developing Countries (CWD) P.O. Box 85, 3800 AB Amersfoort, The Netherlands.

Marier, D. (1985) Windfarm Update. Article in Alternative Sources of Energy, No. 72, March/April 1985. Alternative Sources of Energy Inc., 107 S. Central Avenue, Milaca, MN 56353, U.S.A.

Nelson, V. & Caldera, E.M. (1985) The Latin American Market, A Preliminary Look at Wind Energy Potential. Article in Alternative Sources of Energy, No. 75, Sept/Oct. 1985. Alternative Sources of Energy Inc., 107, S. Central Ave. Milaca, M.N. 56353, U.S.A.

Stephens, H.S. & Goodes, D.H. (1982) Wind Energy Systems. Proceedings of the Fourth International Symposium held at Stockholm, Sweden, September 1982. BHRA Fluid Engineering, Cranfield, Bedford, MK 43 OAJ, U.K.

Sørensen, B. (1980) Renewable Energy. Academic Press, London, U.K.

Twidell, J. et al. (1983) Energy for Rural and Island Communities. Proceedings of the Third International Conference held at Inverness, Scotland, September 1983. Pergamon Press Ltd., Headington Hill Hall, Oxford OX3 OBW, U.K.

U.N.S.O. (1982) Development of the Utilization of Wind Energy in Cape Verde. UNSO/CVI/81/002. United Nations Sudano-Sahelian Office (U.N.S.O.)

Wind Power Digest (1984) Wind Energy Directory 1984. Wind Power Publishing Company, 398 E. Tiffin Street, Bascom, P.O. Box 700, Ohio 44809, U.S.A.

25 WATER PUMPING: AN OVERVIEW

"Safe drinking water and sanitation for all by the year 1990" is the goal of the UN Water Decade. Over half of the population in the developing countries have no access to safe water and three out of four people have inadequate facilities for the disposal of excreta. The WHO estimates that between 10-25 million deaths each year, and 80% of all the world's sickness, can be attributed to inadequate water or sanitation.

According to World Bank calculations, the total investment required to meet the Water Decade target could run up to US$ 600 billion. The United Nations International Children's Education Fund (UNICEF), one of the leading proponents of low-cost appropriate water and excreta disposal technologies, cuts this figure by half to US$ 300 billion, but either figure demands an astronomical input.

There are several techniques to ensure and improve water supplies, for example improving existing water services, or building new dams and reservoirs to provide water for large urban populations. The needs of the rural poor can also be met by installing small, decentralized pumping systems to use available groundwater − a community hand pump may serve up to 5000 people. This overview chapter deals with small-scale water pumping for rural use and, in particular, with the possibilities of using renewable energy as the power source.

Several of the technologies discussed in this Chapter are treated in detail in other Chapters, for example, Draught Animal Powered Pumps in Chapter 10, Biomass Engine Pumps in Chapters 12, 13 and 14, Solar Powered Pumps in Chapters 16 and 17, Wind Pumps in Chapter 26 and Hand Pumps in Chapter 27.

Water Use

In rural areas, the three main uses for water are: domestic demand, livestock demand and irrigation.

The main domestic demands are drinking, cleaning and washing. The arrangement of water distribution points varies due to factors such as the volume of water required, the spacing of the consumers, the degree of reliability that is wanted and what is affordable. Typical demand values are about 10-15 litres per capita/day, for people carrying water a moderate distance of between 500 and 1000 m. In the case of individual tap connections, demand is usually higher,

being in the order of 30-50 litres per capita/day.

The value of water for livestock is high because even quite modest quantities can sustain a sizeable herd. The consumption varies from 5-75 litres per unit/day for sheep and cattle respectively.

Due to the limited value added in terms of crop production/m^3 of water, the volumes of water required for irrigation purposes are of such an order that irrigation water has to be really cheap. To produce 5000 kg of corn over 12 000 m^3 of water are needed. Irrigation water demand also fluctuates considerably with the seasons − peak daily water requirement may be about 6 millimetres (i.e. 60 000 litres/day/ha). The amount of water needed also depends on the irrigation systems themselves which have different efficiences depending on leaching, evaporation and other losses. Open channels have an overall efficiency of 50% while trickle pipes have efficiencies of up to 85%.

Use	Daily requirement
Domestic	
minimum for survival	5 l/person
water carried home from distant communal supply	10 l/person
water carried home from nearby communal supply	30 l/person
one tap in each house	50 l/person
multiple tap connections	200 l/person
Livestock	
cattle	35 l/head
horses, mules and donkeys	20 l/head
sheep and goats	5 l/head
poultry	25 l/100
pigs	15 l/head
Irrigation	
including conveyance and field application losses	5-10 mm or 50-100 m^3/ha

Approximate water requirement for various purposes.

Source: IRC (1983).

Choice of Pump

Pumps vary greatly in performance, depending on their efficiency and output capacity. Correct pump choice is probably the single most important factor in designing a small-scale pumping system. In the case of electric pumps, about 10% of the input energy is lost in the electric motor, 2% in the coupling between motor and pump, and another 20% in the pumping process, even when operating at optimal efficiency. If the water pumping is subject to further, avoidable, energy losses, this will add greatly to the overall pumping cost.

The required water output and the water sources, depth and type of construction will influence the choice of pump. Pumps that rely on suction cannot lift water from depths of more than 6-7 m. If the water is further down than this, the pump works by pressing up the water and thus needs to be positioned below ground level sufficiently close to the water table, or submerged in the water. Seasonal variations in water level and water needs must be taken into account when selecting a pump.

All pumps wear with time, and the amount and rate of wear greatly affects the pump's efficiency. This means that maintenance is very important. Even when there is only a limited choice of pumps available locally, it is generally better to opt for one of these than to import a different type for which few spare parts and little maintenance expertise will be available.

Clearly, the range of available energy options is of decisive importance in the choice of a suitable pump. In the absence of a power grid, factors such as the price and reliability of the supply of diesel fuel, and wind regime characteristics, etc. will greatly influence the overall performance and economy.

Basic Pump Types

Single stage centrifugal pumps have the advantages of compactness and simplicity. They can be constructed for a wide range of flow and head situations, and be combined in series for high head applications. For deep wells, electrically-driven submersible pumps are generally more suitable than pumps directly coupled to a windmill or an engine. In general, centrifugal pumps require a low rate of maintenance and have life times of about 5-10 years. Efficiencies are 80% in large systems and around 50-70% in small set ups.

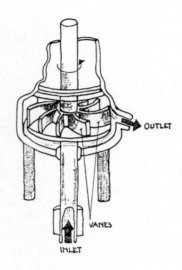

Centrifugal pump.

To choose the right pump for the right situation is essential for a successful water programme.

The displacement pump is the traditional design for water pumping and
most hand and wind pumps are of this
type. For higher heads, displacement
pumps can be more efficient than
centrifugal pumps. Displacement pumps
work well both at low and high speed.
The delivery head can vary over a
wide range, even if they are best
suited for high-head and low-flow
conditions. The main problem is
their high maintenance requirements.
The efficiency of displacement pumps
can be up to 80% if the packing and
valves are in good condition. With
leaking packing or cracked valves
the efficiency will decrease sharply.

Special types of displacement pumps
are the diaphragm pump, the petro
pump and the Vergnet pump. There
have been hopes that diaphragm pumps
might, through low internal friction,
prove a superior type of positive
displacement pump for low head applications. However, this particular
type of pump has not proved to be
competitive with the better centrifugal pumps. Most displacement pumps
run at much lower speeds than standard
engines and thus require (at additional cost) a transmission system
to match the differing speeds.

The rotary pump is the modern version
of the Archimedean screw. The pump
consists of a screw which rotates
inside a drum. It can only be used
to lift water through a few m, but
it is insensitive to water with a
high content of particles or mud.
A modern rotary pump is the Helical
Pump. This can be used in a borehole
submerged in water. These pumps also
have a good tolerance to abrasive
particles in the water and can be
made of corrosion-resistant materials. Helical pumps do not have
a very high efficiency, but their
output is very versatile. For heads
less than 15 m, however, they probably
cannot compete with centrifugal pumps.

In a jet pump, water at high pressure
is passed through a nozzle to endow
it with a high velocity. The resultant jet creates a low-pressure area
causing water to flow in from the
suction entrance. A jet pump has
no moving parts and can be built
compactly; however, it needs an auxiliary pump to pressurize the driving
fluid. A jet pump system is a good
alternative when the suction height
exceeds 5 to 8 metres or when the
bore-hole has a small diameter.
Efficiency is lower than that of
the centrifugal pump and it is also
more sensitive to water with a high
proportion of suspended material.
Jet pumps can also work with gases
as the driving fluid. One variant
of this theme is the air-lift pump.

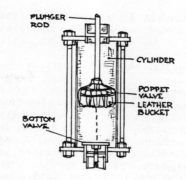

Piston pump.

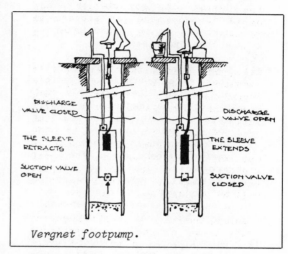

Vergnet footpump.

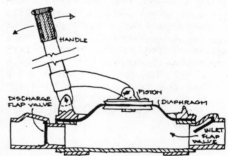

Diaphragm pump.

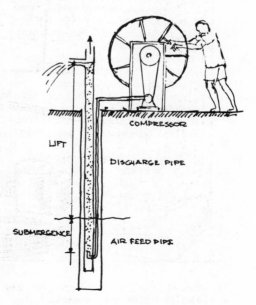

Airlift pump.

The following is a summary of the pumping options presently available:

1 Traditional Pumping Methods

a) <u>Gravity fed</u>. If a water source is available at a higher elevation than its point of use, gravity can supply water via pipes or aqueducts. Horizontal wells can be drilled or dug in order to transfer water by gravity from mountains, for example the Persian qanats. The collection of water during rains from roofs or catchment areas and its storage in cisterns or as ground water is another possibility.

b) <u>Human power</u>. (See Chapter 27). There are many traditional methods of lifting water manually such as a rope and bucket, the Archimedean screw and the counter weight. Modern devices to pump with human power include a great variety of hand pumps. The major problem with these is reliability. More power can be produced with legs than with arms, and several footoperated pumps have been developed.

c) <u>Animal power</u>. (See Chapter 10). There are several traditional methods of lifting water which can incorporate the use of animal power e.g. the sack and rope. Pumps developed specifically to use animal power include the Noria wheel, the chain and washer pump and the Sakia. Modern developments have included the introduction of gear boxes and rotary pumps.

2 "Free" Fuel Pumping Methods

a) <u>Water power</u>. (See Chapter 22). If falling water is available then mechanical pumps such as the hydraulic ram and the Chinese turbine pump are suitable. Water wheels in rivers have been used traditionally to pump water, and river turbines are modern interpretations of this concept. Electricity from generating units (micro- hydro) which are located near falling water can be used to drive pumps.

b) <u>Wind power</u>. (See Chapter 26). A wide variety of modern wind pumps are produced and adapted for different heads and capacities. As irrigation water must be cheap, at present, only simple, locally made wind pumps are economical for irrigation purposes. In many places the water source is in one place and the wind in another - here, electricity generated by wind can run pumps.

c) <u>Solar power</u>. (See Chapters 16 and 17). Small solar photovoltaic pumps are commercially available but are still expensive, although development is proceeding rapidly. Solar thermodynamic pumps are still at the prototype stage. Cheaper photovoltaics might make them uneconomical.

3 Mechanized Pumping Methods

a) <u>Commercial</u>. If a central electrical grid to power pumps by electricity is close, it is usually the most economical solution. However, extension of a grid is costly. For diesel pumps, the cost depends on fuel prices and maintenance possibilities. High heads and large flows prove more economical. Kerosene pumps are cheaper and smaller than diesel pumps but fuel is usually more costly and their lifetime shorter.

b) <u>Biomass in internal combustion engines</u>. (See Chapter 12). Biomass (e.g. charcoal), can be used in producer gas units to yield a gas that can power a diesel or petrol engined pump. Some forms of biomass (e.g. sugar cane) can be used to make alcohol which is then used, either on its own or as a supplement, to fuel petrolengined pumps. Biogas (see Chapter 9) from the digestion of animal wastes can be used to power a petrol (or kerosene) engine and drive a pump.

c) <u>Biomass in external combustion engines</u>. (See Chapter 13). Biomass can be burned in a boiler to produce steam which can be used directly to drive a water pump. Biomass can also be burned in a boiler to produce steam to power a steam engine that can drive a pump. Alternatively biomass can be combusted to produce heat that can power a Stirling engine and so drive a pump or a generator.

Gravity-fed systems

A pump always needs regular maintenance, spare parts and has a limited
life. So if you can distribute water
without pumps you should do so. There
are several methods which are fed
by gravity.

Gravity can be used to move water
downhill from a source at higher
elevation than the village. Water
is collected at an intake and is
fed through a pipeline to a reservoir
above the village and fed to tap-
stands. Tanks may be needed to break
water pressure en route, and a sedi-
mentation tank can be used for purification. This was a method used
formerly by the Romans in their aqueducts.   Recent work on the method
has been carried out by UNICEF in
Nepal.

Horizontal drilling is another reasonably easy and low cost method for
tapping water resources in a way
that would provide water without
pumping, just by gravity flow. This
is also an old method dating back
to 800 B.C. when the Persians began
to dig the famous qanats. A qanat
exploits the natural terrain to transfer water by gravity from the mountains to the plains. In Iran some
270 000 km of qanats still supply
35% of the country's water. With
modern engineering, geology, and
hydrology applied, this principle
could play a role in future water
production in arid lands. UNICEF
has done some work on it recently
in Indonesia.

Rainwater harvesting, collecting
water during rains and storing it
for use during the rest of the year,
usually in underground cisterns,
is an extremely ancient practice. The
rainwater can be harvested from roofs
or from the ground. If the ground
is permeable an artificial catchment
can be made by laying concrete, asphalt or polythene sheeting on the
ground or by treating the soil surface
chemically or biologically to make
it impermeable. Catchment tanks should
be covered to prevent evaporation
so a common practice is to build
the tanks underground. In individual
systems the tank can be placed inside
the house in order to cool the house.

A rain catchment system developed
by the Danish aid organisation DANIDA
in Kenya does use rain catchment
dams of earth or concrete to catch
the water in streams during the rainy
season. This gives the water time
to sink down in the earth, and an
artificial ground water (or subsur-
face dam) is created which can be
tapped through a well for the rest
of the year.

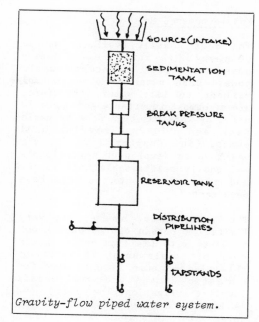

Gravity-flow piped water system.

Horizontal well.

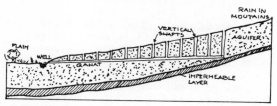

Persian qranat.

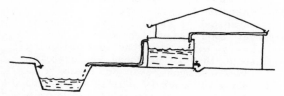

Rainwater catchment.

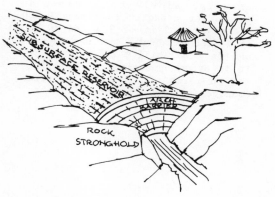

Water catchment dam.

Muscle-powered Methods

The most readily available source of energy in the rural areas of developing countries is human muscle power. Hand pumps or other manual methods to lift water, therefore, are of utmost importance and recently a lot of work has been done by World Bank and UNDP to develop hand pumps. (See Chapter 27.) The aim has been to develop cheap, reliable and easily maintainable hand pumps, and many different concepts have been tested.

The human power, however, is very limited. A person can produce about 50 watts with his arms and with the help of pedals about 75 watts. So this kind of pump is only used for domestic demand and small-scale low-lift irrigation. The energy cost is high, and depends on the actual salary. But this cost is acceptable for rural water supply when people are pumping their own water, but the cost tends to be too high for irrigation purposes. The capacity is low and depends on the lift. However, most hand pumps have outputs below 1 litre/s (e.g. about 1000–3000 litres/hour). Low-lift irrigation devices like the Archimedean screw have capacities of about 10 000–30 000 litres/hour when lifted less than one metre. One-man chain pumps have the capacity of about 10 000 litre/metre at a lift of 3 m.

Because of this limitation in power, man has sought, since ancient itmes, to utilize other sources of power. Once he learned that a pair of oxen have the pull of about 14 men, he pressed these and other animals into his service. Draught animals are a common and vital source of power in many developing countries. (See Chapter 10.) The power output of an animal is usually about 600 watts per 1000 kg of its weight. Thus a heavy horse or a bullock weighing about 1000 kg produces at least ten times as much power as a man. This source of power is usually associated with slowmoving water-lifting devices for irrigation, such as water wheels which can be driven by an animal treading slowly on a circular track. The capacity is about 10 000 litres at a lift of 9 m and 70 000 at a lift of about 1 m. Cost depends on the availability of fodder, but it is often economical at low lifts and on farms of about 1–4 hectares compared to other methods.

Chinese ladder pump.

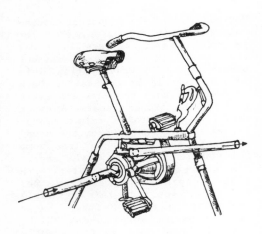

Chinese pedal pump.

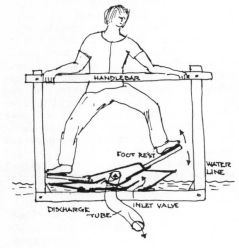

Bellow pump.

Water Pumping Economy

The unit cost of water delivered by a pumping system is probably the crucial factor that determines which type of pumping system should be used for a given application and location.

There are three main uses for water pumps and the economics of two of them, village water supply and live-stock water supply, can be treated in one way whilst water pumping for irrigation uses must be treated separately. Water for irrigation purposes is characterized by a large month-to-month variation in water demand. It is necessary to size the pumping system so that it can satisfy the critical month's water demand. Water for irrigation has a relatively low value because the cost of its supply for irrigation must not be more than the value of the additional crops that can be grown. This means that pumping systems have to be cheap, and accordingly, pumping heads low. On the other hand, however, the need for frequent maintenance is acceptable as the farmer is regularly in the field anyhow.

Water for rural supplies (livestock and villages) is characterized by a constant month-to-month demand and is, therefore, more suited to pumping systems that derive their energy from a diffuse source such as sun or wind. This water has a high value and it is, of course, critical that there is water available on demand. Hence this type of pumping system must include a storage tank, and reliability and low maintenance are of great importance.

There are many factors which influence the cost of water from a pumping system and it is not possible to account for them simultaneously. It is useful to define a baseline model and then look at the effects of changes on different parameters. These are the different calculation stages:

1   For a given daily water demand and a given water table head, the system energy requirements are calculated. This is then used to determine the size of the pumping device.

2   The system capital cost is then found. It comprises three parts: site preparation and transport, the power source, and the pump cost.

3   Maintenance, operating and replacement costs are calculated. All costs are discounted at the present value to give the present worth of life cycle cost.

4   The unit water cost can then be found by calculating the annual equivalent of the life cycle costs and dividing them by the water pumped in one year.

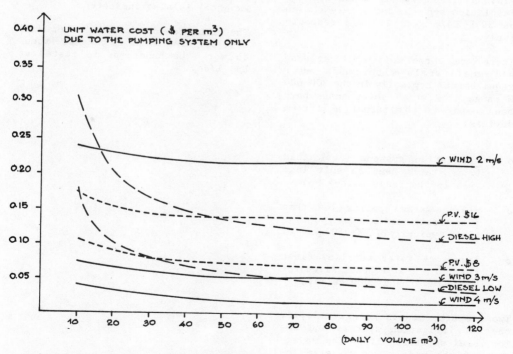

Unit water cost as a function of daily water demand for different pump technologies for windpumps (3 different windspeeds), for photovoltaic pump systems (2 different cost levels) and for diesel pump (2 different cost levels). For further explanation, see text.

Source: Fraenkel et. al., 1983.

RET-L

Water Pump Comparison

As part of the UNDP Project GLO/80/003 on small-scale solar-powered pumps by the World Bank, Sir William Halcrow and Partners in association with I.T. Power compared different small-scale pumping systems (see Chapters 16, 17 and 26). This section summarizes their main results.

Five different pumping systems were compared for each application: solar pumps, wind pumps, diesel pumps, animal-powered pumps and hand pumps. Diesel systems were considered twice due to uncertainties in their operation and maintenance costs. These are referred to as diesel high case and diesel low case and represent a range of operating conditions. For the irrigation studies, kerosene pumps were also considered but these were found to be less cost effective than diesel pumps, so diesel pumps were then considered to be the prime example of petroleum-powered pumps.

Water-driven pumps are very site specific, and do not effectively compete with the other pumping systems. However, where there is a suitable location they are the obvious choice.

The solar and wind pumps are sized to meet the water requirements in the critical month when the ratio of solar or wind energy available to hydraulic energy required is a minimum. The critical month's solar insolation and windspeed were chosen as 20.8 MJ/m^2 and 2.5 m/s respectively.

There are three main applications, all small-scale with shaft power requirements typically in the 200-600 W range, for which solar pumps have been compared with alternative lifting devices:

o irrigation systems (less than 10 m static head to suit land area in the range 0.25-4 ha)

o village water supplies (less than 30 m static head to serve populations typically of 300-1500 people)

o livestock water supplies (less than 30 m static head, sized for herds of around 2000 cattle).

Two computer models were developed, one for irrigation systems and one for rural water supplies. The objective of the models was to size the systems for a particular application and calculate unit water costs. A modular approach was used for both models utilizing components for water source, power source, water-storage unit, conveyance or distribution method and field application method. (The latter only for the irrigation model).

Three basic analyses were carried out:

a) Baseline Studies; chosen to represent typical technical and economic conditions. The technical parameters were based on average conditions for small farms and villages, while the cost and economic data were drawn from world market data.

b) Sensitivity Analysis of Baseline Scenarios; to indicate the relative sensitivity of various parameters and assumptions included in the baseline models.

c) Country Specific Case Studies; to investigate how local conditions in countries selected (Bangladesh, Kenya and Thailand) might influence the competitiveness of solar pumps.

Computer runs were made for all combinations of distribution and pumping methods over supply areas ranging from 0.125 ha to 4 ha. The optimum supply area and distribution method was then selected for each pump. The resulting unit water costs for each pump are shown in Figures below.

The global target for economic water cost delivered to the crop (estimated to be 10 ¢/m^2 by World Bank irrigation experts) is also indicated.

The main reason for the high unit cost for hand pumps is because labour to work the handpumps is costed at US$ 1/day.

Technical Characteristics of Pump Prime Movers - Irrigation and Rural Water Supply Baseline Scenario

Solar pump
Tilt	20 degrees
Critical Radiation level(daily ave)	250 W/m²
PV array efficiency	11% (4)
Array lifetime	15 years
Motor/pump efficiency(ave)	40% (3)
Motor/pump lifetime	20,000 operating hours or 10 years maximum (1)

Windpump
Cut in wind speed	Mean wind speed of least windy month
Rated wind speed	10 m s⁻¹
Furling wind speed	12 m s⁻¹
Efficiency at rated wind speed	9%
Pump lifetime	40,000 operating hours or 10 years maximum (1)
Tower/Rotor lifetime	30 years

Diesel pump (high case)
Size (Rated shaft power)	2.5 kW
Engine efficiency	10%
Engine lifetime	3700 operating hours or 7.5 years maximum (1)
Pump efficiency	60%
Pump lifetime	40,000 operating hours or 10 years maximum (1)

Diesel pump (low case)
Size (rated shaft power)	2.5 kW
Engine efficiency	15%
Engine lifetime	5000 operating hours or 10 years maximum (1)
Pump efficiency	60%
Pump lifetime	40,000 operating hours or 10 years maximum (1)

Kerosene pump (high case)
Size (rated shaft power)	1.1 kW
Engine efficiency	3%
Engine lifetime	2700 operating hours or 5 years maximum (1)
Pump efficiency	60%
Pump Lifetime	40,000 operating hours or 10 years maximum (1)

Kerosene pump (low case)
Size (rated shaft power)	1.1 kW
Lifetime	3700 operating hours or 5 years maximum (1)
Engine efficiency	5%
Pump efficiency	60%
Pump Lifetime	40,000 operating hours or 10 years maximum

Animal pump
Mean animal power rating	350 W
Length of working day per animal	5 hours
Pump efficiency	60%
Pump lifetime	15 years maximum
Animal lifetime	10 years

Hand pump
Human power rating	60W
Length of working day	8 hours social) irrigation only
	5 hours actual)
Pump efficiency	60%
Pump lifetime	20,000 operating hours or 10 years maximum (1)
Peak flow rate (2)	(at 2m static head) = 1.83 lit/s
	(at 7m static head) = 0.52 lit/s
	(at 20m static head) = 0.18 lit/s

NOTES:

(1) : Unit replaced at whichever limit is reached first.
(2) : Extrapolated from data from 'Performance Index for Man Powered Pumps' by A R O'Hea, Appropriate Technology Vol 9 No. 4, March 1983 assuming 36 watt output maintained for duration of 10h.
(3) : Based on ratio of hydraulic energy output to energy input when pump is working
(4) : Based on gross cell area (not array)

Source: Halcrow et. al., 1983.

Illustrations of Pump Costs for Selected Hydraulic Outputs

System	Hydraulic Rating of Pump (watts)	Capital Cost of Pumping Systems	Maintenance Cost $ per yr	$ per 1000 operating hours	Operating Cost
Solar	100-400	$17.1-18.7 Wp of array output(3)	50(2)	12	0(4)
Wind	N/A	$300 per m² of rotor (for 1m <dia <10m)	50(2)	6	0(4)
Diesel - low	1.5 kW	$850 per kW engine shaft output	-	200	40¢ per litre(5)
Diesel - high	1.5 kW	$850 per kW engine shaft output	-	400	80¢ per litre(5)
Kerosene - low	550	$400 per kW engine shaft output	-	200	40¢ per litre(5)
Kerosene - high	550	$400 per kW engine shaft output	-	400	80¢ per litre(5)
Animal	210	$5.2 per hydraulic Watt output(1)	10(2)	-	$2.25 per animal day
Handpump	36	$6-$8 per hydraulic Watt	50(2)	-	$1 per man day

NOTES:

(1) On the basis that one animal generates 350 watts
(2) Probably low for one pump alone, but when groups of pumps are involved total cost is reasonable
(3) See Table 8.4 for details
(4) Values will be small and in absence of firm data costs have been set to zero
(5) Costs of attendance on engine not included

Source: Halcrow et. al., 1983.

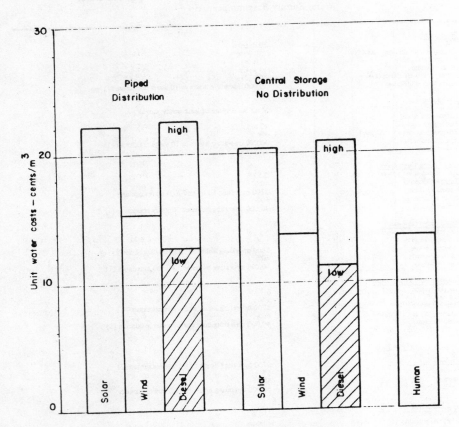

Histogram showing unit water costs from studies
of village water supply baseline scenarios
(population 750 at a 20 metre lift, 30m³ per day)

Source: Halcrow et. al., 1983.

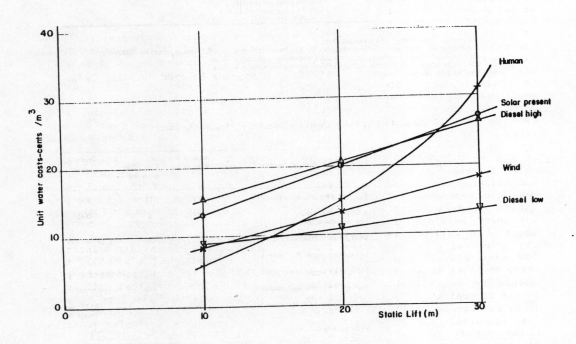

Effect of Static Lift on Cost of Water for Village Water Supplies
(Baseline scenario - population 750 - no distribution)

Source: Halcrow et. al., 1983.

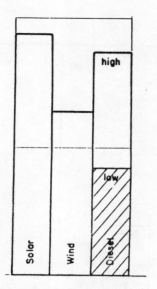

Histogram showing unit water costs from studies of livestock water supply baseline scenarios (2000 cattle at a 20 metre lift, 80m³ per day)

Source: Halcrow et. al., 1983.

Sensitivity Studies

Four groups of parameters were varied: (A) <u>Climate</u>, (B) <u>Agricultural</u> and <u>Technical Factors</u>, (C) <u>Costs</u> and (D) <u>Economic</u>. One of the major parameters of interest was the effect of future cost reductions for photovoltaic (PV) systems and the influence on static lift (Fig.) Three cost scenarios were considered for the solar PV pump: "Present", "Target" and "Potential". These produce systems at price levels of around US$ 18/Wp, US$ 9/Wp and US$ 6/Wp respectively. It is anticipated that when PV arrays reach a level of US$ 5/Wp (say in five years time) the "Target" case will be reached, but it is impossible to predict when systems might be available commercially at US$ 6/Wp. To reach that level array prices would need to fall to the region of US$ 2/Wp.

It can be seen that "Target" case solar systems lie between the high and low case diesel lines. "Target" case Solar systems with channel distribution provide water to the crop for under 10 ¢/m³ for lifts up to about 9 m more economically than any other alternative, other than low case diesel. Thus within five years the best solar pumping systems should be establishing themselves as the economically preferred option in circumstances similar to the baseline scenarios. The "Potential" case solar pumps are competitive with all other systems including low case diesel at heads up to 14 m.

Conclusions of the UNDP Study

Irrigation

- solar pumping systems have an optimum capacity to suit areas of around 1 ha for heads in the 2 m-7 m range.

- at present solar pumping systems are competitive with diesel engine pumps midway between the high and low cases at heads of around 2 m. Anticipated reductions in PV module costs over the next five years will make solar competitive with low case diesel at lifts up to 7 m.

- worst month windspeeds of around 2.5 m/s represent an approximate breakeven requirement for windpumps compared with solar pumps.

- at heads up to around 20 m, handpumps provide cheaper water than any mechanized option, providing human energy requirements are not costed.

- under the baseline model conditions solar pumps are marginally more expensive than either wind or low case diesel pumps. However, the difference is small, and in practice it is thought that the operational advantages of solar pumps even at present prices could outweigh the small cost differences.

- given an adequate wind regime, windpumps generally are the least cost option for livestock water supplies.

- with future cost reductions, solar pumps will become the cheapest mechanized option for heads up to 30 m in regions where windspeeds are low.

It can thus be seen that low case diesel is cheaper than any other option, but the real cost of maintenance and operation is very crucial and needs to be examined very carefully before opting for diesel. Wind pumps could be cheaper in better wind regimes. If the wind regime is very poor and diesel is excluded because of maintenance and fuel supply problems, solar might be the best option. For hand pump options it is necessary to consider maintenance cost, eventual extra borehole costs and charge for the time spent pumping. The most critical influencing factors regarding the type of power generation system installed appear to be climate, i.e. solar radiation and wind speed, the static lift, i.e. head of water, the costs of equipment and discount rates.

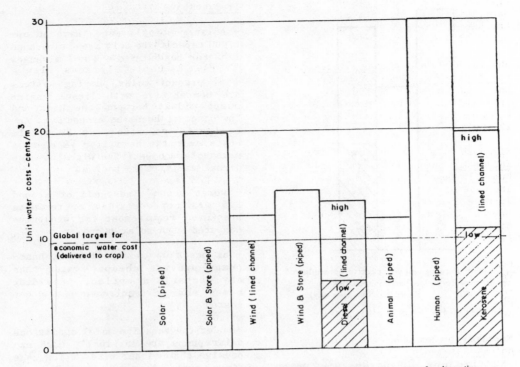

Histogram showing optimum irrigation water costs for alternative
pumping methods to supply 2ha at a 7 metre lift with peak water
requirements of 6mm per day (using baseline models)

Source: Halcrow et. al., 1983.

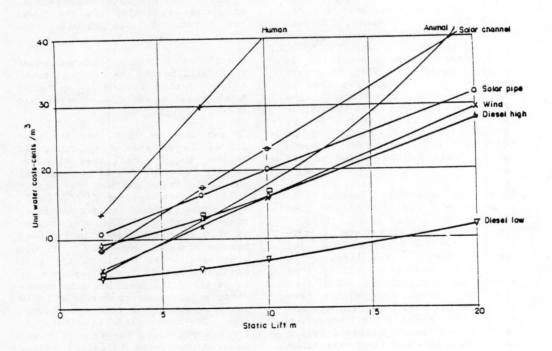

Effect of Static Lift on Unit Water Costs
(Baseline irrigation scenario - no storage)

Source: Halcrow et. al., 1983.

Bibliography

Agarwal, A. et al. (1981) Water, Sanitation, Health - for all? Earthscan, 3 Endsleigh Street, London WC1 ODD, U.K.

Bhatia, R. (1984) Energy Alternatives for Irrigation Pumping: an economic analysis for Northern India. International Labour Office (ILO), 4, route des Morillons, CH-1211 Geneva 22, Switzerland.

Birch, D.R. & Rydzewski, J.R. (1980) Energy Options for Low-Lift Irrigation Pumps in Developing Countries. The Case of Bangladesh and Egypt. ILO Working Paper No. 57, International Labour Office (ILO), 4, route de Morillons, CH-1211 Geneva 22, Switzerland.

Engebak, P. (1980) Horizontal Water Wells - Water Without Pumping. UNICEF Report WS/773/80. UNICEF, 866 United Nations Plaza, New York, N.Y. 10017, U.S.A.

Fraenkel, P. et al. (1983) Wind Technology Assessment Study. UNDP Project GLO/80/003 executed by the World Bank. Intermediate Technology Power Ltd., Mortimer Hill, Mortimer, Reading, Berks, R67 3PG, U.K.

Halcrow, Sir W. and Partners & I.T. Power Ltd. (1983) Small-Scale Solar Pumping Systems: The Technology, its Economics and Advantages. UNDP Project GLO/80/003 executed by the World Bank. United Nations Development Programme (UNDP). 1, United Nations Plaza, New York, N.Y. 10017, U.S.A.

IRC (1983) Alternative Energy Sources for Drinking Water Supply in Developing Countries. The International Reference Centre for Community Water Supply and Sanitation (IRC), P.O. Box 5500, NL 2280 HM Rijswijk, The Netherlands.

Johansson, S. & Nilsson, R. (1985) Renewable Energy Sources in Small-Scale Waterpumping Systems. Swedish International Development Authority (SIDA), 105 25 Stockholm, Sweden.

Jordan, T.D. (1983) Handbook of Gravity-Flow Water Systems. United Nations Children's Fund (UNICEF), Box 1187, Katmandu, Nepal.

Nissen-Petersen, E. (1983) Rural Water Supply. No Pumps. No Pipes. Danish International Development Agency (DANIDA). P.O. Box 40412, Nairobi, Kenya.

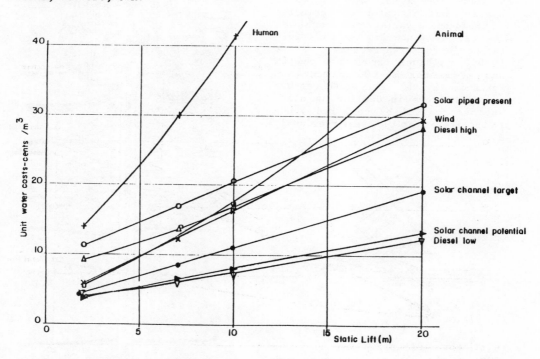

Effect of Reduced PV Solar Pumping System Capital Costs in Comparison with Alternatives as a Function of Static Lift (Baseline irrigation scenario - no storage)

Source: Halcrow et. al., 1983.

26 WIND PUMPS

Introduction

The techniques for harnessing wind to lift water have been known for a great number of years. The Chinese first used windpump systems 2000-3000 years ago and records exist of windmill operation in eastern Iran and western Afghanistan from about 200 BC.

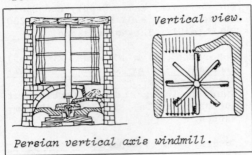

Persian vertical axis windmill.

The first windpumps consisted of crude vertical axis systems and it was not until about 1000 AD that the first Mediterranean horizontal axis windmills with sails were developed. By the 12th century windmills were in regular use throughout Northern Europe not only for water pumping but also for the grinding of grain and a variety of other uses. The use of windmills in this role peaked in the 17th century at about 10 000 machines throughout the UK and a further 12 000 machines in Holland. By the early 19th century windmills were in decline as steam engines fuelled by coal had begun to proliferate in the areas previously dominated by wind power.

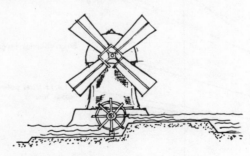

Dutch wind pump.

With the widespread use of steam engines in Europe, concentration on windpump development shifted to the USA and Australia where windpumps had become important devices on remote homesteads for raising sufficient quantities of water for farming livestock herds. By the 1880s the all-steel American windpump had evolved and it is estimated that about six million windpumps were built and installed in the USA from 1890 up to the present day. Many of these are no longer operational due to competition from other water raising systems and through the spread of piped water supplies, but it is estimated that approximately one million windpumps are still working, mainly in Argentina, the USA and Australia.

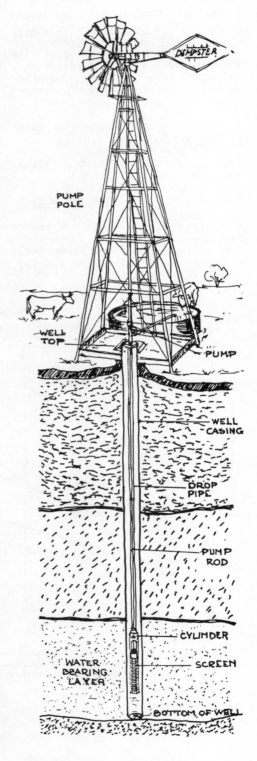

American farm windpump.

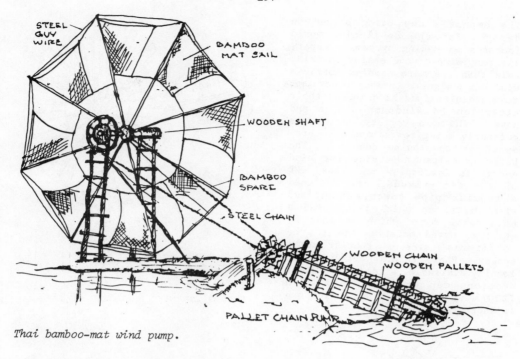

Thai bamboo-mat wind pump.

While the USA was developing its relatively expensive steel machines, lower cost windpumping systems were being developed in other parts of the world. China, for example, developed a wind-powered ladder pump with a maximum lift of 1.5 m for use on irrigation and salt-working projects, and Crete developed a horizontal axis machine with rotors and cloth sails for farm irrigation. In Thailand in the early 1950s, more than 20 000 simple wooden windmills are reported to have been in use lifting water for irrigation, and in northern Peru similar machines, made from local materials, were constructed for the same purpose.

Most traditional American windpump designs are based on the use of galvanized steel components bolted together using a large number of nuts and bolts to connect the rotor, the tower and tail assemblies. A transmission system is used to convert the rotary action of the wind mill to the reciprocating motion of the pump, and is normally oil-bath lubricated.

Most current windpumps use between four and twenty blades rotating on a horizontal axis to capture wind energy and transfer it to pumping mechanisms. The most common machine in use is the American multi-bladed fan-type rotor windpump. The machine has changed very little since its inception in the 19th century due to the original design's durability and reliability under extreme conditions.

On the other hand, the sailwing, or Cretan, windmill consists of a three to six m horizontal axis rotor, comprised of several metal or wooden blades in the form of spokes on which are mounted cloth sails which can be furled or removed during high wind conditions. Although the cloth has to be replaced every two or three years, the remaining windmill components usually last between 10-15 years if well-constructed. Thousands of these machines still provide irrigation water for farms around the Mediterranean. There are also several thousand working machines in Thailand which are built on similar lines and use bamboo mats for sails.

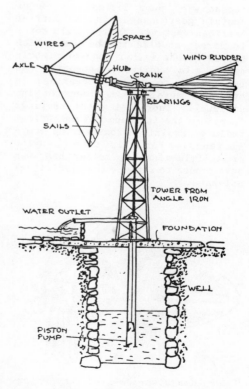

Cretan sail wind pump.

RET-L•

The emphasis now being placed on designs for the developing world focusses on making systems cheaper, lighter, simpler and easier to maintain than the more complex American windpump designs. A number of development organizations have turned their attention to windpump design and have produced a variety of systems currently being tested and also produced in developing countries. The Dutch consultancy service for wind energy in developing countries, CWD of the Netherlands, has designed a tubular pipe tower, stabilized with wires, on which is mounted a horizontal rotor which is made by welding curved metal sheets to the "sheltered" side of tubular metal spokes. Where tubular metal is not locally available, welded angle steel sections can be used. The British ITDG work on a deep-well windpump is another example.

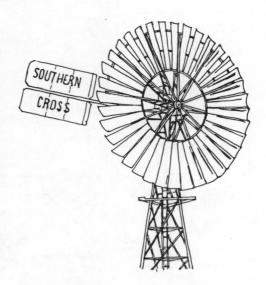

Australian wind pump.

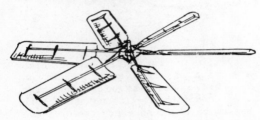

Sheet-metal rotor of the CWD-design.

Besides adapting the components for local conditions, the agencies involved in windpump design have concentrated on reducing the number of components in the systems. Most recent wind pump designs no longer contain gearboxes as in the earlier American systems. Instead, the rotor is direct-coupled to the pump which and is grease rather than oil-lubricated. In addition, attention has been given to windpump bearings in order to develop some means of individually lubricating the bearings to reduce the incidence of mechanical failure resulting from bearing inefficiency. For this reason, nylon or self-lubricating roller bearings have tended to replace the traditional oil-bath.

Work by the various design groups has resulted in the development of a wide array of different windpump types:

1 Traditional Multi-bladed Farm Windpumps

This design was developed in the 1880-1930 period in the USA and Australia. The Aeromotor, Dempster, Climax, Comet and Southern Cross designs are all horizontal axis machines having many fan-type blades and requiring a transmission system. Components in these designs include galvanized rolled steel sections bolted together with the transmission being made of large castings or forgings lubricated by an oil-bath. Such machines are inherently extremely heavy and material-intensive and are complicated to assemble; however, they compensate for this with their high robustness, reliability and ease of maintenance. They require servicing only on an annual basis – usually just a change of lubricant.

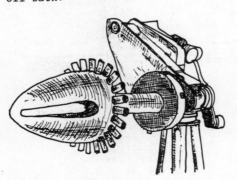

Modern direct-drive wind pump rotor.

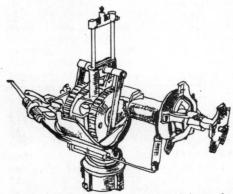

Traditional wind pump rotor with gearbox in oil-bath.

Kijito wind pump.

Organization for Rural Poor (ORP)/TOOL machine developed jointly by India and the Netherlands, the WEU1/3 model developed jointly by Sri Lanka and CWD and the Colombian Gaviotas system are examples of such machines which are reaching the stage of full maturity.

4 Modern, Lightweight, Lowlift, Small Windpumps

These are smaller direct-drive units suitable for locations having relatively good wind regimes, where pumping heads are generally less than 10 m and where pumped water demand is modest. Such systems are generally produced in Europe and include the Finnish Wind-Acrobat, the West German Windkraft, Denmark's Sparco and the French Humblot machines.

2 Modern, Lightweight, Deep-well Windpumps

These incorporate a concept that is similar to the earlier, traditional windpump designs but which uses modern manufacturing techniques including welded modular sub-assemblies, in place of the rolled steel sections, roller bearings and grease lubrication. Such designs are easier to assemble and erect than the traditional designs as they generally consist of prefabricated welded sections with lighter transmission units located at the top of the tower. Their modern composition causes no loss of operating efficiency and in many cases actually improves performance. While the traditional Australian Southern Cross with a 7.6 m diameter rotor on a 12 m high tower weighs a total of 3950 kg, the modern Kenyan Kijito design, developed by The Intermediate Technology Development Group (ITDG) in Britain and built by Kenyan company, Bob Harris Engineering Ltd., is of similar size but weighs just 1100 kg.

3 Modern, Lightweight Medium-depth, High-capacity Wind-pumps

These also utilize lighter welded steel components and are simpler to manufacture and assemble than the traditional designs, but they are not suitable for such heavy duties as the traditional or modern deep-well systems since they tend to run at faster speeds and have lower pumping power. Such systems as the (Indian)

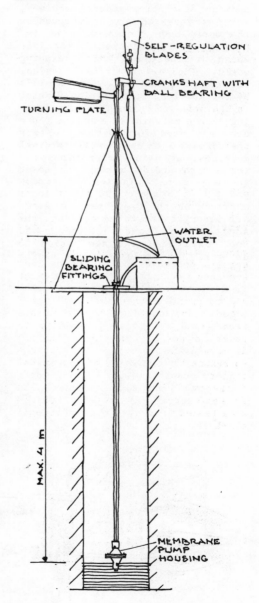

Modern light-weight low lift wind pump.

5   Wind-electric Pumps

Although a wide variety of wind
powered generators are now manufactured around the world, only a handful
of manufacturers have experience
in matching wind generator output
to pumping demand, and the manufacture
of suitable components is complex.
As a result, experience with this
relatively high technology (in comparison with direct-drive windpumps)
is limited, despite the advantage
of being able to site the wind turbine
away from the pump site in an area
of better wind regime. Companies
involved in this field include
France´s Aerowatt, India´s Jyoti
and the USA´s North Wind Power Company.

6   Locally Manufactured Windpumps

Besides developing modern machines,
various research groups throughout
the world have concentrated on improving the traditional machines
which originated in the developing
world. The Cretan windmill design,
for example, was taken up in 1975
by the American Presbyterian Mission
in Ethiopia and refined by substituting dacron material (a man made fibre
commonly used for boat sails) for
the cotton cloth used on the original
machines, by using water pipes for
the spokes and by making the tower
as a single unit. The refined version
was put into production at a local
workshop on the Omo river but since
all material had to be brought from
Addis Ababa, construction proved
too costly for the new design to
do well and production was short-
lived.

The importance of using local materials in windmill construction
is now being stressed by most research
organizations involved in this field
and work on systems like the Chinese
and Thai bamboo mat windmills – which
drive a wooden pallet chain pump
(the dragon bone pump) – is continuing. In Thailand, as well as in India,
work is concentrating on developing
more efficient locally produced windmills. India has developed the Poghil
windpump and Thailand has produced
a high-speed, four-bladed, 6 m diameter, wooden rotor machine. VITA
has developed an improved version
of the Thai machine incorporating
eight laminated wooden blades as
well as split bamboo blades mounted
on a metal tower.

The Sahores windpump from Mali offers
a different approach to the same
problem. This machine has a rotor
consisting of a plywood disc with
16 bamboo blade spears bolted to
it. Galvanized steel sheets are
attached to the spokes of the rotor
in such a way as to allow them to
be feathered. The tower is constructed from locally-available poles
and all materials, with the exception
of the steel ball bearings and the
piston pump, can be obtained locally.

In addition, a large number of designs
and prototypes of locally made windpumps are constantly being produced
at many development centres in the
developing countries.

Windpump Applications

Windpumps are used to supply water
for livestock, human requirements
and for irrigation.

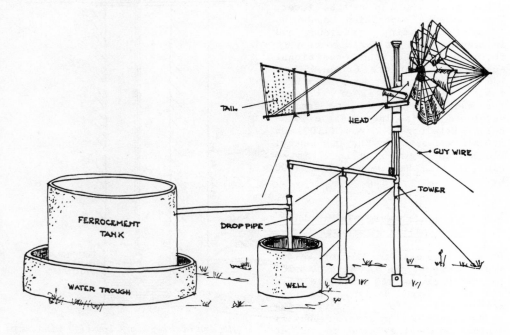

Sahores wind pump.

Livestock   The primary application in developing countries of windpower at present is water pumping for live-stock, and this has also been the case historically.   The value of water for livestock is high because quite modest quantities of water can sustain a sizeable herd.   For example, in 3.8 m/second winds, a typical 12 foot (4.88 m) rotor farm windpump pumping through a typical depth of 35 m will sustain a herd of around 70 steers (the value of which, even if of bush-cattle quality in Africa, is typically in excess of US$ 10 000), or 500 sheep. A wind-pump of that size will cost approxi-mately US$ 1500 - 3000 and will have a life of about 20 years with once-a-year maintenance.   Windpumps are very suitable as prime suppliers of ranch water: they need little attention, they can be seen to be working from a considerable distance, they are reliable and long lasting, cannot easily be stolen and are robust enough for cattle to scratch them-selves against the tower legs without damaging the windmill. Ranch water supply is probably of considerable future importance in arid regions where about the only method for ex-ploiting the land, is to graze herds of hardy livestock.   However, one must be aware of the problem of over-grazing caused by excessive concen-trations of animals in one area, due to lack of water everywhere else.

Human water supplies   Many of the requirements that make traditional farm windpumps convenient for supply-ing water to livestock also apply to human water supplies; i.e. the general lack of attention demanded by the windpump, the high value that

can be assigned to the water and the large number of people who can be sustained by quite a modest and relatively inexpensive machine. So far as developing countries are concerned, it is still very rare to find windpumps used for human water supplies although it is clear that there is potentially a very large market for this purpose. Wind-pumps are probably the most cost-ef-fective water pump in any location with a mean wind-speed exceeding 3.5 m/s - a scenario which includes quite substantial areas of the devel-oping world.   The scale of isolated human water supplies is such that pumps should deliver 10-15 litres/person and day, for people carrying water a distance of 500-1000 m and 20-50 litres/person and day where standpipes are provided within 250 m. A typical well-designed 12 foot (4.88 m) farm windpump will deliver 500 000 litres per meter head and day in a mean windspeed of 3.8 m/s, i.e. enough water for a community of 600 people if pumping through 50 m head or 1200 people at 25 m head. The capital investment required for the windpump will be about US$ 3.50 per capita at a 50 m head or 1.75 per capita at a 25 m head and the life of the system can be expected to exceed 20 years if regularly main-tained.

Irrigation Traditional industrially-manufactured farm windpumps are not ideal for irrigation and are rarely used for this purpose. This is be-cause the techno-economic requirements are rather different to those for drinking water supply. Unlike such water supplies, which are normally all-the-year-round applications with

Small wind pump for irrigation.

a fairly constant daily demand, irrigation water demands fluctuate considerably with the seasons and the peak demand can be over twice the mean. The consequence of this is that the pump has to be sized according to the month of maximum irrigation demand, using the wind resource available at that time (which may well not be at the maximum level at that time, although fortunately there is a reasonable correlation in many areas between water demand and windspeed). As a result most irrigation windpumps will be grossly oversized with a rather poor utilization factor. Consequently it is important to keep the capital cost of an irrigation pump as low as possible. A common result of this is that most windpumps actually used for irrigation are locally produced and crude. Windpumps of this nature can be cheap and simple, as they only operate a few months every year, pump through very small heads and are virtually constantly attended when in operation should anything go wrong. Also, irrigation water must be cheap because of the small added value it generates in terms of crop production/m^3 of water. While human water supplies can be viable at levels running at over US$ 1/m^3 and figures of US¢ 20 are fairly normal, irrigation water must cost under US¢ 10/m^3 and preferably less than US¢ 5.

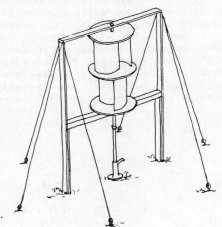

A Savonius wind rotor is simple to build, but is inefficient and heavy.

The Economics of Windpump

Windpumps are very difficult to compare with each other and with other types of pumps, because they are made to meet different conditions. They can be designed to work at different windspeeds and pump differing volumes of water through various heads. Or they can be built with different amounts of technical sophistication (efficiency and mass), require different amounts of maintenance

(quality) and can be of different sizes (rotor diameter).

One way to compare windpumps is to look at the capital cost/m^2 of rotor area. However, the problem with this is that the figures do not account for windpumps of different sophistication and quality. On one hand there are products dating from the 1930s that are fairly reliable but heavy and expensive to ship. An American farm windpump with a 3-4 m diameter rotor and the capacity to pump through a 30-60 m head costs about US$ 500-550/m^2. The same kind of windpump made in Australia costs only US$ 150-300/m^2. This probably reflects both the higher manufacturing costs and higher profit margins that are typical of the American market.

On the other hand there are a number of more modern designs, none of which, however, have yet been able to operate for the 10-20 years necessary for a proper evolution. Some of the designs which show promise for the future are: the Dutch-designed 3 m windpump produced in Sri Lanka costing US$ 130/m^2 and the bigger deepwell Kenyan Kijito design at US$ 260 to 350/m^2.

Locally-manufactured designs can work out to be extremely low in cost. There are Indian, and Thai designs in the US$ 25-50/m^2 price range. This is mainly because they are not of the same standards of strength and durability as traditional windpumps and they tend to be constructed with much smaller towers. Most are intended for irrigation pumping where continuous attendance is possible and reliability need not be so important.

As the different manufacturers performance data relates to different heads and windspeeds, one crude way to draw comparison between designs is to look at the specific capital cost (SCC) for a given water supply, calculated for the wind speed of 3.9 m/s (14 km/h). This is the specific energy output in kJs, i.e. $/kilojoule/day, and gives a measure of the investment required to provide a given hydraulic output at a windspeed of 3.9 m/s during a day. This method has been used for UNDP Project GLO/80/003 (see Chapters 16, 17 and 25).

Many windpumps attain an SCC in the US$ 0.2-0.5/kilojoule/day range. For solar-pumps in a standardized solar day of 5 kW/m^2 of solar irradiation, the SCC would be about US$ 2/kJ/d. Mass production could possibly bring this down to US$ 1/kJ/d. The design and size of a windmill, and hence to a certain extent the costs for the power it delivers, depends to a large extent on the mean windspeed for the critical month. The critical (or "worst") month is the month for which the windpump must be designed to provide the required output. It is generally the month with the worst mean windspeed, but also depends on the water or power demand in that month, e.g. for irrigation during the growing season which has the highest water demand.

In the study referred to above, it is argued that for mean windspeeds in the critical month of 2 m/s or less, a windpump is not competitive with other systems. Windpumps in locations with a worst month mean windspeed of over 3 m/s are usually competitive with current solar pumps or high case diesel pumps (see above). Even if the future cost reduction of PV arrays is realized, windpumps are a cheaper option at this head and for a worst month windspeed greater that 3 m/s. When compared with the low case diesel, the windpump is more economical at demands less than 60 m^3/d, and at a worst month windspeed of over 4 m/s, windpumps are often the cheapest option of all.

The study thus indicates that windpumps are definitely an interesting technology, and that they may be an economically sound investment under a wide range of conditions.

If wind pump costs were to drop as much as 50% due to mass production of modern designs, then, in this case, the mean wind speed in the critical month would probably drop to nearer 2 m/s for windpump economy.

Environment

The main environmental problems concerning wind pumps arise in connection with irrigation. The impacts of irrigation must be fully understood and carefully studied before an extensive irrigation system is introduced. Typical problems include salination of the land, waterlogging and the removal of nutrients from the soil due to soaking.

Problems in connection with cattle can be overgrazing and desertification close to the well, if too many animals are dependent on the water supply. Other problems may be the curiosity value of windmills to people and the resulting dangers if they are climbed. These can be overcome with suitable barriers and signs.

Production

Attempts to introduce windpumps into developing countries have run into problems. The reason for this is mainly connected with the high cost of importing and erecting windpumps

in remote areas, and the serious difficulties in providing spare parts and necessary maintenance. Another important factor is that there seems to be no great attraction in finding a substitute for expensive, imported diesel engine pumps and diesel fuel in the form of yet another imported and less well understood energy system. However, if windpumps are locally manufactured within a developing country they gain a number of advantages. These are principally:

1 Import substitution and hence a saving of foreign currency.

2 Shorter supply lines from manufacturer to customer, resulting in more locally adapted systems, reduced transport costs, elimination of import formalities, fewer "middle-men", and improved spare-part and after-sales service.

3 Enhancement of developing country capability.

4 Job creation within the local economy.

5 Provision of more dependable power source than with diesel in remote areas.

6 Wider use of pumped water, leading to improved agricultural outputs.

There appear to be two lines of development proceeding, designing new windpumps suited for production in factories in the developing countries, and developing windpumps suitable for do-it-yourself builders. Varying approaches will probably work best in different societies and in different economic situations, and it is impossible to prescribe a single strategy that will work everywhere.

The Introduction of Windpumps

A windpump is a comparatively complicated machine: it rotates, experiences severe dynamic loads with the inherent wear and tear of bushes, bearings, washers and joints. It is exposed to the rigours of the weather, not only storms but also the damaging effects of corrosion. A windpump operating 3000 hours a year for 10 years at a 1 rotation/s makes approximately 1 hundred million cycles, a fact which clearly exposes the fatigue problems involved. Well designed windpumps have certain advantages: in a good wind regime (i.e. more than 3 m/s in the most critical month) they are usually the most economic way to pump water. They give independence from fuel supplies, have a long life, high reliability and low maintenance, and are less prone to theft than diesel pumps. Importing windpumps creates a number of difficulties: foreign currency is required, delivery time can be long and shipping costs high, also it is difficult to get service and spare parts. Thus the ideal conditions for dissemination are a local production of cheap high quality windpumps. Unfortunately a developing country usually lacks the technical expertise necessary to develop an indigenous design. There is also often a lack of wind data, capital and official governmental support.

It would thus seem that this is a situation that is well suited for cooperation between a developed country with knowledge of windpump technology and manufacture, and a developing country where the biggest needs for windpumps exist. First, it is necessary to understand the local wind regime and the local need for windpumps in order to specify the size of the local market and the size of an adequate windpump and its specific requirements. Secondly, a reliable and cost effective design must be developed, tested and put into production. Government support is essential so that existing subsidies on other energy sources (i.e. diesel or kerosene), or taxes on new, imported products, do not obstruct the dissemination of windmills.

Dissemination

Put briefly, some general aspects regarding dissemination are these: Large numbers of windpumps will be needed. The hardware should be first class. The programme should aim at educating local engineers, technicians, etc. The local organizations should become independent of foreign experts. The local organization should be the only body responsible for dissemination. The technology should be economically sound. If an external agency is involved, there should be a long-term involvement. There should be sufficient continuity of man-power and money for an adequately long period. Dissemination should always be started in regions where conditions are the most favourable. Finally, it must be remembered that there are no standard solutions to dissemination, it all depends on the local circumstances and the programme's ability to understand and adapt to these.

The important thing to consider with the dissemination of wind pumps is that the whole chain, from the wind to the weeding of the extra crop, must be dealt with. Questions that must be raised over the suitability of windpumps to a particular area

include: is wind speed sufficient throughout the critical months? Do the local farmers think that the availability of more water would serve any useful purpose? Can wells be made and is water available? Will the local culture accept windpumps and an eventual extra crop?

If the results of such an analysis are positive, a wind pump that is suited to the local conditions should be selected from existing designs, or a new one should be developed. One way to do this is by cooperation between a technical institution with knowledge on windpumps, and a potential producer in the developing country concerned. This is usually a trial and error process that will take some years, during which several improved models will evolve.

Other essential factors which must be considered are; the financing of the windpumps for the farmers, entailing cooperation between the government and the banks; the provision of education for new farming and irrigation methods and the utilization and maintenance of the windpumps; the availability of maintenance and spare parts in the country; the training of workers to transport and install the windpumps, and the generation of a firm commitment to windpumps across the spectrum from the government, banks and producers to the farmer, serviceman and public.

At the same time it is necessary to study the new farming and irrigation systems made possible by wind pumps. This can be done by cooperation between an agricultural university and local farmers and agricultural schools. An aspect not to be forgotten is the methods of water-tank and irrigation system construction - the material and skills for these have to be easily available and not too costly. Finally, everything must be tied together and coordinated in order to get the production of the windpumps going. This means that materials, skills and machinery have to be available in the country, or else easily imported - often entailing that the modification of the design is necessary to suit the locally available materials.

Groups Working with Windpumps

As noted earlier, many of the new concepts and improved versions of traditional windmill design have yet to reach their full performance potential. Work is proceeding to optimize machine designs to suit local conditions prevailing in the developing world. Among the foremost research agencies involved in this field are the CWD of the Netherlands,

Intermediate Technology Power (a branch of the ITDG) of the UK, the TOOL Foundation of the Netherlands and VITA of the USA.

CWD in the Netherlands was established by the Dutch Ministry for Development Cooperation. This organisation operated under the name Steering Committee Wind Energy Developing Countries (CWD) until 1984. In that year the name of the organisation was changed into Consultancy Services Wind Energy Developing Countries (CWD). CWD promotes interest in wind energy in developing countries and aims to help governments institutions and private parties in the Third World with their efforts to utilize wind energy. The parties in the Netherlands that are currently participating in CWD are these: DHV Consulting Engineers in Amersfort are project coordinators; the Wind Energy Group of Eindhoven University and the Windmill Group of Twente University do the technical work; the Institute of Land Reclamation and Improvement (ILRI) in Wacheningen does the work on agriculture and irrigation. Close contacts also exist with the Working Group on Development Technology (WOT) at the Twente University and the Dutch National Wind Energy Research Programme.

Among CWD activities in developing countries are: <u>Sri Lanka</u> – By 1984 150 3m diameter windpumps were installed for irrigation by small farmers. Development of a 2 m diameter and a 5 m diameter windpump are underway. <u>Tunisia</u>-designing of a 5 m diameter windpump and testing of prototypes to start production. <u>Cape Verde</u> – training of windpump staff and development of a 5 m diameter high windpump. <u>Pakistan</u> – to start production of a windpump similar to the Sri Lankan one with the local company Merin Ltd. <u>Tanzania</u> – the Ujizi Leo industries in Arusha have been supported in improving their windpump. <u>Peru</u> – a windpump prototype has been built for an agricultural project. An interesting approach has been made in Sri Lanka where CWD has built a network around the Wind Energy Unit (WEU), which reports to the Water Resources Board (WRB) that is part of the Ministry of Lands and Land Development. The unit manages a design group, a workshop, and an agricultural extension organization. It also keeps in contact with local banks to provide financing to the farmer who wants to buy a windpump.

IT Power, with assistance from Britains's Overseas Development Admi-

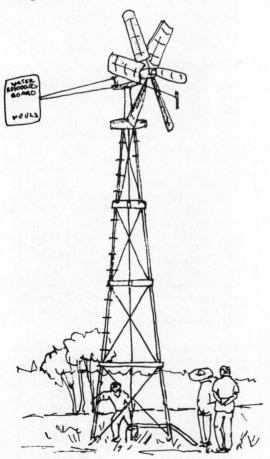

CWD wind pump in Sri Lanka.

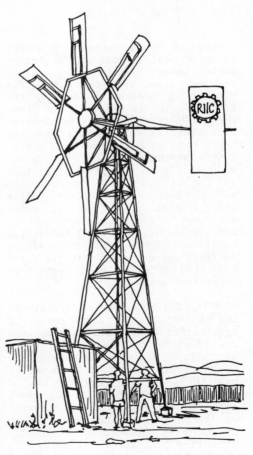

ITDG windpump in Botswana.

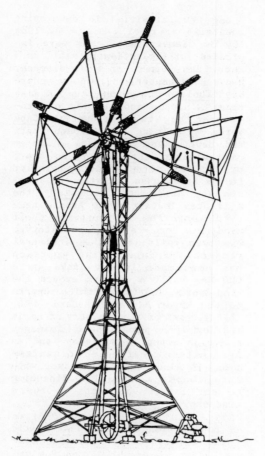

VITA windpump in Thailand.

nistration, has embarked on a number
of projects throughout the developing
world, among them: a Kenyan project
in cooperation with the Bob Harris
Engineering Company to develop the
Kijito design of windmill and windpump; a Pakistani project in cooperation with the Karachi-based company
Merin Ltd. to build and test a prototype windpump and to later start
up production of the optimized design;
an Indian project in cooperation
with the Bombay-based concern, Voltas
Ltd., to develop a 6 m diameter rotor
windpump; a project in Botswana in
cooperation with the Rural Industries
Innovation Centre to improve windpump
designs; and a project with the World
Bank to conduct a wind technology
assessment study.

TOOL is a Dutch foundation participating in the renewal, development and
application of socially appropriate
technolgies. TOOL is mainly funded
by the Dutch Ministry for Development
Cooperation. The TOOL Foundation
has assisted the Organization of
the Rural Poor (ORP) in Ghazipur,
India in developing a 5 m diameter
windpump which is now being mass-
produced by the Institute for Engineering and Rural Technology in
Allahabad. The 500 units produced
so far have been sold for about US$
1000 each to be used for irrigation.

VITA in Washington, funded by the
US Government, have several wind
projects in Thailand in conjunction
with the National Research Council
of the Royal Thai Government. Current
work is in progress to improve a
version of the traditional windpump
by using bamboo or laminated wooden
blades. The windpump has a 6 m diameter and is used for irrigation. The
VITA also supports windpump projects
in Mexico, Honduras and India.

Conclusions

Wind pumps are a good and cheap way
to pump water if windspeeds of about
3-4 m/s prevail during the time of
the year when water is needed. The
technology is well matched to the
task of providing water, as during
a brief windless period the user
can simply draw on water previously
pumped into a storage tank.

The UN Conference on New and Renewable
Sources of Energy (Nairobi 1981)
made the following recommendation:
"Socio-economic system studies indicate that small, wind-powered waterpumping systems are quite appropriate
for irrigation in small farms, even
in comparison with other energy sources. The panel strongly recommends
inclusion of wind-powered pumps in
rural development programmes wherever
local conditions are favourable.
An important positive aspect of small
windmills is the likelihood of additional employment in rural areas".

This appears justified but only after
the inclusion of several provisos,
viz., windmills have to be produced
in the country of intended use in
order to reduce the price and to
ease access to spare parts and maintenance; the windpump has to be
adapted to local conditions (this
is not an easy task and should be
done by someone with vast experience
of windpumps in developing countries
and in cooperation with the local
producer). It should also be realized
that the windpump is just one part
of a new agricultural practice and
that the local farmers and agricultural experts have to be involved
in the process, while the funding
of the projects must be assumed by
the government and banks.

A second generation of windpumps
is now being developed in countries
such as India, Sri Lanka, Kenya and
Colombia in cooperation with agencies
such as the CWD, ITDG and UNDP. Given
the overall prospects for windpump
introduction, it is very likely that
those new improved designs will be
the beginning of a new wave of windpumps that will be seen in many countries all over the world in the years
to come.

Bibliography

Both, D. & van der Stelt, L.E.R. (1984) Catalogue of Windmachines. Consultancy Services Wind Energy Developing Countries (CWD), P.O. Box 85, 3800 AB Amersfort, The Netherlands.

CWD (1985) Report of activities of CWD 1983. Consultancy Services Wind Energy Developing Countries (CWD). P.O. Box 85, 3800 AB Amersfort, The Netherlands.

Fraenkel, P. et al. (1983) Wind Technology Assessment Study. UNDP Project GLO/80/003 executed by The World Bank. Intermediate Technology Power Ltd., Mortimer Hill, Mortimer, Reading, Berks, RG7 3PG, U.K.

Goedhart, P. (1982) Prospects for Small and Marginal Farmers in Trichy District (Tamil Nadu, India) to use Water Pumping Windmills for Irrigation. Technology and Development Group, Twente University of Technolgy, P.O. Box 217, 7500 AE Enschede, The Netherlands.

Lysen, E.H. (1982) Introduction to Wind Energy. Consultancy Services Wind Energy Developing Countries (CWD), P.O. Box 85, 3800 AB Amersfoort, The Netherlands.

Mueller, A.M. & DHV (1984) Windmills in the Lift. The results of a CWD-project. Consultant Services Wind Energy Developing Countries (CWD). P.O. Box 85, 3800 AB AMERSFOORT, The Netherlands.

Mueller, A.M. (1984). Farm Economics of Water Lifting Windmills. Consultancy Services Wind Energy Developing Countries (CWD). P.O. Box 85, 3800 AB Amersfoort, The Netherlands.

Rastogi, T. (1982) Windpump Handbook. Tata Energy Research Institute, Documentation Centre, Bombay House, 24 Homi Mody Street, Bombay 400 023, India.

Smulders, P.T. (1982) Experiences of Dissemination of Windmill Technology. Wind Energy Group, Eindhoven University of Technology, P.O. Box 513, 5600 MB Eindhoven, The Netherlands.

Sørensen, B. (1980) Renewable Energy. Academic Press, London, U.K.

Vilsteren, A. v. (1981) Aspects of Irrigation with Windmills. Consultancy Services Wind Energy Developing Countries (CWD). P.O. Box 85, 3800 AB Amersfoort, The Netherlands.

VITA (1983) An Integrated Program of Technical Assistance to the Private Sector for the Manufacture and Dissemination of Low-Lift Wind Powered Pumps for the Small Scale Farmer/Entrepreneurial Sector in Thailand. Volunteers In Technical Assistance (VITA), 1815 N. Lynn Street, Suite 200, Arlington, Virginia 22209-2079, U.S.A.

WINDPUMPS	Rotor diameter m	Cost $	Pumping head m	Wind speed m	Water quant. m^3/h	Specific Cost $/m^3$	Specific Cost $/kg$	Spec. Capital Cost (SCC) $/kJ/day$
Traditional, multiblade								
Aeromotor, USA	3.66	5482	47	30	1.8	521	5.78	2.20
Southern Cross, Australia	3.66	1548	37	29	1.64	147	1.29	0.96
Modern, multiblade								
BHEL, Kijito, Kenya ITDG	3.66	3874	70	12	0.37	368	7.04	0.40
Climax No 12, South Africa	3.66	2171	61	11.7	0.21	206	2.64	0.40
Modern, shallow well								
WEU 1/3, Sri Lanka/CWD/NL	3.00	922	10	13	2.0	130	2.53	0.16
ORP/TOOL 12PU500 India/NL	5.00	1060	6	16	7.2	54	2.65	0.16
Modern, small								
Lubing MO15-6-6, W Germany	1.5	637	6	14.4	0.84	360	8.61	0.56
C&S Varcoe, Australia	1.83	1259	22	14.6	0.5	478	-	0.55
Low cost, locally built								
Anila 1, India	3.66	159	8	25	2.0	15	-	0.24
Village built, Thailand	7.00	1200	0.9	22	90.0	31	-	0.24

Comparison of different types of windpumps. Source: Fraenkel, 1983.

27 HAND PUMPS

Introduction

Although it is generally agreed that hand pumps are a cheap and appropriate solution to the problem of supplying adequate drinking water in remote rural areas of the developing world, in the past the technology has been plagued by frequent system failures due to ineffective design, poor system quality, poor installation and incorrect matching of pump systems to the desired use. For example, some early hand pump installation projects in the developing world used fragile pumps designed for infrequent use in European gardens and totally unsuited to often remote and hostile environments. As a result it came as little surprise to some that these projects failed miserably leaving broken hand pumps littering the landscape.

In some areas, as many as 80% of the pumps installed were abandoned shortly after installation, due to breakdowns and lack of spare parts. It is now being recognized that the successful use of water pumps depends, to a large extent, on non-technical factors. The involvement of the users in maintaining their pumping units, the availability of spare parts, and the possibilities of manufacturing the pumps locally are examples.

The main problem with a well is how to protect it from being contaminated. This can be done by lining the well with a watertight concrete seal in such a way that surface water cannot enter. This necessitates that it must then be provided with a pump. Selecting a hand pump for a particular site is difficult, partly because little objective information is available to the purchaser.

Testing and evaluation of different hand pumps is being undertaken in the UNDP/World Bank programmes called "Rural Water Supply Project for the Testing and Technological Development of Handpumps" (INT/81/026). The scheme is intended as the first real attempt to not only test and compare different pump designs for rigorous use in the developing world, but also to stimulate the development of new systems better suited to these conditions. Part of the approach centres on the selection of suitable designs for particular uses and locations, and on the production of low

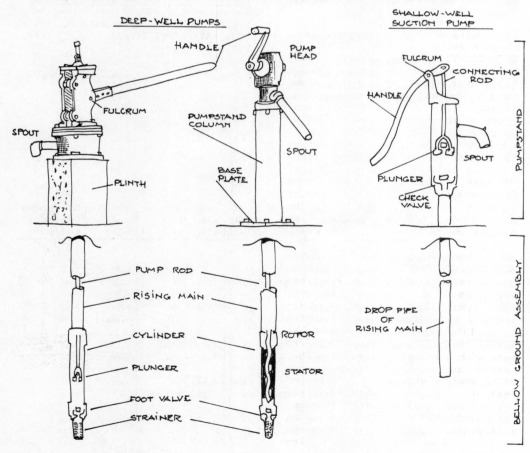

Basic pump nomenclature. Source: Arlosoroff, 1983.

308

Hand Pump Development

As with all other pumping systems, current work on hand pumps aims to produce simpler, cheaper, more sturdy and more reliable pumps. The aim is to be able to fabricate and assemble the pumps and spare parts in the country of use, rather than having to import handpumps and components.

In terms of technical improvements in this field, a great deal of research has been concentrated on designing systems in which the connecting rods, pistons, foot valve and rising main can be removed for inspection and repair by hand through the pump head without having to resort to the use of a winch and tripod. This is already possible with the Maldev pump in Malawi.

Reducing the cost of maintenance is also an important part of the research work underway, and one possible solution may be to replace certain metal components with locally-produced plastic parts. If this is possible, it could substantially reduce overhaul costs which, on the Indian Mark II pump, for example, can range from US$ 35 to 2000 per year. One idea is to use locally-produced plastic materials for those parts of the pump that must be regularly replaced (by injection moulding).

For shallow well pumps, several cheaper designs are under way - the Blair PVC-VLOM, developed in Zimbabwe, has no lever and no leather seal. It uses standard PVC water pipes and fittings and non-return ball-type valves. it is easy to make and operate and requires little maintenance. It costs between US$ 75 and 150. Another new cheaper design is the Rover pump in Bangladesh. The pump cylinder, a 5 cm diameter PVC pipe, is inclined at 30 degrees from horizontal, enabling the operator to pull and push directly on the piston rod. It costs only 60% as much as the Bangladesh new No 6 pump.

Other pumping mechanisms under development include the Mono and Moyno pump which feature a spiral stainless steel rotor in a rubber housing; the "hydropump Vergnet", a pedal-operated pump using hydraulic action on an inflatable silicon rubber bag inside the pump cylinder; the Petro pump with a rubber hose section that narrows and expands, displacing the water upwards with a simple arrangement of two ball valves; and the Pulsa-3 that uses a diaphragm pump. The potential of air-lift pumps (using a compressor) and centrifugal pumps is also being investigated. (See Figures in Chapter 25).

The modified India Mark II pump is easy to maintain and repair.

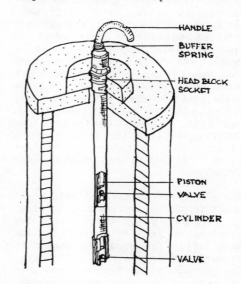

The Blair pump.

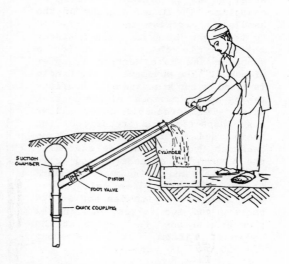

The Rover pump.

cost systems most applicable to individual end-uses. Testing is being
carried out at Britian´s Consumer
Association Testing and Research
Laboratories at Harpenden, and many
hand pumps, manufactured by different
suppliers, are being examined. Laboratory tests of a number of typical
hand pumps is under way. The first
interim report was issued in March
1982 and the fourth report in February
1985. The laboratory testing is to
be followed by extensive field trials,
but the testing and developing of
better handpumps is also continuing.

Hand Pump Technologies

Hand pump technologies can be divided,
according to use, into shallow well
(less than 7 m) and deep well (greater
than 7 m) systems. The most commonly
used models work by the reciprocating
principle. This can be used in
suction mode, with the plunger above
ground (which makes maintenance easy)
in shallow wells, or in lift mode
when the cylinder and plunger are
located below the water level in
the well (the pump can be used as
deep as 180 m). It is generally agreed
that one of the best hand pumps yet
developed, using the reciprocating
principle for deep well operation,
is the Indian Mark II pump which
was first introducced in 1966 and
is often referred to as one of the
first pumps suitable for village
level operation and maintenance
(VLOM).

The pumpstand is made of mild steel
which is not as brittle as cast iron.
It can be easily taken apart for
maintenance and has three legs for
stability. The spout is angled to
prevent the introduction of stones.
The entrance of the handle is designed
to prevent fingers being mauled.
The handle is of square cross-cut
steel so as to prevent breaking or
snapping. The inner segment of the
handle is circular and is attached
to the connecting rod with a sturdy
motor cycle chain to provide a
smoother and more centred transfer
of power. The movement of the handle
is eased by two standard ball-bearings
taking up the movement of the central
axle. The detachable cover allows
easy inspection and greasing of the
chain, and is designed so that no
backwash from the outside can enter
the pump. The connecting rod (circular steel) and cylinder are standard
items. The brass-line cylinder is
made from steel with leather or PVC
buckets and two valves, one in the
plunger and one check valve at the
cylinder bottom. It costs about
US$ 350 (1983)

Among the shallow well pumps the
cast iron Bangladesh New No. 6 is
another success story.

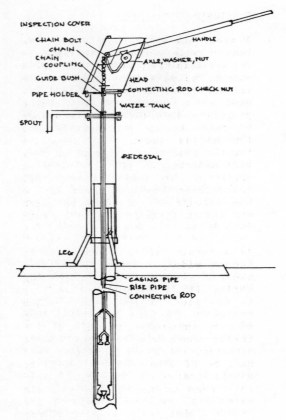

India Mark II handpump.

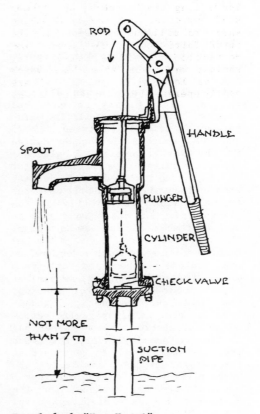

Bangladesh "New No. 6".

Conclusion

A great number of hand pumps have been installed throughout the developing world in the past and, to a large extent, their installation and successful operation has been short-lived. Hand pumps installed in development programmes have failed for a number of reasons, but primarily because the designs chosen were inadequate for the conditions under which they were expected to give trouble-free operation. In essence, in the past, insufficient thought was given to the rigours of operation to which the chosen systems were to be subjected.

As a result, if hand pumps are to become widely accepted and disseminated throughout the developing world, it is essential that they be designed to cope with the job for which they are intended. They must be extremely durable and virtually unbreakable if they are to withstand the harsh conditions, and must be of simple enough design to enable local manufacturers to fabricate them using locally-available materials wherever possible. The UNDP´s Global Project for Hand Pump Testing and Development has rigorous tests in order to identify and develop those hand pumps best-suited to operation in the developing world.

For a long time to come, hand pumps will continue to provide drinking water to millions of people in the rural Third World. It is thus important to find durable, trouble-free, sanitary and inexpensive pumps, which can be locally maintained and are easily operated by women and children.

Bibliography

Arlosoroff, S. et al. (1983) Rural Water Supply Handpump Project.
Laboratory Evaluation of Hand-Operated Water Pumps for Use in Developing Countries.
World Bank technical paper No 6, A UNDP Project Management Report.
The World Bank, 1818 H. Street N.W. Washington, D.C. 20433, U.S.A.

Arlosoroff, S. et al. (1984) Rural Water Supply Handpump Project.
Laboratory Testing of Handpumps for Developing Countries: Final Technical Report.
World Bank technical paper No 19. A UNDP Project Management Report.
The World Bank, 1818 H. Street N.W. Washington, D.C. 20433, U.S.A.

Arlosoroff, S. et al. (1985) Rural Water Supply Handpumps Project.
Handpump Testing and Development: Progress Report on Field and Laboratory Testing.
World Bank Technical Paper Number 29, A UNDP Project Management Report.
The World Bank, 1818 H Street N.W., Washington, D.C. 20433, U.S.A.

Beyer, M.G. (1983) Water-Social Barriers to the Efficient Utilization of Water Resources.
United Nations International Children´s Emergency Fund (UNICEF).
866 United Nations Plaza, New York, N.Y. 100 17, U.S.A.

Beyer, M.G. (1983) the India Mark-II Handpump.
A Case study.
United Nations International Children´s Emergency Fund (UNICEF), 866 United Nations Plaza, New York, N.Y. 100 17, U.S.A.

Blair, The Blair Pump.
Articles in Blair Research Bulletin, Blair Research Laboratory, Ministry of Health, P.O. Box 8105, Causeway, Harare, Zimbabwe.

Pacey, A (1976) Hand-pump Maintenance.
Intermediate Technology Publications Ltd., 9 King Street, London WC2E 8HN, U.K.

Appendices

Appendices

1 LIST OF ABBREVIATIONS, UNITS AND ACRONYMS

We have not tried to be completely consistent in our choice of units in the different chapters of this book, but rather to be practical. Depending on circumstances and on the source of data, different sets of units have sometimes been used in different chapters, although in most cases metric units (MKSA) are given.

A	ampere		kWh	kilowatthour
AC	alternate current		l	litre
b	bar		lb	pound
bl	barrel (159 litres)		LCZ	lower convection zone
BTU	British Thermal Unit		m	meter
C	Carbon		m^2	square meter
^{o}C	degree Celcius (centigrade)		m^2.d	square meter and day
cal	calorie		m^3	cubic meter
cap	capita		M	mega (prefix: million = 10^6)
cd	candela		Mtoe	million ton oil equivalent
CH_4	methane		MW	Megawatt
cm	centimeter		MWd	megawattday
cm^2	square centimeter		MWhr	megawatthour
cm^3	cubic centimeter		N_2	nitrogen
CO	carbon monoxide		NER	net energy(consumption)ratio
CO_2	carbon dioxide		O_2	oxygen
C/N	carbon to nitrogen ratio		ORC	Organic Rankine Cycle
d	day		OTEC	ocean thermal energy conversion
DC	direct current		p	peak
dm	dry matter		P	Peta (prefix: million billion=10^{15})
^{o}F	degree Fahrenheit		psi	pound per square inch
g	gram		PVC	polyvinylchloride
G	Giga (prefix: billion=10^9)		rpm	revolutions per minute
GJ	gigajoule		R&D	research and development
GW	gigawatt		s	second
GWh	giga watthour		SCC	specific capital cost
h	hour		t	ton
ha	hectare		T	Tera (prefix: thousand billion=10^{12})
hp	horsepower		tce	ton coal equivalent
hr	hour		toe	ton oil equivalent
H_2	hydrogen		twe	ton wood equivalent
H_2S	hydrogen sulphide		UCZ	upper convection zone
IRR	internal rate of return		USD	US Dollar
IRS	Indian rupees		US$	US Dollar
J	Joule		V	volt
K	kilo (prefix: thousand)=10^3		VLOM	village level operation and main-tenance
kCal	kilo calories			
kg	kilogram (kilo)		W	watt
kJ	kilo Joule		Wh	watthour
kJ.d	kilo Joule and day		Wp	peakwatt
km	kilometer		y	year
km^2	square kilometer		yr	year
kW	kilowatt			

List of Acronyms (not explained in text)

AIT   Asian Institute of Technology; Bangkok

ASTRA   Application of Science & Technology to Rural Areas; Bangalore, India.

BRET   Botswana Renewable Energy Technology Centre; Gaborone

CIDA   Canadian International Development Agency; Ottawa

CILSS   Interstate Committee for Draught Control in the Sahel; Burkina Faso

CWD   Consultancy Services Wind Energy Developing Countries; the Netherlands

DANIDA   Danish International Development Agency; Copenhagen

FAO	Food and Agricultural Organization, UN; Rome
FINNIDA	Finnish International Development Agency; Helsinki
GATE	German Appropriate Technology Exchange, in Deutsche Gesellschaft für Technische Zusammenarbeit (GTZ); F.R.G.
IBRD	International Bank for Reconstruction and Development (World Bank); Washington D.C.
ICRAF	International Council for Research in Agroforestry; Nairobi
IDRC	International Development Research Centre; Canada
IIED	International Institute for Environment and Development; U.K.
ILCA	International Livestock Centre for Africa; Nairobi and Addis Ababa
INE	Institute Nicaguense de Energia; Managua
ITDG	Intermediate Technology Development Group; U.K.
ITIS	Intermediate Technology Industrial Services Ltd.; U.K.
ITP	Intermediate Technology Power Ltd; U.K.
KVIC	Khadi and Village Industry Commission; India
MBZ	Royal Ministry of Foreign Affairs; the Netherlands
NORAD	Norwegian Agency for International Development; Oslo
ODA	Overseas Development Administration; U.K.
ORP	Organization for the Rural Poor; India
RECAST	Research Centre for Applied Science and Technology; Kathmandu, Nepal
SAREC	Swedish Agency for Research Cooperation with Developing Countries; Stockholm
SIDA	Swedish International Development Authority; Stockholm
SWD	Steering Committe Wind Energy Developing Countries; the Netherlands (now CWD).
TPI	Tropical Products Institute; London (now TDRI)
TDRI	Tropical Development and Research Institute; London
UN	United Nations
UNDP	United Nations Development Programme; New York
UNEP	United Nations Environment Programme; Nairobi
UNERG	United Nations Conference on New and Renewable sources of Energy; Nairobi, 1981
UNICEF	United Nations Children Fund; New York
UNSO	United Nations Sudano-Sahelian Office
USAID	United States Agency for International Development; Washington D.C.
VITA	Volunteers in Technical Assistance; U.S.
WHO	World Health Organization; Geneva
WMO	World Meteorological Organization; Geneva

2 CONVERSION TABLES

To convert from	into MTOE	Energy conversion factors* MBOE	MTCE	Bm³gas	TWh	Pj
Million tonnes oil equivalent (MTOE)	1	7.33	1.55	1.15	12.60	45.37
Million barrels oil equivalent (MBOE)	0.136	1	0.21	0.156	1.71	6.17
Million tonnes coal equivalent (MTCE)	0.65	4.74	1	0.74	8.14	29.31
Million tonnes wood+ equivalent (MTWE)	0.39	2.8	0.6	0.44	4.9	17.5
Billion (10^9) cubic metres natural gas (Bm3 gas)	0.87	6.41	1.34	1	10.93	39.26
Terawatt-hours (10^{12} watt-hours, Twh)	0.079	0.58	0.12	0.091	1	3.6
Peta joules (10^{15} joule, PJ)	0.022	0.16	0.034	0.025	0.28	1

* e.g. , 1 million barrels oil (MBOE)=0.136 million tonnes oil (MTOE)
+ Wood varies depending upon species water content, and density, but an average figure given here refers to equilibrium air-dried wood (i.e. containing c.15 per cent water). Note: 1 tonne wood averages 1.4 m³.

Power Equivalents

To Convert From	into Mtoe/yr	Mbd	Mtce/yr	GWth	PJ/yr
Mtoe/year (Mtoe/yr)	1	0.02	1.55	1.43	45
Mb/day (Mbd)	50	1	77	71	2235
Mtce/year (Mtce/yr)	0.65	0.013	1	0.92	29
Giga watt (10 watt) thermal(GW$_{th}$)	0.70	0.014	1.09	1	32
PJ/ year (PJ/yr)	0.02	4.5×10^{-4}	0.034	0.031	1

Other conversion factors

1 cal   = 4.19 J = 1.16 · 10^{-3} Wh
1 kcal = 4.19 kJ = 1.16 Wh
1 BTU   = 1.06 kJ = 0.29 Wh
1 toe   = 10^{10} cal = 10 G cal (approximately)
1 kg   = 2.2 lbs
1 acre = 0.405 ha
1 year = 8760 hours
1 day   = 86 400 seconds

The heat value in 1 m³ (1000 liters) of fuel oil (diesel) corresponds approximately to:

0.84	ton	fuel oil (diesel)
1.2-1.4	ton	coal
3.2-3.5	ton	urban refuse
4.3	ton	fresh wood (50% humidity)
2.4	ton	air dried wood (20% humidity)
5	m³	fine wood (30% humidity)
2.5	ton	straw (10% humidity)
4	ton	peat (50% humidity)
10	m³	peat (50% humidity)
2.3	m³	methanol
1.7	m³	ethanol

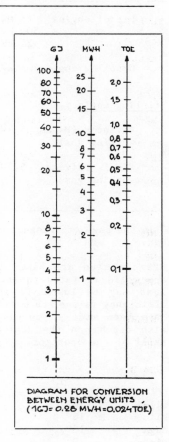

DIAGRAM FOR CONVERSION BETWEEN ENERGY UNITS ,
(1GJ= 0.28 MWH =0.024TOE)

Energy value of selected fuels: MWh

1 ton coal   7-8.5
1 ton crude oil   12   (approx.)
1 ton diesel oil   11.8
1 ton kerosene   9.5
1 ton gasoline   11.9
1 ton ethanol   7.4
1 ton methanol   5.5
1 ton urban refuse   3   (approx.)
1 ton wood (20% humidity)   5   (approx.)
1 ton straw (15% humidity)   4   (approx.)
1 ton peat (50% humidity)   2.5   (approx.)
1 ton dung (15 % humidity)   3.5   (approx.)
1 ton crop residue (15% humidity)   4.0   (approx.)
1 ton peat (50% humidity)   2.5   (approx.)

Additonal data on energy contents of biomass fuels are given in Chapter 11.

Typical efficiencies for different types of power plants.

Power plant technology		Fuel	
		Oil	Wood
Steam:			
Steam turbine	200 MW	41%	-
	15 MW	23-25%	22-24%
	1 MW	20-22%	19-21%
Steam engine	500 kW	12-14%	11-13%
	25 kW	8-10%	7-9%
Internal combustion:			
Diesel engine	15 MW	38-40%	30-32%*
	500 kW	37-39%	29-31%*
	25 kW	26-28%	19-21%*
Gasoline engine	5 kW	15-22%	10-16%*
Stirling engine:			
Advanced (He as working gas)	5 kW	about 20%	about 16%
Simple (air as working gas)	5 kW	about 5%	about 4%

* In the case of Internal combustion engines, wood is converted to Producer Gas.

Efficiencies are based on the lower heating value of the fuel. They are valid at rated power. Average efficiency for operation with varying load may be lower. It should be observed that both higher and lower efficiency values are possible depending on the extent to which investments have been made in energy efficient equipment and systems. Efficiencies also depend on the degree of maintenance of the equipment and on the skill of the operator.

Prefixes:

milli	10^{-3}	one thousands (1/1000)
kilo	10^{3}	thousand (1.000)
mega	10^{6}	million (1.000.000)
giga	10^{9}	billion (1.000.000.000)
tera	10^{12}	million million (1.000.000.000.000)
peta	10^{15}	billion million (1.000.000.000.000.000)

3 LIST OF BACKGROUND REPORTS COMMISSIONED BY THE BEIJER INSTITUTE

Report title:	Author:
1. Forestry	Jan Erik Nylund
2. The Potential Role of Agroforestry in Fuelwood Production	B. Lundgren B. van Gelder
3. Marine Biomass as an Energy Source	Torgny von Wachenfeldt
4. Alcohol Production from Biomass in Developing Countries	Britt Sahleström
5. Vegetable Oil Fuels	Per Johan Svenningsson
6. Combustion of Biomass Fuels for Industrial Applications in Developing Countries	Jackie Bergman
7. Producer Gas	Björn Kjellström
8. Biomass Fuelled Stirling Engine Systems	Stig Gummesson
9. Acetylene Lamps for Rural Areas of Developing Countries	Intermediate Technology Industrial Services (ITIS)
10. Wind Energy for Electricity Production in Developing Countries	Staffan Engström
11. A Project on Solar Crop Drying	M.S. Sodha N.K. Bansal
12. Prospects for Solar Industrial Process Heat Systems in Developing Countries	Dennis Costello Kathyn Lawrence
13. Passive Cooling	Christer Nordström
14. Renewable Energy Technologies in Kenya: A Case Study	Tom Harris
15. Energy Flows on Smallhold Farms in the Ethiopian Highlands	Guido Gryseels Michael R. Goe
16. Prospects for New and Renewable Energy Sources in Developing Countries: Social Assessment Component	Kirsten Johnson Cindi Katz